Springer Tracts in Mechanical Engineering

Springer Tracts in Mechanical Engineering (STME) publishes the latest developments in Mechanical Engineering - quickly, informally and with high quality. The intent is to cover all the main branches of mechanical engineering, both theoretical and applied, including:

- Engineering Design
- Machinery and Machine Elements
- Mechanical structures and Stress Analysis
- Automotive Engineering
- Engine Technology
- Aerospace Technology and Astronautics
- Nanotechnology and Microengineering
- Control, Robotics, Mechatronics
- MEMS
- Theoretical and Applied Mechanics
- Dynamical Systems, Control
- Fluids mechanics
- Engineering Thermodynamics, Heat and Mass Transfer
- Manufacturing
- Precision engineering, Instrumentation, Measurement
- Materials Engineering
- Tribology and surface technology

Within the scopes of the series are monographs, professional books or graduate textbooks, edited volumes as well as outstanding PhD theses and books purposely devoted to support education in mechanical engineering at graduate and post-graduate levels.

Indexed by SCOPUS and Springerlink. The books of the series are submitted for indexing to Web of Science.

To submit a proposal or request further information, please contact: Dr. Leontina Di Cecco Leontina.dicecco@springer.com or Li Shen Li.shen@springer.com.

Please check our Lecture Notes in Mechanical Engineering at http://www.springer.com/series/11236 if you are interested in conference proceedings. To submit a proposal, please contact Leontina.dicecco@springer.com and Li.shen@springer.com.

More information about this series at http://www.springer.com/series/11693

Panfeng Huang · Fan Zhang

Theory and Applications of Multi-Tethers in Space

Springer

Panfeng Huang
National Key Laboratory of Aerospace
Flight Dynamics, School of Astronautics,
Research Center for Intelligent Robotics
Northwestern Polytechnical University
Xi'an, Shaanxi, China

Fan Zhang
National Key Laboratory of Aerospace
Flight Dynamics, School of Astronautics,
Research Center for Intelligent Robotics
Northwestern Polytechnical University
Xi'an, Shaanxi, China

ISSN 2195-9862 ISSN 2195-9870 (electronic)
Springer Tracts in Mechanical Engineering
ISBN 978-981-15-0386-3 ISBN 978-981-15-0387-0 (eBook)
https://doi.org/10.1007/978-981-15-0387-0

This Springer imprint is published by the registered company Springer Nature Singapore Pte Ltd.
The registered company address is: 152 Beach Road, #21-01/04 Gateway East, Singapore 189721, Singapore

Preface

Theory and Applications of Multi-Tethers in Space provides a comprehensive overview of the recently developed space multi-tethers, including maneuverable space tethered net and space tethered formation, with detailed system description, dynamics modeling and analysis, and controller design. For each application of space multi-tethered system, detailed derivatives are given to describe and analyze the mathematical model of the system, and then, different control schemes are designed and proved for different problems of the application. In the textbook, Newton and Lagrangian mechanics are used for dynamics modeling, Hamilton mechanics and Poincare surface of section are introduced for dynamics analysis, and both of centralized and distributed controllers are employed to figure out the formation question of the multi-tethered system. Besides the equations and words, 3D design drawing, schematic diagram, control scheme block, Table et al. is used for easy reading and understanding. The graduate students in related research area can systematically learn space multi-tethered system and its applications, and we hope other researchers could be inspired by this technical book and make much more contribution to this topic.

This book is not prepared as a collection of existing papers on dynamics and control of multi-space tethers; rather, we first establish the structure that the theory and applications of multi-space tethers should follow, and then fill in the details. The book is organized and presented in a widely accessible fashion. For example, the related control problems of maneuverable space tethered net are introduced under a space mission sequence, which has been described at the very beginning of the part. In order to help the reader who is interested about the field and wants to study the results by himself or to exploit the space tether's applications for his own needs, the calculations and derived equations are detailed explained in this text.

This book provides the readers with extensive material from the first concept of space tether up to the cutting-edge art in the area of the applications of multi-tethers in space presented in a pedagogical fashion. In addition to the emphasis on practicality, many theoretical equations and theorems (which may have practical relevance and importance) are numerically verified using MATLAB/Simulink software. This book, which is based on a major developed at the Northwestern

Polytechnical University of Space Science and Technology, is appropriate as a textbook for graduate and senior students, or as a self-study reference book for practicing engineers, who are with a basic knowledge of theoretical mechanics, classical control theory, and some knowledge of spacecraft orbital dynamics. This book is intended to be intelligently challenging for students.

Organization

This book consists of nine chapters and two appendixes which consisted of three parts including the Introduction (Chap. 1), Maneuverable Space Tethered Net (Chaps. 2–5), and Space Tethered Formation (Chaps. 6–9), in which two parts represent two classic applications of multi-tethers in space.

Chapter 1 starts with a brief historical overview of space tether. The motivations of the space tether's study and previous research including the classic single space tether and multi-tethers have been introduced.

Chapter 2 opens the second part of the book, which is focused on the space tethered net. This part is organized based on a space mission of active debris removal. In Chap. 2, a complex space tethered net system is described in detail and modeled by mathematical equations. Then in Chaps. 3–5, it is organized by the mission sequence, namely, folding and storing, releasing, and approaching. The folding criteria of the flexible net are introduced in Chap. 3, and corresponding unfolding characteristics are studied. In Chap. 4, the control scheme during approaching for both configuration-keeping and configuration-maneuvering are studied. Different from the centralized control scheme studied in Chap. 4, distributed control is addressed to solve the same approaching questions. The readers can learn and study further control strategies based these two chapters.

Chapter 6 is the first of last four chapters, which compose the third part of the book. In this part, another important application of multi-space tether, namely space tethered formation is researched. Two different formations, closed-chain and hub-spoke formations, are modeled and analyzed in Chap. 6. The control schemes of these two formations are presented in the following chapters. In Chap. 7, the formation-keeping control scheme is studied. Besides the traditional thruster-based control scheme, a novel thruster and tether-tension-based controller is also proposed and analyzed. Chapter 8 presents the deployment and retrieval control of hub-spoke system based on natural motion without any controller. In Chap. 9, advantage control schemes of the hub-spoke system for formation-keeping are proposed including error-based and time-based coordinated control.

Xi'an, China
2019

Panfeng Huang
Fan Zhang

Acknowledgements

This book is based on the research supported by the National Natural Science Foundation of China. For the facilities and daily-operation supporting, thanks to the National Key Laboratory of Aerospace Flight Dynamics, Research Center for Intelligent Robotics, and School of Astronautics of Northwestern Polytechnical University.

We owe special thanks to Mr. Konstantin Eduardovich Tsiolkovsky for his much fruitful work on the creation of the theory of jet aircraft, multistage rockets, gas turbine engine, and of course, the ***SPACE TETHER***.

Sincere thanks to our friends for their sagacious companionship, colleagues for scientific and friendly support and in particular Zhongjie Meng, Yizhai Zhang, Zhengxiong Liu, Gangqi Dong, and Zhiqiang Ma. Thanks as well to Yakun Zhao, Ya Liu, Yongxin Hu, and other students in the Research Center for Intelligent Robotics, who helped in proofreading and providing valuable feedback. We are grateful to all their help.

Finally, many thanks to our families for their encouragements and tender supports.

Contents

Chapter 1
Introduction

Ever since the concept of the space tether was first proposed by Tsiolkovsky in 1895, it has generated a multitude of possibilities for use in space exploration [1]. In the beginning, applications for the space tether seemed science fictions than reality. Most traditional space tether applications (e.g., orbit transfer or artificial gravity) are impossible or excessively costly with the existing space technology and engineering capacity, and the research of the space tether had fallen to the wayside for a number of years. Space tethers were imagined for use within artificial gravity, as a space elevator, for orbit transfer, and for similar, seemingly otherworldly tasks. The space missions of the 1960s saw a turning point for the application of the space tether. In the later decades, many of publications were released and proposed applications became great in number and in variety. Scholars begun to focus on the space tether as a new approach to space exploration again. In the following decades, space tether started its golden age.

1.1 Form Single Tether to Multi-tethers

Tethers are commonly considered to be as useful in their performance in space as they have been used on the Earth [2]. The space tether concept has been studied for over one hundred years, and applications in various capacities have been proposed by many different researchers. Artificial gravity, orbit transfer, attitude stabilization, and space debris capture are typical applications of space tether, which will be introduced as follows.

Artificial gravity is highly desirable for long manned space flights since even small fractions of gravity will improve living conditions aboard a space station. It was precisely for this task that the space tether was first proposed. The centrifugal force of inertia can be used to create artificial gravity on Earth or in space. Tsiolkovsky thought to implement this idea in 1895 wherein two spacecraft were

P. Huang and F. Zhang, *Theory and Applications of Multi-Tethers in Space*,
Springer Tracts in Mechanical Engineering,
https://doi.org/10.1007/978-981-15-0387-0_1

connected by a tether chain and the whole system was rotated to create artificial gravity [3]. The length of the chain is a key factor in the magnitude of the force generated as well as the square of the angular velocity of the mechanical system's rotation.

There are advantageous and far-reaching applications associated with the space tether for transportation. Traditionally, thrusters are mounted on a spacecraft as a reactive mass for maneuvering in orbit. When the working medium is exhausted, however, this process will fail. For tethered satellites on opposite sides, the use of a space tether system utilizes a pure exchange of energy and angular momentum between them. Since there is no fuel consumed, this kind of orbit transfer system promises a sizable reduction in fuel usage. Hence, it is a viable alternative to the traditional approach.

The orientation of a spacecraft mission is usually oriented toward the Earth's direction. When in space, it may be necessary to maintain this orientation for a long time. Active stabilization systems can assist in satisfying this requirement. Traditionally, these systems use jet engines with small thrust for attitude adjustment. However, these kinds of systems have the same disadvantage as the requirement for the use of a propellant. The passive stabilization system was developed as a solution to this problem; a long flexible tether with the load is one such approach.

The Tethered Space Robot (TSR) (Fig. 1.1) is composed of a space platform, a space tether, and a gripper. Because of its flexibility and large workspace, the TSR is a promising future solution for On-Orbit Servicing such as in-orbit maintenance and repair, in-orbit refueling, orbit maneuvering, and space debris removal [4, 5, 6].

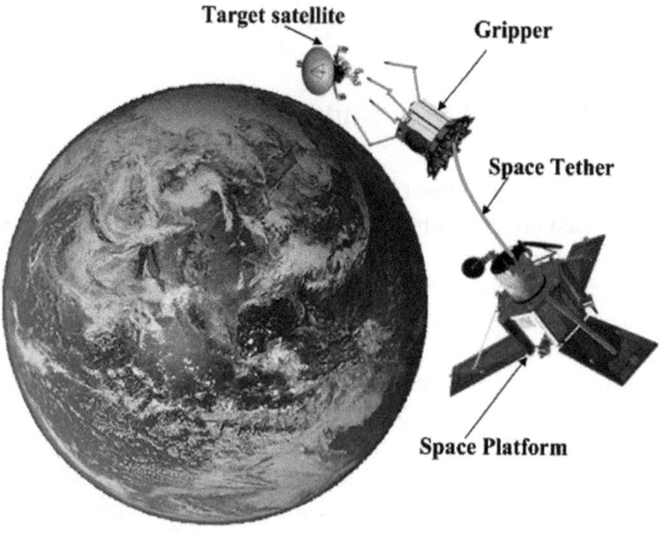

Fig. 1.1 Tethered space robot [66]

Some other tether applications have been also suggested by Levin [7], including upper atmosphere exploration, interplanetary transfers, space escalator, creation of a traffic artery linking Earth and the moon, auxiliary structural members, space harpoons, and so on. This list can be extended. More so, any discussion on tether applications often brings about new proposals—in short, there is much further potential for the space tether concept.

With several space experiments in the 90s, this concept was widely studied by researchers. Although some people think space tether is a radical attempt, the pace of scientists never stops. Based on the research on single space tether, multi-tethers in space is daringly proposed for advanced space missions, such as active debris removal and spacecraft formation. As a classic application, space net is first completely proposed in the project of RObotic GEostationary orbit Restorer (ROGER) by ESA. A tether knitted square net with four mass bullets is released from a platform satellite and fly toward the target space debris. The net is connected to the platform satellite via a single tether for retrieval or deorbit, and this design is considered as an improved means of the TSR.

Another research focus of the multi-tethers in space is space tethered formation. It stems from the spacecraft formation, which is used for the Earth and deep space observation. The precise relative position and attitude orientation require higher fuel consumption. Because the neighboring spacecrafts of the space tethered formation system are connected via space tethers, it seems to be a good solution due to its exact relative position. As the research is developing, more advantages of this system are found such as the spatial orientation property.

In the rest of this chapter, these two proposals of multi-tethers in space will be reviewed in detail.

1.2 The Research on Space Tethered Net

1.2.1 Tethered Space Net in Early Applications

Net shape and mesh geometric topology were not the focus in early studies on this subject [8]. Because centrifugal forces are used for release and deployment in a traditional TSN [9], the centrifugal force field leads to the nonuniform pre-stress distribution of the net. The entire net must have sufficient tension. Pre-stress in the net causes stiffness in the out-of-plane angle, which relies on the geometric topology of the mesh [10]. Stiffness and spinning of the net in the out-of-plane angle generate unsatisfactory altitude and positional motions of the net, namely, oscillation transfer. These additional motions cause a reduction in the net's performance [11]. An exact dynamics model and analysis for net shape and mesh geometric topology is critical, to this effect.

Maximum permissible sag-to-span ratios of different corner masses were investigated by Tibert and Gärdsback [12], who concluded that a large sag-to-span ratio produces undesirable pointier vertices. They also identified optimal dynamics and numerical simulations of triangular mesh [13, 14]. Schuerch and Hedgepeth [15] proposed a quadrangular mesh that is able to withstand grand shearing deformations; this attribute is important for package design. Kyser [16] analyzed relationships of mesh geometry with uniform pre-stress. Since triangular, square, and hexagonal are the three prime topologies, many researchers have compared the structure and stresses among them to find that equilibrium geometry is determined by the force distribution [17–21]. Another important conclusion about the eigenfrequency and eigenmode of the TSN is that eigenfrequencies and free-vibration modes are important for the effects of traveling waves, out-of-plane damping requirements, and other phenomena such as orbital maneuvering [22–24].

Structure design of the net is also the focus of research for TSN, and the information gathered is the foundation of the space net. MacNeal [25] proposed a criterion for the structure design of the space net. The optimal design of the net is governed via: (a) pre-stressability, (b) manufacturability, (c) mass, (d) stiffness normal to orbit plane, and (e) eigenfrequencies. Based on the aforementioned requirements, only square mesh is pre-stressable under centripetal force. A square web with a square mesh is the best option based on feasibility. Additionally, both the releasing velocity and bullet masses are determining factors of stiffness normal to the orbit plane and eigenfrequencies.

According to the special deployment required for space net, namely, centrifugal deployment, a good folding structure design is crucial for a proper release. Many researchers have explored folding patterns based on the basic dynamics of the flexible tether [26–30]. Scheel [27] developed a pattern with straight folding lines, while McInnes [28] proposed another folding pattern with 36 radial spars emanating from a central hub. Schuerch and Hedgepeth [31] later derived a star-like shape folding pattern which allows the net to be folded into the central hub in the releasing subsystem.

1.2.2 *Tethered Space Net*

(a) Structure and Configuration

The TSN is an extended application of the TSR shown in Fig. 1.2, which is another solution to Active Space Debris Removal. This kind of TSN is released from a platform satellite via a flexible tether [32]. Normally, the TSN is used for space debris or uncooperative target capture, and other in-orbit service. The various applications of the TSN make its structure and configuration design different from that of the TSN.

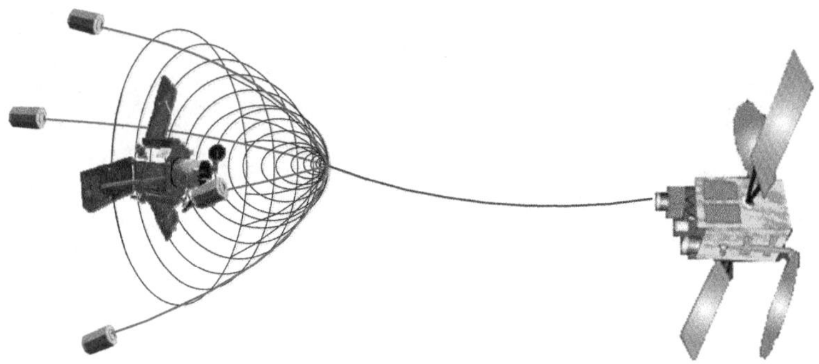

Fig. 1.2 Tethered space net

After the project, Robotic Geostationary Orbit Restorer (ROGER) from ESA [32] was proposed for in-orbit space debris and uncooperative satellite capture and retrieval, the TSN emerged as a good option because of its safety and large operation radius. The TSN is a good trade-off in Active Debris Removal (ADR) in particular. Generally, the TSN is catapulted through the release subsystem and the corner bullet masses propel the net forward. Benvenuto [33] gave a detailed design proposal for the TSN, including the net material, net size and configuration, mesh size and topology, and the closure device system. Zhai [34, 35] gave a very detailed design description of the TSN including the mechanism of net ejection, configuration of the net, and configuration of the capture element.

(b) Releasing Dynamics and Control of TSN

Zhai [36] presented a control strategy to compensate for error through tether tension. They [35] also studied the performance of the motion equations in the deployment phase including both free and non-free motions in circular orbit as well as the motion dynamics. A later study investigated the disturbance of orbital dynamics which may lead an altitude vibration [37]. Liu et al. [38] derived a sophisticated mathematical model for the TSN after the capture. They studied the orbital motion of the system, relative altitude motion of the net and base satellite, and the dynamics of re-orbit after net capture. Benvenuto et al. [39] discussed the possible problems of TSN, namely the GNC issues, in the capture and removal phase. The collision detection, contact dynamics in capture, and tether tension in retrieval were all studied in this paper.

1.2.3 Maneuverable Tethered Space Net

Huang et al. proposed a new configuration as an improvement to the traditional TSN called the Maneuverable Tethered Space Net (MTSN), which is shown in Fig. 1.3. The MTSN is comprised of a main net and four maneuverable units at the

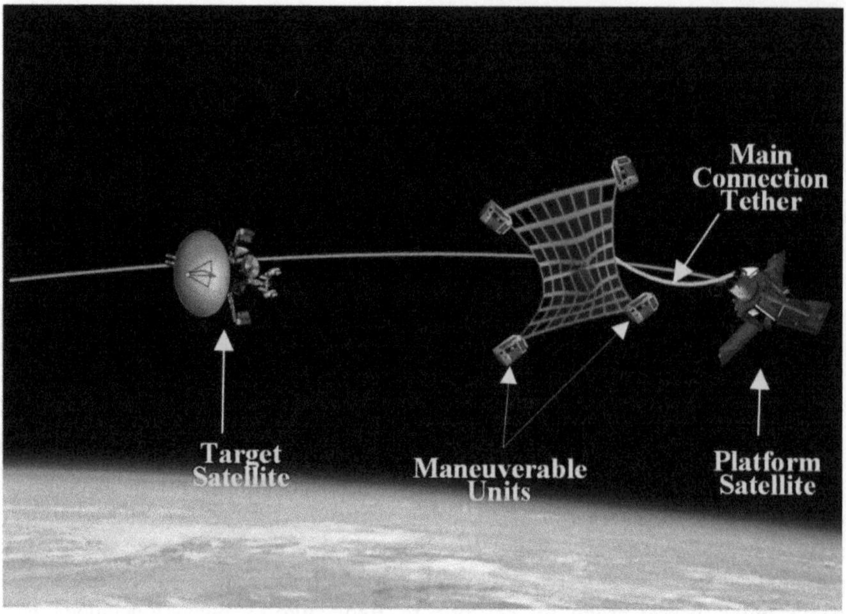

Fig. 1.3 Maneuvering tethered space net

four corners of the net that guide mass bullets in the TSN [40]. Since the corners are maneuverable units are rigid and maneuverable bodies and the net is flexible, the dynamics of the MTSN are fairly complicated. Maneuvering may lead to the net being asymmetrical, which generates more complicated vibration motions of the flexible net. The issues concerning coupled dynamics modeling and shape maintenance in the deployment phase were examined by the same research team [41]. Gao et al. [42] also studied the kinematics and dynamics of a formation-based space net system and derived an improved dynamics model by employing the extended Hamilton's principle. In this paper, the inverse dynamics were transformed into a double-level optimization problem.

1.3 The Research on Space Tethered Formation

Formation flying has been considered as an efficient means to reduce expenses and add new potential missions for space-based programs [43]. Large spacecraft could be replaced by many smaller and less complicated satellites that make up a particular spatial configuration, and the satellites would communicate to share information. Compared to single large spacecraft, satellite formations are more flexible and reliable. Satellite formation technology has become the focus for developing future space mission technology accordingly.

Currently, satellite formation has been used in synthetic aperture radar satellite formation, distributed meteorological satellite stereoscopic imaging, high-resolution synthetic aperture optical interferometry, electronic surveillance, and more [44]. Spatial formation flight is complicated due to a number of factors, however, especially space environmental disturbance such as interference by gravity, air resistance, sunlight pressure, electromagnetism, and modeled force. This makes spatial formations difficult to employ for formation stability in the long term. Further, satellites have to consume large quantities of fuel to maintain stable configurations, so the service lives of the satellites in formation are greatly reduced.

Tethers can be applied to the satellites, connected to the spacecraft, remain taut, and the relative distances between spacecraft can be maintained accurately. A satellite formation connected by tethers is a "tethered satellite formation", which rotates along an axis or with thrusters. Gravity gradient or air resistance keeps the tether in tension and maintains the shape of the formation. The primary advantages of the usage of space tethers for space formation missions were summarized by Fedi [45]. The tether also has many disadvantages in space formation, however, including more complex deployment in tether dynamics which complicate controller design and practice, as well as the possibility of environmental deterioration and risk of collisions [46]. Tethered satellite formation has become a popular research object accordingly.

1.3.1 Structure and Configuration

Many researchers have focused on two-body systems, since the initial applications of tethered spacecraft formation techniques involved only two bodies. Sarychev [47] investigated the equilibrium of a dumbbell system in circular Keplerian orbit in relation to the orbital reference frame. Lorenzini [48] investigated a satellite system with three end satellites: The Space Station, an end mass fixed 10-km away from the station via a Kevlar tether, and a gravity laboratory which is able to crawl along the tether. Pizarro [49] was dedicated to the dynamics of multi-tethered satellite formations. This system consists of two configurations. One is a central hub configuration containing a central body for the tether revolving spindle and a satellite fixed to the end of tether [50]. The other is a closed hub-and-spoke configuration, which is a series of satellites connected by tethers that add to the aforementioned configuration [51]. When spinning in the orbital plane, the hub-and-spoke configuration can stabilize the end bodies [52]. The feasibility of a rotating formation of satellites using flexible tethers was investigated by previous researchers. The system consisted of three spacecraft connected by tethers in a triangle-like formation [53]. The open-loop control scheme for deployment and retrieval was investigated in the same paper. Guerman [54] studied the equilibrium of the tetrahedral configurations, i.e., four rigid bodies attached by three massless rods. The dynamic stability of a spinning triangular formation during the deployment and retrieval stages was also investigated [55].

1.3.2 Dynamics and Control

The dynamics and control of tethered satellite formations in flight have been studied extensively. Avanzini [51] explored the multi-tethered satellite formation by modeling the tether as a sequence of point-masses and massless springs. The results showed that the dynamics model using a massless link to represent the tether is precise enough for parameter estimation and controller design during capture. A control scheme for tracking via the θ-D technique was proposed by Liu [56] and shown to satisfy the TSS working on halo orbits. Another formation near libration points in a spinning triangular formation was investigated by Lorenzini [48] including deriving dynamic motion equations and analyzing the dynamic characteristics of the system. Zhao [57] investigated formations composed by multi-satellites and tethers and derived corresponding full coupled dynamics models; the parent satellite was working on a large halo in 3D and the center point surrounded the second liberation point with respect to the Sun–Earth system. The orbit dynamics performed well even when initial spin rate was increased; similar results were obtained for other tether liberations. Pizarro examined the dynamics model of a hub-spoke formation [54]. Zhao [58] presented the dynamics of multi-tethered satellite formations in which subsatellites surround a parent satellite attached by varying-length tethers in a hub-spoke formation. Chung [59] investigated the multi-vehicle tethered spacecraft and established a corresponding nonlinear dynamics model; they found that both reshaping and spin-up are controllable only by the flying wheels and tether tension.

1.3.3 Formation Control

It is necessary for the satellite formation to change the rotation direction in many missions such as interferometry observation [60]. Nakaya [61] discussed altitude maneuvers of a spinning tethered configuration system and presented a feedback maneuver control scheme based on the virtual configuration method as proposed. Other researchers proposed that the configuration of a tethered satellite formation can be maintained or changed by using an active controller of both tether tension and length; a coordinate controller comprised of thrusters, fling wheels, and tether tension was also addressed [62]. Menon [63] investigated a tethered formation system consisted of two platforms linked by a flexible tether a few hundred meters long. A decentralized coordinated altitude control strategy was derived under the behavior-based control approach in another study [64]. Existing controllers can assure globally asymptotical reachability for a reference trajectory in the presence of model uncertainties and external perturbation [65].

1.4 Scope and Organization

This book gives a comprehensive overview of the recently developed space multi-tethers, including maneuverable space tethered net and space tethered formation, with detailed system description, dynamics modeling and analysis, and controller design. For each application of space multi-tethered system, detailed derivatives are given to describe and analyze the mathematical model of the system, and then, different control schemes are designed and proved for different problems of the application. In the textbook, Newton and Lagrangian mechanics are used for dynamics modeling, Hamilton mechanics and Poincare surface of the section are introduced for dynamics analysis, and both of centralized and distributed controllers are employed to figure out the formation question of the multi-tethered system. Besides the equations and words, 3D design drawing, schematic diagram, control scheme block, Table et al. is used for easy reading and understand. The graduate students in related research area can systematically learn space multi-tethered system and its applications, and we hope other researchers could be inspired by this technical book and make much more contribution to this topic.

This book is organized into nine chapters and has two appendixes.

This chapter starts with a brief historical overview of space tether. The motivations of the space tether's study and previous research including the classic single space tether and multi-tethers have been introduced.

Chapter 2 opens the second part of the book, which is focused on the space-tethered net. This part is organized based on a space mission of active debris removal. In Chap. 2, a complex space tethered net system is detailed described and modeled by mathematical equations. Then in Chaps. 3–5, it is organized by the mission sequence, namely, folding and storing, releasing, and approaching. The folding criteria of the flexible net are introduced in Chap. 3, and corresponding unfolding characteristics are studied. In Chap. 4, the control scheme during approaching for both configuration-keeping and configuration-maneuvering are studied. Differently from the centralized control scheme studied in Chap. 4, distributed control is addressed to solve the same approaching questions. The readers can learn and study further control strategies based on these two chapters.

Chapter 6 is the first of the last four chapters, which compose the third part of the book. In this part, another important application of multi-space tether, namely space tethered formation is researched. Two different formations, closed-chain and hub-spoke formations, are modeled and analyzed in Chap. 6. The control schemes of these two formations are presented in the following chapters. In Chap. 7, the formation-keeping control scheme is studied. Besides the traditional thruster-based control scheme, a novel thruster and tether tension based controller is also proposed and analyzed. Chapter 8 presents the deployment and retrieval control of hub-spoke system based on natural motion without any controller. In Chap. 9, advantage control schemes of the hub-spoke system for formation-keeping are proposed including error-based and time-based coordinated control.

References

1. Tsiolkovsky KE (1895) Speculations between Earth and Sky. Isd-voAN-SSSR, p 35 (reprinted in 1959)
2. Beletsky VV, Levin EV (1993) Dynamics of space tether systems. Univelt Incorporated
3. Pearson J (2005) Konstantin Tsiolkovski and the origin of the space elevator. AAS Hist Ser 26:17–24
4. Janeski JA (2013) Dynamics of an electrodynamic tether system in a varying space-plasma environment. PhD thesis, Virginia Polytechnic Institute and State University
5. Cartmell MP, McKenzie DJ (2008) A review of space tether research. Prog Aerosp Sci 44 (1):1–21
6. Kumar KD (2006) Review on dynamics and control of nonelectrodynamic tethered satellite systems. J Spacecr Rocket 43(4):705–720
7. Levin EM (2007) Dynamic analysis of space tether missions. Univelt Incorporated
8. Grossman J (2000) Solar sailing: the next space craze? Eng Sci 63(4):18–29
9. Kaya N, Iwashita M, Nakasuka S et al (2005) Crawling robots on large web in rocket experiment on Furoshiki deployment. J Br Interplanet Soc 58(11–12):403–406
10. Tibert G (2002) Deployable tensegrity structures for space applications. Royal Institute of Technology
11. Nakasuka S, Kaya N. Quick release on experiment results of mesh deployment and phased array antenna by S-310-36. The Forefront of Space Science
12. Tibert G, Gardsback M (2006) Space webs final report. ESA/ACT, Adriana ID: 05, 2006, 4109
13. William M (1967) Spinning paraboloidal tension network. National Aeronautics and Space Administration
14. Robbins Jr WM (1967) The feasibility of an orbiting 1500-meter radiotelescope. National Aeronautics and Space Administration
15. Schürch HU, Hedgepath JM (1968) Large low-frequency orbiting radio telescope. National Aeronautics and Space Administration, Washington, DC. NASA contractor report, NASA CR-1201, 1968, 1
16. Kyser AC (1965) Uniform-stress spinning filamentary disk. AIAA J 3(7):1313–1316
17. Schek HJ (1974) The force density method for form finding and computation of general networks. Comput Methods Appl Mech Eng 3(1):115–134
18. Gunnar T (1999) Numerical analyses of cable roof structures. Department of Structural Engineering, Royal Institute of Technology, Stockholm
19. Lai C, You Z, Pellegrino S (1998) Shape of deployable membrane reflectors. J Aerosp Eng 11 (3):73–80
20. Young W, Budynas RG (2002) Roark's formulas for stress and strain. McGraw-Hill, New York
21. Pickett WL, Pratt WD, Larson ML et al (2002) Testing of centrifugally deployed membrane dynamics in an ambient ground environment. In: 43rd AIAA/ASME/ASCE/AHS/ASC structures, structural dynamics, and materials conference, Denver, CO
22. Cook RD (2007) Concepts and applications of finite element analysis. Wiley
23. Eversman W (1968) Some equilibrium and free vibration problems associated with centrifugally stabilized disk and shell structures. National Aeronautics and Space Administration
24. Guest S (2006) The stiffness of prestressed frameworks: a unifying approach. Int J Solids Struct 43(3):842–854
25. MacNeal RH (1967) Meteoroid damage to filamentary structures. National Aeronautics and Space Administration
26. Guest S, Pellegrino S. Inextensional wrapping of flat membranes. In: Proceedings of the first international seminar on structural morphology, pp 203–215
27. Scheel H (1974) Space-saving storage of flexible sheets: US Patent 3,848,821, 1974 Nov 19

28. McInnes CR (2013) Solar sailing: technology, dynamics and mission applications. Springer
29. Koshelev VA, Melnikov VM (1998) Large space structures formed by centrifugal forces. CRC Press
30. Denavit J (1955) A kinematic notation for lower-pair mechanisms based on matrices. J Appl Mech 22:215–221
31. Schuerch HU (1964) Deployable centrifugally stabilized structures for atmospheric entry from space. National Aeronautics and Space Administration
32. Bischof B, Kerstein L, Starke J et al (2004) ROGER-robotic geostationary orbit restorer. Sci Technol Ser 109:183–193
33. Benvenuto R (2012) Implementation of a net device test bed for space debris active removal feasibility demonstration. MS thesis, Politecnico di Milano
34. Zhai G, Zhang J (2012) Space tether net system for debris capture and removal. In: 2012 4th International conference on intelligent human-machine systems and cybernetics (IHMSC). IEEE, pp 257–261
35. Zhai G, Zhang J, Yao Z (2013) Circular orbit target capture using space tether-net system. Math Probl Eng
36. Zhai G, Qiu Y, Liang B et al (2009) System dynamics and feedforward control for tether-net space robot system. Int J Adv Robot Syst 6(2):137–144
37. Zhai G, Qiu Y, Liang B et al (2009) On-orbit capture with flexible tether–net system. Acta Astronaut 65(5):613–623
38. Liu H, Zhang Q, Yang L et al (2014) Dynamics of tether-tugging reorbiting with net capture. Sci China Technol Sci 57(12):2407–2417
39. Benvenuto R, Salvi S, Lavagna M (2015) Dynamics analysis and gnc design of flexible systems for space debris active removal. Acta Astronaut 110:247–265
40. Huang P, Zhang F, Ma J et al (2015) Dynamics and configuration control of the maneuvering-net space robot system. Adv Space Res 55(4):1004–1014
41. Huang P, Hu Z, Zhang F (2016) Dynamic modelling and coordinated controller designing for the manoeuvrable tether-net space robot system. Multibody Syst Dyn 36(2):115–141
42. Gao Z, Yang D, Min H et al (2012) Kinematics analysis and simulation on formation-based space net dragging process. In: 2012 International conference on control engineering and communication technology (ICCECT). IEEE, pp 455–460
43. Guo J, Gill E (2013) DelFFi: formation flying of two CubeSats for technology, education and science. Int J Space Sci Eng 1:113–127
44. Huang P, Liu B, Zhang F (2016) Configuration maintaining control of three-body ring tethered system based on thrust compensation. Acta Astronaut 123:37–50
45. Fedi Casas M (2015) Dynamics and control of tethered satellite formations in low-Earth orbits. PhD thesis, Universitat Politecnica de Catalunya
46. Williams T, Moore K (2002) Dynamics of tethered satellite formations. Adv Astronaut Sci 112:1219–1235
47. Sarychev VA (1967) Positions of relative equilibrium for two bodies connected by a spherical hinge in a circular orbit. Cosmic Res 5:314
48. Lorenzini EC (1987) A three-mass tethered system for micro-g/variable-g applications. J Guid Control Dynam 10(3):242–249
49. Pizarro-Chong A, Misra AK. Dynamics of a multi-tethered satellite formation, In: Proceedings of the AIAA/AAS astrodynamics specialist conference and exhibit, Providence, Rhode Island, AIAA [AIAA 04-5308]
50. Kumar KD, Yasaka T (2004) Rotation formation flying of three satellites using tethers. J Spacecr Rocket 41(6):973–985
51. Avanzini G, Fedi M (2013) Refined dynamical analysis of multi-tethered satellite formations. Acta Astronaut 84:36–48
52. Cai Z, Li X, Zhou H (2015) Nonlinear dynamics of a rotating triangular tethered satellite formation near libration points. Aerosp Sci Technol 42:384–391
53. Pizarro-Chong A, Misra AK (2008) Dynamics of multi-tethered satellite formations containing a parent body. Acta Astronaut 63(11):1188–1202

54. Guerman AD, Smirnov GV, Paglione P et al (2008) Stationary configurations of a tetrahedral tethered satellite formation. J Guid Control Dyn 31(2):424–428
55. Misra AK, Modi VJ (1987) A survey on the dynamics and control of tethered satellite systems. Tethers in Space 62:667–719
56. Liu G, Huang J, Ma G et al (2013) Nonlinear dynamics and station-keeping control of a rotating tethered satellite system in halo orbits. Chin J Aeronaut 26(5):1227–1237
57. Zhao J, Cai Z (2008) Nonlinear dynamics and simulation of multi-tethered satellite formations in Halo orbits. Acta Astronaut 63(5):673–681
58. Zhao J, Cai Z, Qi Z (2010) Dynamics of variable-length tethered formations near libration points. J Guid Control Dyn 33(4):1172–1183
59. Chung SJ, Kong EM, Miller DW (2005) Dynamics and control of tethered formation flight spacecraft using the SPHERES testbed, In: Proceedings of the AIAA guidance, navigation and control conference, San Francisco, AIAA [AIAA 05-6088]
60. Quadrelli M (2001) Modeling and dynamics analysis of tethered formations for space interferometry. In: Proceedings of the AAS/AIAA space flight mechanics meeting, AAS [AAS 01-231]
61. Nakaya K, Matunaga S (2005) On attitude maneuver of spinning tethered formation flying based on virtual structure method. In: Proceedings of the AIAA guidance, navigation, and control conference and exhibit, San Francisco, AIAA [AIAA 05-6088]
62. Mori O, Matunaga S (2007) Formation and attitude control for rotational tethered satellite clusters. J Spacecr Rocket 44(1):211–220
63. Menon C, Bombardelli C (2007) Self-stabilising attitude control for spinning tethered formations. Acta Astronaut 60(10):828–833
64. Liang H, Wang J, Sun Z (2011) Robust decentralized coordinated attitude control of spacecraft formation. Acta Astronaut 69(5):280–288
65. Cai Z, Li X, Wu Z (2014) Deployment and retrieval of a rotating triangular tethered satellite formation near libration points. Acta Astronaut 98:37–49
66. Wang D, Huang P, Cai J et al (2014) Coordinated control of tethered space robot using mobile tether attachment point in approaching phase. Adv Space Res 54(6):1077–1091

Part I
Maneuverable Space Tethered Net

Chapter 2
Dynamics Modeling and Analysis

Since on-orbit service and active space debris removal become popular research subjects in the space technology field, the mechanism design and dynamics and control of space capture operations have seen a great deal of research. There are many teams studying the capturing and removing methods in different countries, and some achievements have been achieved.

2.1 Mission and System Description

2.1.1 Mission Design

We propose a new conceptual device called the Maneuverable Space Tethered Net (MSTN) for uncooperative space target capture and removal (shown in Fig. 2.1) [2, 3]. The MSTN is effectually a combination of the Space Tethered Net (STN). It is composed of a space net and four Maneuverable Units (MUs) located on the four net corners. The proposed device inherits the maneuverability and long working range from the STN, as well as the easy capture capability from the Space Tethered Net.

Because the MSTN was inspired by the STN, a brief literature review of the STN is appropriate. We discuss scientific progress related to the STN in three parts: Mechanics design and hardware layout, releasing and deployment, and capture and impact analysis.

In this chapter, the dynamic derivation of a complete Maneuverable Space Tethered Net(MSTN) is given. However, due to the flexible weaving structure of the net model, its real physical model is extremely complex, so a classical spatial single tether model will be given first; under given reasonable assumptions, the net model will be simplified as a "11 × 11" net. Then, based on the development of a single tether model, the dynamic model of the Maneuverable Space Tethered Net is

P. Huang and F. Zhang, *Theory and Applications of Multi-Tethers in Space*,
Springer Tracts in Mechanical Engineering,
https://doi.org/10.1007/978-981-15-0387-0_2

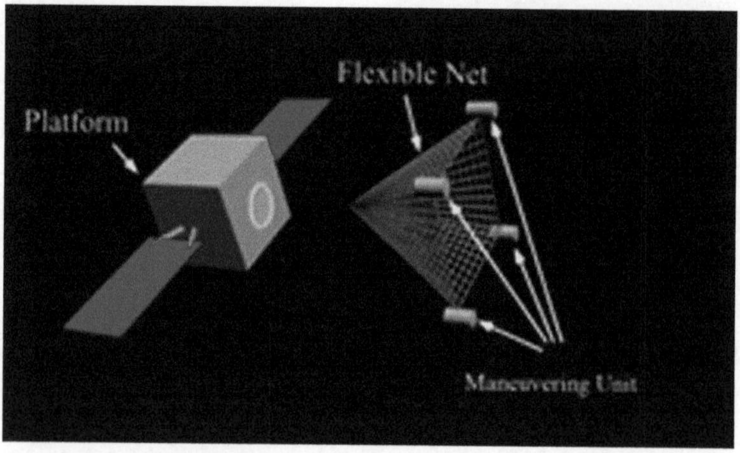

Fig. 2.1 Maneuverable tethered space net robot

eventually extended. In the end, the symmetrical and asymmetrical configurations of the net are respectively analyzed.

2.1.2 System Description

The Maneuverable Space Tethered Net is roughly composed of two parts: The net and four-corner Maneuverable Units(MUs), as shown in Fig. 2.2. The net is a Kevlar knit which is lightweight, easily folded, and stored but strong enough for space missions. The mesh of the net is quadrangular, which represents a trade-off between total mass and stiffness. The net itself is similar in structure to the traditional Tethered Space Net. The side length of the net (L) is 5 m and the nominal side length of the mesh (l) is 5 cm, which satisfies the conclusion $l/L \in [1-5]\%$.

Three primary assumptions are utilized here.

A1. The net is assumed to be a serial of mass points connected via elastic massless links.
A2. Four-corner MUs are assumed to be mass points coinciding w
A3. ith four-corner mass points on the net.
A4. The center of mass of the MSTN system lies on a circular Keplerian reference orbit.

By now, mass–spring model and Absolute Nodal Coordinate Formulation (ANCF) model are the most widely used methods to model the space net. In a mass–spring model, the interaction knots are treated as serial mass points, and the tethers between these knots are regarded as massless spring damping elements. References [2–4] are all based on this mass–spring model, although they focused on

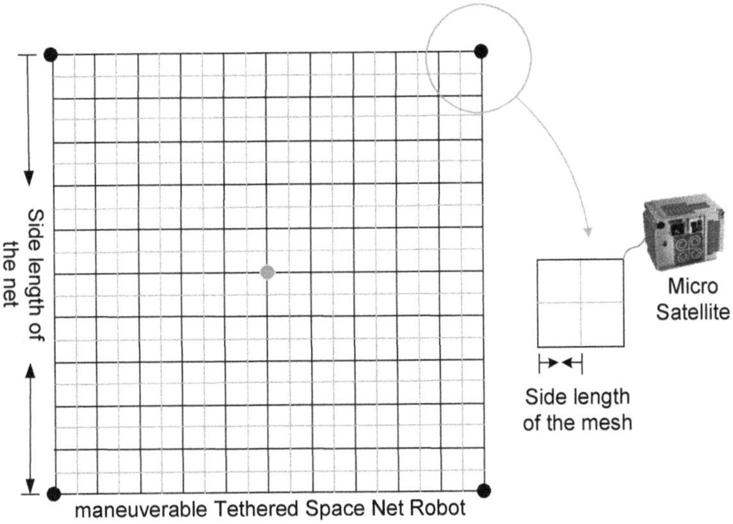

Fig. 2.2 Maneuverable tethered space net

different dynamics characteristics. Shan et al. [5] used ANCF method to model the space net. Compared with the mass–spring model, ANCF model is closer to the real net, and the flexibility of the net can be better simulated. However, the ANCF method originates from the finite element method, which is a salient weakness that it is too complex for controller design.

Therefore, the mass–spring model is used for dynamics modeling in this research. Compared with paper [1], we update the coordinates of the system, so the simulation is more concise. The detailed derivation of the dynamics modeling will be given in this section.

2.2 Dynamics Modeling

2.2.1 Coordinate Description

The generalized coordinates and a schematic diagram of the mathematic model used to describe the MSTN's motion are shown in Fig. 2.3. Under A3, system centroid C is on the circular Keplerian orbit defined by a true anomaly γ, declination δ, and radial coordinate $\mathbf{R}_C$. Rotating coordinate systems $(C - x_O y_O z_O)$ is used with its origin at the center of the system's mass. The coordinate system $x_O - y_O - z_O$ has the x_O-axis radially outward from the Earth along the local vertical, the.y_O.-axis along the local horizontal, and the z_O-axis along the orbit normal completing the right-hand triad.

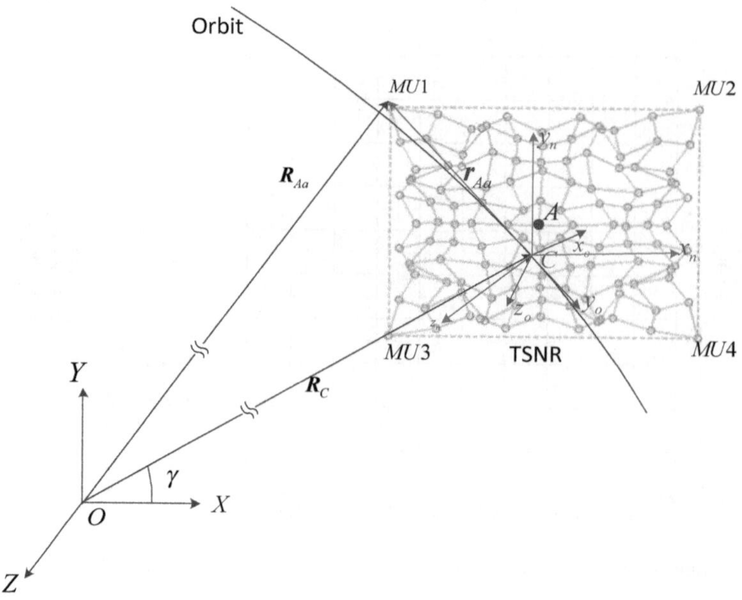

Fig. 2.3 Coordinate of the system

The rotating coordinate system. $C - x_n y_n z_n$. is defined as the body coordinate system of MSTN. It is necessary to introduce a reference plane before any detailed description of this coordinate system. The reference $x_i - y_i - z_i$ plane is parallel to the launch plane defined by four MUs, and the geometrical center A (centroid when MSTN is totally spread out) is also in this plane (Fig. 2.3). The origin of the body frame $C - x_n y_n z_n$ is located at the point A. The x_n-axis is vertical to the line of $MU2 - MU4$ and the y_n-axis is vertical to the line of $MU1 - MU2$. The rotation angles from the orbital frame to the body frame are $(\psi \quad \varphi \quad \theta)$ in sequence; these are defined as the configuration attitude angle of the MSTN.

The rotating coordinate systems $i - x_i y_i z_i$ $(i = Aa, \ldots, Ak, \ldots, Kd)$ are defined as body coordinate systems $MU1 - MU2$ of each mass point (namely, the knitted cross-point of the net) with the origins at each mass point. The coordinate system is obtained via the rotation α_i (in-plane angle) about the x_{i-1}-axis., yielding the axes x'_i, y'_i, and z'_i as well as the rotation β_i (out-of-plane angle) about the z_{i-1}-axis.

2.2.2 Dynamics Modeling

In this chapter, a mesh length and relative attitude angles are used to locate the position of each cross mass point, so the net coordinates are different from the other mass–spring models. As shown in Fig. 2.4, the coordinate of each cross mass point

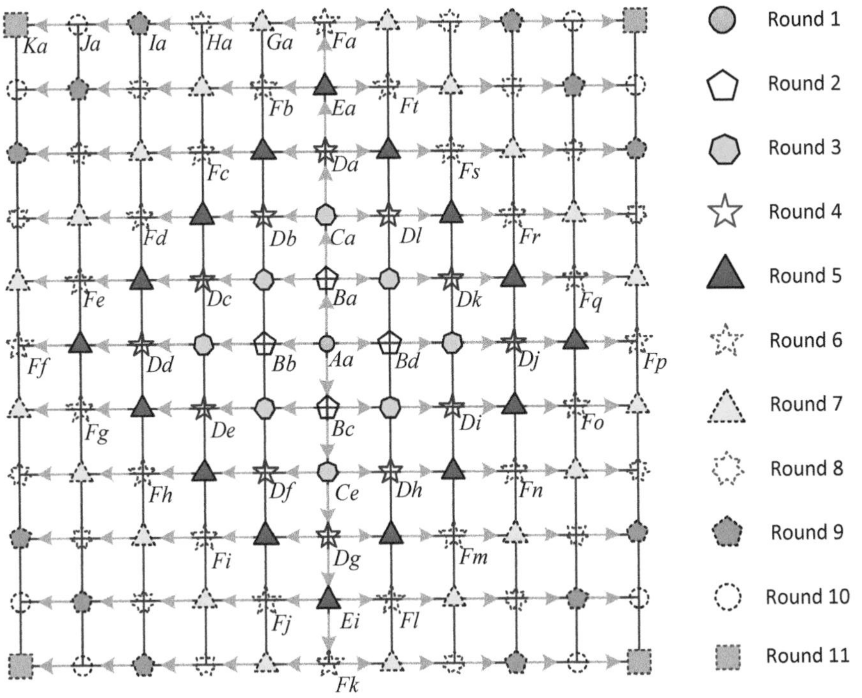

Fig. 2.4 The sequence of coordinates

is obtained by the rotation from the last one; this generates the sequence of coordinates shown in Fig. 2.4. There is only one coordinate in round 1, so its origin coincides with the geometry center A. In round 2, the four gray coordinates are obtained from round 1 (pink) via two rotations (α_i and β_i). The arrowheads in Fig. 2.4 mark the source coordinate of each generated coordinate.

The position of mass point *Aa* with respect to inertia frame is expressed as

$$^{I}\mathbf{R}_{Aa} = {}^{I}\mathbf{R}_{C} + {}^{I}\mathbf{r}_{\Delta} \tag{2.1}$$

where $^{I}\mathbf{R}_{C}$ denotes the vector of centroid of system; $^{I}\mathbf{r}_{\Delta}$ is the vector from centroid point to point *Aa* (geometry center when the net is fully expanded as shown in Fig. 2.3), which can be exactly written as

$$^{I}\mathbf{r}_{\Delta} = \boldsymbol{\Psi}_{o}^{I}\boldsymbol{\Psi}_{n}^{on}\mathbf{r}_{\Delta} \tag{2.2}$$

where $\boldsymbol{\Psi}_{o}^{I}$ and $\boldsymbol{\Psi}_{n}^{o}$ are the rotation matrices from inertia to orbit coordinate, and from orbit to body coordinate, respectively:

$$\boldsymbol{\Psi}_o^I = \begin{pmatrix} \cos\gamma & \sin\gamma & 0 \\ -\sin\gamma & \cos\gamma & 0 \\ 0 & 0 & 1 \end{pmatrix} \tag{2.3}$$

$$\boldsymbol{\Psi}_n^o = \begin{pmatrix} 1 & 0 & 0 \\ 0 & \sin\psi & \cos\psi \\ 0 & \cos\psi & -\sin\psi \end{pmatrix} \begin{pmatrix} \cos\varphi & 0 & -\sin\varphi \\ 0 & 1 & 0 \\ \sin\beta & 0 & \cos\varphi \end{pmatrix} \begin{pmatrix} \cos\theta & \sin\theta & 0 \\ -\sin\theta & \cos\theta & 0 \\ 0 & 0 & 1 \end{pmatrix} \tag{2.4}$$

where γ is the true anomaly.

The positions of other mass points except Aa with respect to inertia frame are expressed as

$${}^I\mathbf{R}_i = {}^I\mathbf{R}_C + \boldsymbol{\Psi}_o^I \boldsymbol{\Psi}_n^o \left({}^n\mathbf{r}_\Delta + \sum_{j=1}^{r(i)-1} {}^n\mathbf{r}_{j\_j+1} \right) \tag{2.5}$$

where $i = Aa, \ldots, Ft, \ldots, Kd$; both the expression j and $r(\cdot)$ here denote the round number shown in Fig. 2.4. The exact vector of ${}^n\mathbf{r}_{j\_j+1}$ has been given in Fig. 2.4.

With the definition of each mass point of the system, the following relation holds:

$$\sum_{i=Aa}^{Kd} m_i \left[{}^I\mathbf{R}_C + \boldsymbol{\Psi}_o^I \boldsymbol{\Psi}_n^o \left({}^n\mathbf{r}_\Delta + \sum_{j=1}^{r(i)-1} {}^n\mathbf{r}_{j\_j+1} \right) \right] = M^I\mathbf{R}_C \tag{2.6}$$

where m_i is the mass of each mass point of the net (when $i \neq Ka, Kb, Kc, Kd$), and the mass of corner MUs (when $i = Ka, Kb, Kc, Kd$), respectively; ${}^I\mathbf{R}_C$ is the position vector of the center of mass of the system with respect to the center of the Earth, which is expressed as

$${}^I\mathbf{R}_C = R\cos\delta\cos\gamma\boldsymbol{i} + R\cos\delta\sin\gamma\boldsymbol{j} + R\sin\delta\boldsymbol{k} \tag{2.7}$$

where $\boldsymbol{i}, \boldsymbol{j}$, and $\boldsymbol{k}$ are the unit vector with respect to the inertial frame, respectively.

According to Eq. (2.7), we derived the relative position with respect to the center of mass of the MSTN system:

$${}^I\mathbf{r}_\Delta = \boldsymbol{\Psi}_o^I \boldsymbol{\Psi}_n^{o\,n}\mathbf{r}_\Delta = -\frac{1}{M}\sum_{i=Aa}^{Kd} m_i \boldsymbol{\Psi}_o^I \boldsymbol{\Psi}_n^o \sum_{j=1}^{r(i)-1} {}^n\mathbf{r}_{j\_j+1} \tag{2.8}$$

So Eq. (2.1) is rewritten as

$$ {}^{I}\mathbf{R}_{Aa} = {}^{I}\mathbf{R}_{C} - \frac{1}{M}\sum_{i=Aa}^{Kd} m_i \boldsymbol{\Psi}_{o}^{I} \boldsymbol{\Psi}_{n}^{o} \sum_{j=1}^{r(i)-1} {}^{n}\mathbf{r}_{j\_j+1} \tag{2.9} $$

As an important conclusion from the single tether's research, a lumped mass–spring model is selected for the dynamics modeling, which is a better trade-off of precision and computability. Therefore, the tension–strain relation can be written as

$$ {}^{I}\boldsymbol{F}_{ij} = \begin{cases} (-EA|\xi| - a\dot{r}_{ij})^{I}\hat{\boldsymbol{r}}_{ij} & \xi > 0 \\ 0 & \xi \leq 0 \end{cases} \tag{2.10} $$

where $\xi = r_{ij} - l$, l is the nominal side length of net mesh; r_{ij} is the spatial distance of adjacent point i and j; α is the damping coefficient; A is the sectional area of the cord; E is the Young's modulus of the cord; ${}^{I}\hat{\boldsymbol{r}}_{ij}$ is the unit vector from i to j.

The sum of internal force is

$$ \sum_{i=1}^{n}\sum_{j=1}^{n} \mathbf{F}_{ij} = 0 \tag{2.11} $$

The momentum of the whole system can be derived as

$$ \mathbf{p} = \sum_{i=1}^{n} m_i {}^{I}\dot{\mathbf{R}}_i \tag{2.12} $$

where $i = 1, \ldots, n$ is a concise expression of $i = Aa, \ldots, Ft, \ldots, Kd$; $\mathbf{p}$ is the momentum of the MSTN, namely a particle system; ${}^{I}\dot{\mathbf{R}}_i$ is the velocity of each mass point derived in Eq. (2.5). Then the moment of momentum of this particle system can be expressed as

$$ \mathbf{L} = \sum_{i=1}^{n} \left(\mathbf{r}_i \times m_i {}^{I}\dot{\mathbf{R}}_i\right) = \boldsymbol{J}\boldsymbol{\omega} \tag{2.13} $$

where $\boldsymbol{\omega}$ is an equivalent angular velocity of the whole net. We define this equivalent angular velocity to describe the equivalent angular motion, which is important for further controller design when the configuration maneuvering is considered.

So according to Eq. (2.10), the motion equations of the system can be expressed as

$$ \dot{\mathbf{p}}_n = \sum_{i=1}^{\zeta} \boldsymbol{f}_i + \boldsymbol{G}_n + \boldsymbol{F}_{ne} \tag{2.14} $$

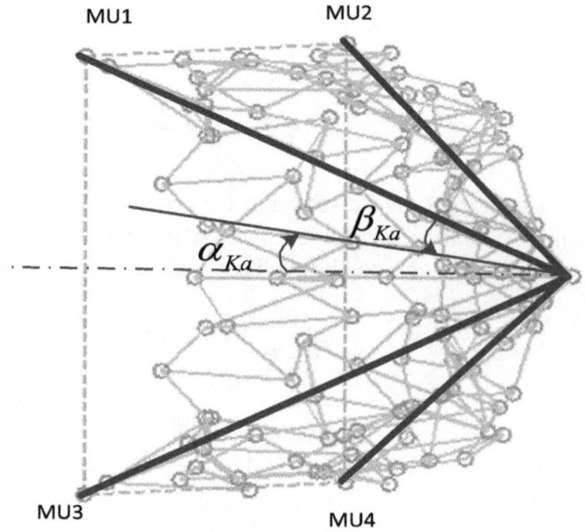

Fig. 2.5 Practical attitude angles of the MUs

where $i = 1, \ldots, \zeta$ is a concise statement of adjacent subscripts of the nth mass point, which can be easily acquired in Fig. 2.4; $\boldsymbol{f}_i$ is the tension–strain working on the ith point mass; $\boldsymbol{G}_n$ is the gravity force of the nth mass point; and $\boldsymbol{F}_{ne}$ is the external force such as the orbital perturbation and the thrust from the MU. In this paper, all the perturbations are assumed to be disturbances, which will be discussed in control scheme.

Before the dynamics analysis and controller design, several free-flying tests have been simulated to prove the mathematical models. A remark here is that, because of the tiny gravity in Geostationary Orbit (GEO), the momentum with respect to the gravity direction is not conserved. The simulation results verify that the momentum is conserved in two directions when all the external forces are not involved.

The in-plane and out-of-plane angle defined in ${}^{n}\mathbf{r}_{\Delta}$ is a relative attitude relationship between the coordinates. In the dynamics analysis and controller design, the real in-plane and out-of-plane angle of the four-corner MUs (as defined in Fig. 2.5) are more applicable to our analysis; they are noted as $\tilde{\alpha}_i$ and $\tilde{\beta}_i$ $(i = Ka, Kb, Kc, Kd)$.

The real in-plane and out-of-plane angle of the MU1 is obtained as

$$\begin{cases} \tilde{\alpha}_{Ka} = \arcsin \dfrac{{}^{n}z_{Ka}}{\sqrt{{}^{n}y_{Ka}^{2} + {}^{n}z_{Ka}^{2}}} \\ \tilde{\beta}_{Ka} = \arcsin \dfrac{{}^{n}x_{Ka}}{\sqrt{{}^{n}x_{Ka}^{2} + {}^{n}y_{Ka}^{2} + {}^{n}z_{Ka}^{2}}} \end{cases} \tag{2.15}$$

where the position of MU1 $({}^{n}x_{Ka} \quad {}^{n}y_{Ka} \quad {}^{n}z_{Ka})$ with respect to the net body frame is derived as

$$\begin{cases} {}^{n}x_{Ka} = \sum\limits_{r(Aa)}^{r(Fa)} (-l\sin\beta_i) + \sum\limits_{r(Aa)}^{r(Fa)} (-l\cos\beta_i) \\ {}^{n}y_{Ka} = \sum\limits_{r(Aa)}^{r(Fa)} (-l\cos\alpha_i\cos\beta_i) + \sum\limits_{r(Aa)}^{r(Fa)} (-l\sin\beta_i) \\ {}^{n}z_{Ka} = \sum\limits_{r(Aa)}^{r(Fa)} (l\sin\alpha_i) + \sum\limits_{r(Aa)}^{r(Fa)} (0) \end{cases} \tag{2.16}$$

2.3 Dynamics Analysis of the Flexible Net

2.3.1 Dynamics Simplification

The real in-plane and out-of-plane angles of each corner MU are highly complex, as evidenced by Eqs. (2.15) and (2.16). To acquire a direct and simple analysis result for the motion of each MU, a simplified motion equation can be built from the aforementioned full dynamics. The simplified model is only used for dynamics analysis in this section. We assume that there are only five mass points (*Ka*, *Kb*, *Kc*, *Kd*) to compose the MSTN system, which can be exactly found in Fig. 2.4. Besides, the elasticity of tether is ignored as well. The longitudinal oscillation makes the analysis far more completed. If the system without elasticity is stable, it is still stable when the elasticity is involved, because a stable state indicates that the positions of all the mass points are constant. In this case, the elasticity of tether is certainly a zero. On the other hand, if the system is not stable, it is still unstable when the elasticity is involved.

From the classic dynamics analysis in single tether case, it is known that the notable dynamics behavior is the oscillation, which is always studied by in-plane and out-of-plane angles of the tether, but not the positions [6, 7]. So in this paper, the same method is used to analysis the MSTN, a multi-tethers case. The assumptions and coordinates used in Lagrangian method is exactly the same as ones in Newton method. So, the motion equations derived by Lagrangian mechanics are exactly the same as ones derived by Newton mechanics in Eq. (2.14) [8]. For dynamics analysis, newly generalized coordinates are defined. A brief derivation of the dimensional equations of motion in generalized coordinates is given in Appendix A.

Based on dimensional equations, nondimensional equations are further studied for dynamics analysis. It is a standard practice in tethered system to formulate the equations of motion in nondimensional form. According to [9] and [10], the orbit radius to the center of mass R_C, and orbit angular velocity $\dot{\gamma}$ are expressed as

$$R_C = \frac{a(1-e^2)}{k} \tag{2.17}$$

$$\dot{\gamma} = \sqrt{\frac{\mu}{a^3(1-e^2)^3}}k^2 \tag{2.18}$$

where $k = 1 + e\cos\gamma$. When in a circular orbit, we get $e = 0$. Then (2.17) and (2.18) retrograde as $\dot{\gamma} = \sqrt{\mu/R_C^3}$. Equations of motion with respect to time are transformed to ones with respect to true anomaly angle, which can be derived as

$$\frac{\mathrm{d}()}{\mathrm{d}t} = \frac{\mathrm{d}()}{\mathrm{d}\theta}\frac{\mathrm{d}\theta}{\mathrm{d}t} = \theta'\Omega \tag{2.19}$$

where $\Omega = \dot{\gamma}$ is the true anomaly angular velocity. Meanwhile, the length of mesh is also transformed via $\Lambda = l/L_r$, where L_r is the reference length of a tether (make up of mesh), Λ is the nondimensional tether length.

The nondimensional equations are then given by

$$\begin{aligned}
&\frac{m_i}{M}\left(\sum_{\substack{j=1\\ j\neq i}}^{4} m_j\right)L_r^2\alpha_i''\cos^2\beta_i - m_iL_r^2\cos\beta_i\left[\sum_{\substack{j=1\\ j\neq i}}^{4}\alpha_j''m_j\cos\beta_j\right]\\
&-2\frac{m_i}{M}\left(\sum_{j=1 j\neq i}^{4} m_j\right)L_r^2(1+\alpha_i')\beta_i'\sin\beta_i\cos\beta_i\\
&+m_iL_r^2\left\{\begin{array}{l}\beta_i'\sin\beta_i\left[\sum\limits_{\substack{j=1\\ j\neq i}}^{4} m_j(1+\alpha_j')\cos\beta_j\right]\\ +\cos\beta_i\left[\sum\limits_{\substack{j=1\\ j\neq i}}^{4} m_j(1+\alpha_j')\beta_j'\sin\beta_j\right]\end{array}\right\}\\
&+2\frac{m_i}{M}\left\{\begin{array}{l}\sin\alpha_i\left[\sum\limits_{\substack{j=1\\ j\neq i}}^{4} m_j\cos\alpha_j\cos(\beta_i-\beta_j)\right]\\ -\cos\alpha_i\left[\sum\limits_{\substack{j=1\\ j\neq i}}^{4} m_j\sin\alpha_j\right]\end{array}\right\} = Q_{\alpha_i}/\Omega^2
\end{aligned} \tag{2.20}$$

$$\begin{aligned}
&\frac{m_i}{M}\left(\sum_{\substack{j=1\\ j\neq i}}^{4} m_j\right) L_r^2\beta_i'' - m_i L_r^2\left[\sum_{\substack{j=1\\ j\neq i}}^{4} m_j\beta_j''\right] \\
&+\frac{m_i}{M}\left(\sum_{\substack{j=1\\ j\neq i}}^{4} m_j\right) L_r^2(1+\alpha_i')^2\sin\beta_i\cos\beta_i \\
&- m_i L_r^2(1+\alpha_i')\sin\beta_i\left[\sum_{\substack{j=1\\ j\neq i}}^{4} m_j(1+\alpha_j')\cos\beta_j\right] \\
&-2\tfrac{m_i}{M}\cos\alpha_i\left[\sum_{\substack{j=1\\ j\neq i}}^{4} m_j\cos\alpha_j\sin\left(\beta_i-\beta_j\right)\right] = Q_{\beta_i}/\Omega^2
\end{aligned} \tag{2.21}$$

where $i = 1, 2, 3, 4$ denotes the four MUs on MSTN; Ω is the angular rate of true anomaly.

A planar unperturbed case is studied as a beginning of the analysis. The in-plane angle motion equations are given as

$$\begin{aligned}
\alpha_i'' = {}&\Phi_1\left(\sin\alpha_i\sum_{\substack{j=1\\ j\neq i}}^{4}\cos\alpha_j - \cos\alpha_i\sum_{\substack{j=1\\ j\neq i}}^{4}\sin\alpha_j\right) \\
&+\Phi_2\sum_{k=1}^{4}\left(\sin\alpha_k\sum_{\substack{j=1\\ j\neq k}}^{4}\cos\alpha_j - \cos\alpha_k\sum_{\substack{j=1\\ j\neq k}}^{4}\sin\alpha_j\right) i = 1, 2, 3, 4
\end{aligned} \tag{2.22}$$

where $\Phi_1 = \dfrac{2(m+2m_l)^2}{m(4m+\frac{1}{4}m_l)}$ and $\Phi_2 = \dfrac{2(m+2m_l)^2}{\frac{1}{4}m_l(4m+\frac{1}{4}m_l)}$.

2.3.2 Dynamics Analysis of the Symmetrical Flexible Net

As defined in [11], a chaotic system is one critically dependent on the initial conditions. The system performs differently with different initial conditions, so the motion of the space tether is a classic chaotic question [6, 12].

We investigated a symmetrical configuration of the MSTN prior to the asymmetrical configuration. The initial states $(a_1\ \ a_1'\ \ a_2\ \ a_2'\ \ a_3\ \ a_3'\ \ a_4\ \ a_4')$

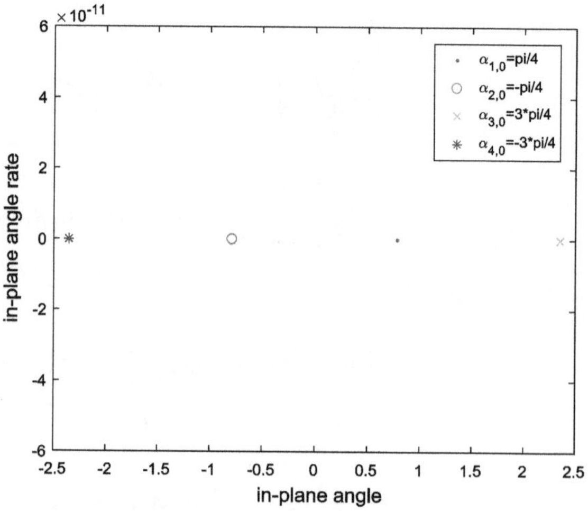

Fig. 2.6 Symmetric initial conditions

are given as $(\pi/4 \quad 0 \quad -\pi/4 \quad 0 \quad 3\pi/4 \quad 0 \quad -3\pi/4 \quad 0)$, which are obviously symmetrical; the phase trajectory for this initialization has been shown in Fig. 2.6. The MSTN system is inherently stable. From a series of phase trajectory analysis, it is clear that a symmetrical initial condition (with no initial angular rate) results in a stable configuration of the MSTN. This can be expressed as

$$\alpha_{1,0} = \hat{\alpha}, \quad \alpha_{2,0} = -\hat{\alpha}, \quad \alpha_{3,0} = \pi - \hat{\alpha}, \quad \alpha_{4,0} = -(\pi - \hat{\alpha}), \quad \alpha'_{1,0} = \alpha'_{2,0} = \alpha'_{3,0} = \alpha'_{4,0} = 0 \tag{2.23}$$

where $\hat{\alpha}$ is an arbitrary angle.

To proof this conclusion, a differential equation of the phase trajectory is derived as

$$\frac{\partial \alpha'_i}{\partial \alpha_i} = \frac{\alpha''_i}{\alpha'_i} \quad i = 1, 2, 3, 4 \tag{2.24}$$

Give zero to $\partial \alpha'_i$ and $\partial \alpha_i$. With the consideration of MSTN's configuration, a specific solution is given a:

$$\begin{cases} \alpha_1 + \alpha_2 = 0 \\ \alpha_1 + \alpha_3 = \pi \\ \alpha_1 - \alpha_4 = \pi \end{cases} \tag{2.25}$$

The solution is called the singularity of differential equation of the phase trajectory, which is also the stable point of the system.

2.3.3 Dynamics Analysis of the Asymmetrical Flexible Net

The initial states of $\left(\alpha'_{1,0} \quad \alpha_{2,0} \quad \alpha'_{2,0} \quad \alpha_{3,0} \quad \alpha'_{3,0} \quad \alpha_{4,0} \quad \alpha'_{4,0}\right)$ in the asymmetrical case are identical to those in the symmetrical case; a slight deviation is added to MU1 for the latter. In this case, $\alpha_{1,0}$ is initialized as $\pi/4.001$, $\pi/4.003$, $\pi/4.005$, $\pi/4.007$, $\pi/4.009$, $\pi/4.011$, $\pi/4.013$, and $\pi/4.015$, respectively. The other three initializations are the same as the symmetrical case.

According to the phase plane in Fig. 2.7, some interesting characteristics are noted here. The first is that MU1's motion is periodic, where each periodic trajectory is slightly different than the previous one. Second, all the phase trajectories seem to pass one point, i.e., a stable point in Fig. 2.7. However, a zoomed-in local drawing indicates that all the trajectories pass next to, rather than exactly through, this point. It can thus be inferred that the motions of MUs are gradually divergent.

To proof this deduction, a simulation with 200 NT (nondimensional time) is plotted in Figs. 2.8 and 2.9, where the periodic motion of the MU1 indeed gradually diverge with time. The three intersection points in Fig. 2.8, namely point A, point B, and point C, correspond to the three points in Fig. 2.10. All these intersection points exist at $\alpha'_1 = 0$. In these two figures, it is obvious that the motion of MU1 is periodic. The simulation step is 0.01 NT and the motion period is 23 steps in our simulations, so the motion period is approximately 0.23 NT.

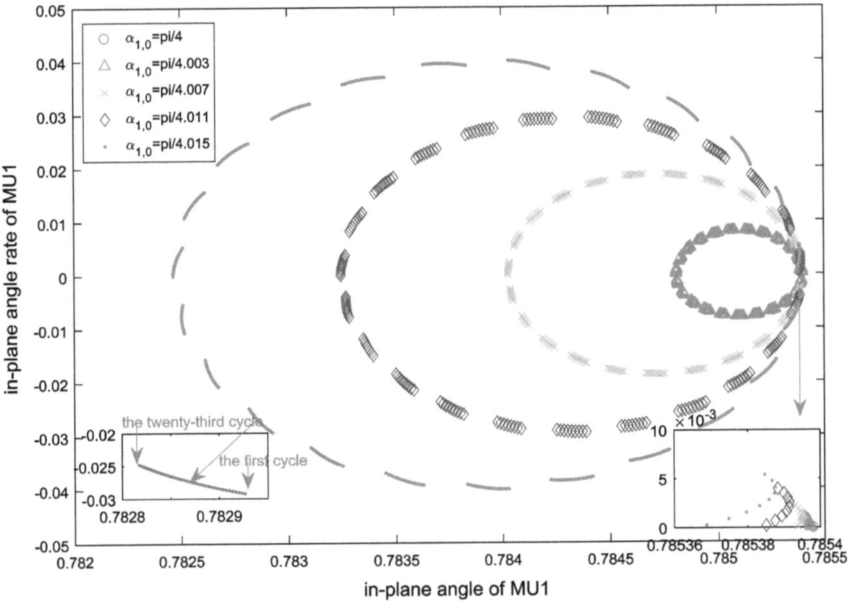

Fig. 2.7 Phase plane of MU1

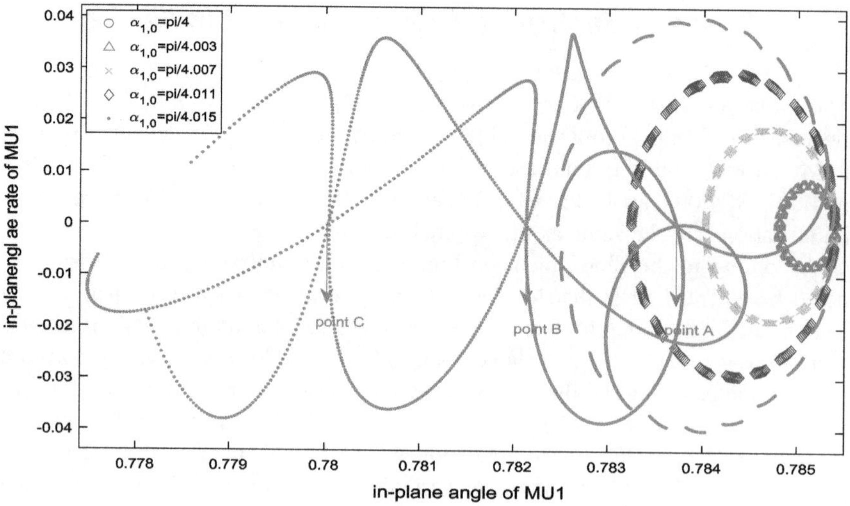

Fig. 2.8 Phase plane of MU1 with 200NT

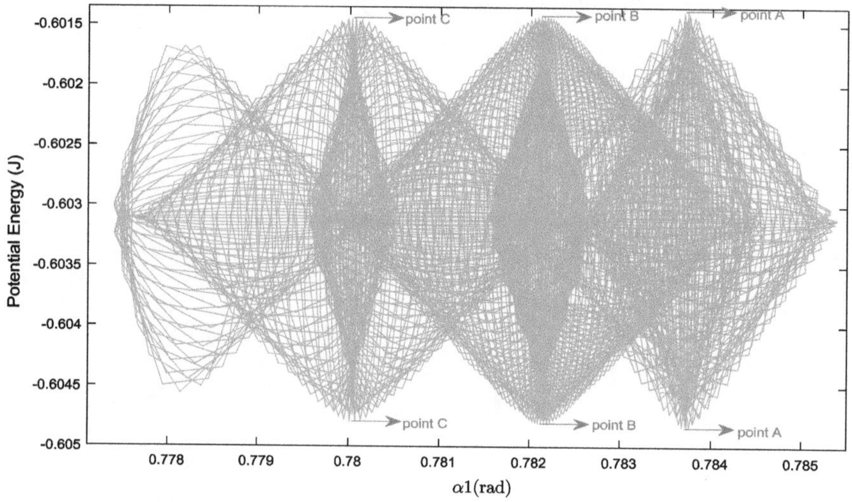

Fig. 2.9 Potential energy of MU1

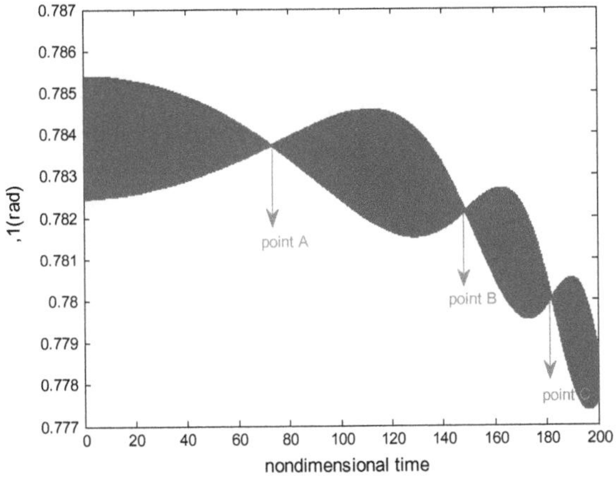

Fig. 2.10 In-plane angle of MU1

For further analysis, an energy integral of Eq. (2.22) is derived as

$$\frac{\alpha_i'^2}{2} + W_{\alpha_i}(\alpha_1, \alpha_2, \alpha_3, \alpha_4) = E_i \quad i = 1, 2, 3, 4 \tag{2.26}$$

where $W_{\alpha_i}(\alpha_1, \alpha_2, \alpha_3, \alpha_4) = \sum_{j=1, j \neq i}^{4} \cos(\alpha_i - \alpha_j)$ is the potential energy of the MUi, and the total energy of each MU is expressed as E_i. The potential energy of MU1 has been plotted in Fig. 2.9 with initial conditions as $(\pi/4.015 \quad 0 \quad -\pi/4 \quad 0 \quad 3\pi/4 \quad 0 \quad -3\pi/4 \quad 0)$, which are same as the last set in Fig. 2.8.

The phase trajectory of each in-plane angle is actually dependent on the corresponding potential energy W_{α_i}. Theoretically, the minimum potential energy corresponds to the center point in the phase portrait while the maximums correspond to the saddle points [7, 13]. Our case is a bit more complicated than this, however. There are three pairs of minimums and maximums, each of which corresponds to an intersection point in the phase portrait. In other words, the maximums no longer correspond to the saddles. The motion of MU1 can be divided into three conditions: $E_i > W_{\alpha_i}$, $E_i < W_{\alpha_i}$, and $E_i = W_{\alpha_i}$. The total energy is dependent on initial conditions.

As aforementioned, the initial conditions of Fig. 2.10 and Fig. 2.11 are $(\pi/4.015 \quad 0 \quad -\pi/4 \quad 0 \quad 3\pi/4 \quad 0 \quad -3\pi/4 \quad 0)$. Compared with symmetric case, it is obvious that a slight deviation is added to $\alpha_{1,0}$. Besides the angular motion of MU1 in Fig. 2.10, the corresponding angular motions of other three MUs are plotted in Fig. 2.11. A visualized conclusion is that the vibrations of α_1 and α_4 are inherent, while the oscillations of α_3 and α_2 are excited. To verify this conclusion, an additional three simulations are done. The same deviation is added to $\alpha_{2,0}$, $\alpha_{3,0}$, and $\alpha_{4,0}$, respectively. The results demonstrate that a deviation on $\alpha_{1,0}$ (or $\alpha_{4,0}$) will immediately cause oscillations of α_4 (or α_1) with the same amplitude, and excites

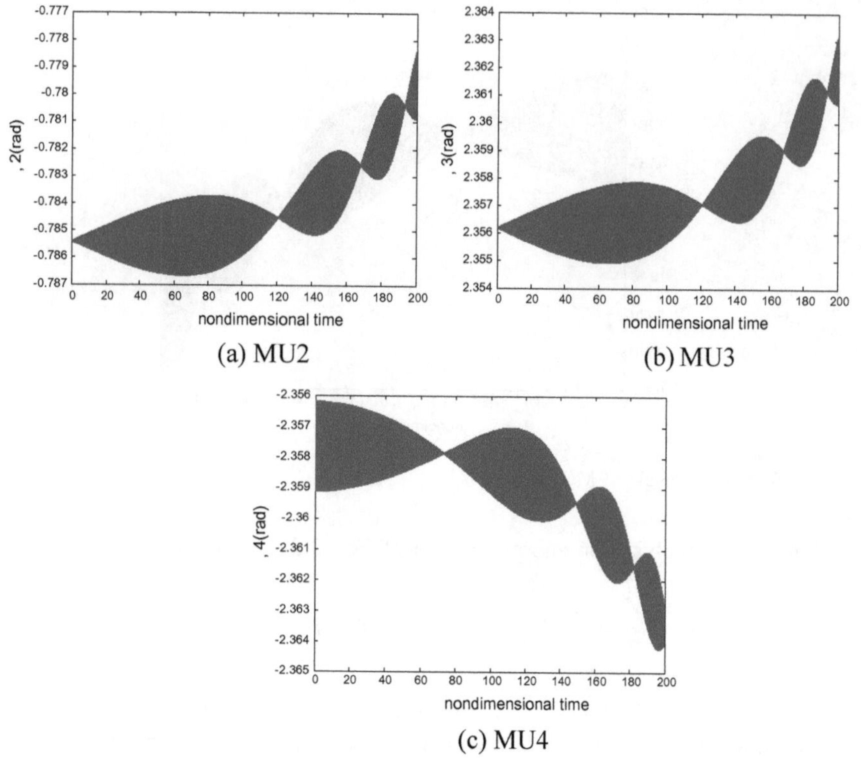

Fig. 2.11 In-plane angles

oscillations of α_2 and α_3 from zeros. Similarly, a deviation on $\alpha_{2,0}$ (or $\alpha_{3,0}$) will immediately cause oscillations of α_3 (or α_2) with the same amplitude, and excites oscillations of α_1 and α_4 from zeros.

The motion of MSTN in an asymmetrical configuration is chaotic. This chaos is inevitable due to shooting error, unpredictable perturbance, and/or installation error. The chaotic oscillation of the space tether may result in net shrinkage if the control scheme is not employed appropriately, which, as discussed above, is a fatal shortcoming of the traditional, unmaneuverable space net. We developed a control strategy to remedy this, as discussed in the next section.

In Figs. 2.7, 2.8, and 2.9, phase trajectories of MU1 are plotted with different initial conditions. These phase portraits indicate that angular motion of each MU is irregular and chaotic, if the initializations are asymmetric. However, this conclusion is based on the long-term simulation. The simulation time of Figs. 2.7 and 2.10 are both 200*NT*. In our design, the MSTN is applied on GEO orbit. According to the nondimensionalization, the nondimensional time is $\tau = \dot{\theta}t$, and one orbit is $2\pi\tau$. However, in our project, the maximum flying distance of the MSTN is 200 meters. Accordingly, the in-plane angle motion of each MU can be assumed to have regular periodic motions.

References

1. Benvenuto R, Salvi S, Lavagna M (2015) Dynamics analysis and GNC design of flexible systems for space debris active removal. Acta Astronaut 110:247–265. https://doi.org/10.1016/j.actaastro.2015.01.014
2. Zhang F, Huang P, Releasing dynamics and stability control of maneuverable tethered space net. IEEE/ASME Trans Mechatron. https://doi.org/10.1109/tmech.2016.2628052
3. Huang P, Hu Z, Zhang F (2016) Dynamic modelling and coordinated controller designing for the manoeuvrable tether-net space robot system. Multibody Syst Dyn 36(2):115–141. https://doi.org/10.1007/s11044-015-9478-3
4. Botta EM, Sharf I, Misra AK et al (2016) On the simulation of tether-nets for space debris capture with vortex dynamics. Acta Astronaut 123:91–102. https://doi.org/10.1016/j.actaastro.2016.02.012
5. Shan M, Guo J, Gill E (2017) Deployment dynamics of tethered-net for space debris removal. Acta Astronaut 132:293–302. https://doi.org/10.2016/j.actaastro.2017.01.001
6. Misra AK (2008) Dynamics and control of tethered satellite systems. Acta Astronaut 63:1169–1177. https://doi.org/10.1016/j.actaastro.2008.06.020
7. Aslanov VS, Ledkov AS (2012) Dynamics of tethered satellite systems. Elsevier, Cambridge, UK
8. Spiegel MR (1982) Theory and problems of theoretical mechanics. McGraw Hill, pp 188–189
9. Williams P, Chris B, Pavel T (2003) Tethered planetary capture: controlled maneuvers. Acta Astronaut 53(4):681–708. https://doi.org/10.1016/S0094-5765(03)00111-5
10. Aslanov VS (2010) The effect of the elasticity of an orbital tether system on the oscillations of a satellite. J Appl Math Mech 74(4):416–424. https://doi.org/10.1016/j.jappmathmech.2010.09.007
11. Poincare H (1912) Calcul des Probabilite as (in French). Gauthier-Villars, Paris
12. Aslanov VS (2009) Chaotic behavior of the biharmonic dynamics system. Int J Math Math Sci 2009:1–18. https://doi.org/10.1155/2009/319179
13. Aslanov VS (2016) Chaos behavior of space debris during tethered tow. J Guid Control Dyn 39(10):2398–2404. https://doi.org/10.2514/1.G001460

Chapter 3
Folding Pattern and Releasing Characteristics

The shooting parameters of the TSNR, including the folding pattern, the direction of shooting, and the value of shooting are very important, which influence the performance of system during free-flying phase. To find an appropriate releasing scheme, we will discuss these, respectively. The releasing dynamics is studied and an optimal selecting scheme for the TSNR releasing is given based on a full analysis of net folding scheme, MUs' shooting conditions. First, the motion dynamics of a TSNR when it is totally deployed is studied, which can be considered as a standard free flying of the system. Then, three-candidate folding schemes are proposed based on the classic origami dynamics and analysis. Criterion are presented to evaluate the free flying of the TSNR after releasing. So, three-candidate folding schemes are compared and analysis is based on this criterion. Based on the decided folding scheme, initial shooting conditions of the four MUs are discussed later, including the initial shooting angles and velocities.

3.1 Folding Pattern of the Stored Flexible Net

3.1.1 Selection Criteria of the Folding Pattern

Three-candidate folding schemes are proposed based on the classic origami dynamics and analysis in this section. First, we will give some definitions about the folding process [1]:

(1) Peak-valley: The peak refers to the outward folding, while the valley refers to the inward folding used in the folding process;
(2) Crease: All the peaks and valleys will have a crease. Each fold in the folding process means that the orientation of any side of the folded object has changed;

P. Huang and F. Zhang, *Theory and Applications of Multi-Tethers in Space*, Springer Tracts in Mechanical Engineering,
https://doi.org/10.1007/978-981-15-0387-0_3

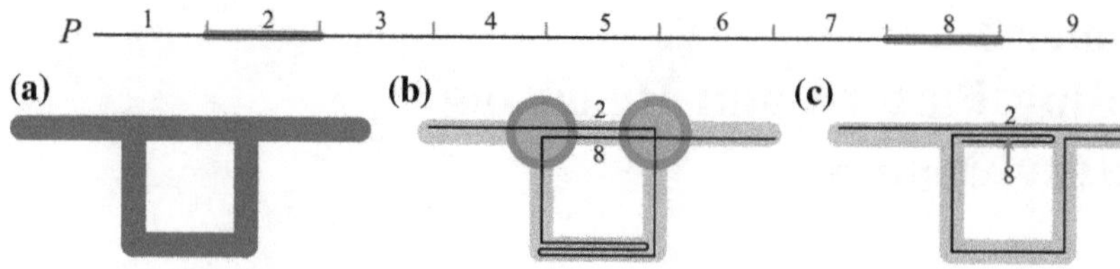

Fig. 3.1 Folding status

(3) Vertices: In the folding process, every two nonparallel creases will produce a vertex, the two vertical creases will produce planar vertices, and non-vertical two creases will produce nonplanar vertices;

All folding process and folded states must be physically meaningful. For example, a piece of white paper is divided into nine equal parts, and paragraphs 2 and 8 are overlapped, as shown in (a) of Fig. 3.1. (b) and (c) denote two folding methods respectively, and the folding pattern in (b) is not practically meaningful. Different paper layers cannot cross each other during the folding process, so only the method (c) is feasible.

Some other mathematical definitions:

(1) The N-division point of the line graph P

Knowing the start point (x_1, y_1) and endpoint (x_2, y_2) of a straight line segment, and the coordinates (x_0, y_0) of the N bisector, we can draw the following conclusions:

If $x_1 \neq x_2$ and $y_1 \neq y_2$, thus

$$x_0 = \frac{(n-1)x_1 + x_2}{n}, y_0 = \frac{(n-1)y_1 + y_2}{n} \tag{3.1}$$

If $x_1 \neq x_2$ and $y_1 = y_2$, thus

$$x_0 = \frac{(n-1)x_1 + x_2}{n}, y_0 = y_1 \tag{3.2}$$

If $x_1 = x_2$ and $y_1 = y_2$, thus

$$x_0 = x_1, y_0 = \frac{(n-1)y_1 + y_2}{n} \tag{3.3}$$

(2) Plane moving geometric transformation

When the point $p(x, y)$ on the plane P is moved to $p(x', y')$, the corresponding projection along x-axis is dx, and the projection along y-axis is dy, then there is a translation matrix transformation:

$$[x' \quad y' \quad 1] = [x \quad y \quad 1]\begin{bmatrix} 1 & 0 & 0 \\ 0 & 1 & 0 \\ dx & dy & 1 \end{bmatrix} \tag{3.4}$$

(3) Plane expansion geometry transformation

Through the expansion and contraction of the plane P, the point $p(x, y)$ is moved to $p(x', y')$, the corresponding stretching (or compression) multiple is s_x in x-axis, and the corresponding stretching (or compression) multiple is s_y in y-axis, then the stretching matrix is transformed into

$$[x' \quad y' \quad 1] = [x \quad y \quad 1]\begin{bmatrix} s_x & 0 & 0 \\ 0 & s_y & 0 \\ 0 & 0 & 1 \end{bmatrix} \tag{3.5}$$

(4) Plane rotation geometry transformation

By performing a rotational transformation θ on the plane P and moving the point $p(x, y)$ to $p(x', y')$, there is a rotation matrix transformation:

$$[x' \quad y' \quad 1] = [x \quad y \quad 1]\begin{bmatrix} \cos\theta & \sin\theta & 0 \\ -\sin\theta & \cos\theta & 0 \\ 0 & 0 & 1 \end{bmatrix} \tag{3.6}$$

Reference [2] has given a very detailed research about how to choose the mesh topology for different requirements, and how to fold the web. However, the typical folding patterns used in space web cannot be applied here, because space web is released by rotation, while the TSNR is shooting directly.

After a simulation of 10 possible folding candidates, 3 basic requirements for the folding scheme are concluded as follows:

(1) The folded net should be exactly symmetric.
(2) The four-corner MUs should be on the top of the net package.
(3) The net package should be easily unfolded.
(4) The folding shape could be 3D, but cannot be a closed hollow.

After a preliminary screening, two folding patterns are chosen as final candidates, namely square and star patterns. The folding processes of these two have been drawn in the following subsections.

3.1.2 Square Pattern

The folding process of the square pattern has been drawn in Fig. 3.2. The full line in the figure is defined as outward folding, that is, creases projecting toward the outside of the plane based on the plane of the net. The dash line is the inward folding reversed with the outward folding. And four maneuverable units have been marked on the way with a blue regular hexagon. As shown in (a) of Fig. 3.2, the quadrilateral net completes the first longitudinal fold in accordance with the "Peak-valley" crease and becomes the appearance in (b). And then, according to the "Peak-valley" crease, the first transverse direction fold is completed in (c). The final version of the net in square pattern is shown in (d). The four MUs are all on the top of the final shape. The net is finally stored in a plane in square case.

Remark 1 All the research contents of the folded part of the fly net, in this paper, are intended to explain the folding method, and then select the appropriate folding method through the comparison of various data indicators in the unfolding process. Therefore, in order to give the folding step more clearly, the folded size in this paper is not compressed to a minimum. In reality, according to the effective volume of the launch tube, the net can be further folded and compressed in the folding manner described in this section.

According to the description of the folding process, the entire folding process is clear and simple, and the folding creases can be arbitrarily increased according to the size of the final launch tube. The folded net wrap is still a fully symmetrical regular quadrilateral, with four maneuverable units at four corners. With the final package size (0.5 m $\times$ 0.5 m, for comparison, the final net package under the latter two folding methods is with the size), the number of the process creases is $4 \times 2 \times 2 = 16$ according to Eqs. (3.1)–(3.6) and the total flip area in grids is

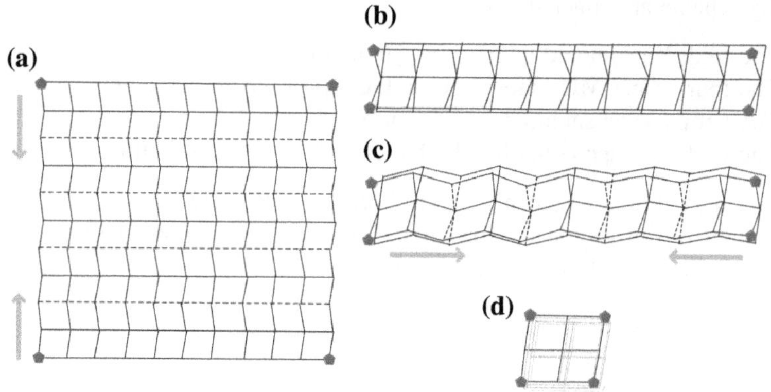

Fig. 3.2 Square pattern

$\sum_{i=1}^{100} 180°\zeta(i)$, where i represents the ith mesh of the net, $\zeta(i)$ is the turn number of mesh i in the whole process.

In our simulations, all of these folding schemes perform well, as long as with suitable other initial conditions. The initial velocities of four MUs are $(0.1 \quad 0.1 \quad 0.1)$ m/s, $(0.1 \quad 0.1 \quad -0.1)$ m/s, $(0.1 \quad -0.1 \quad 0.1)$ m/s, and $(0.1 \quad -0.1 \quad -0.1)$ m/s, respectively. The spread process of the square pattern is shown in Fig. 3.3.

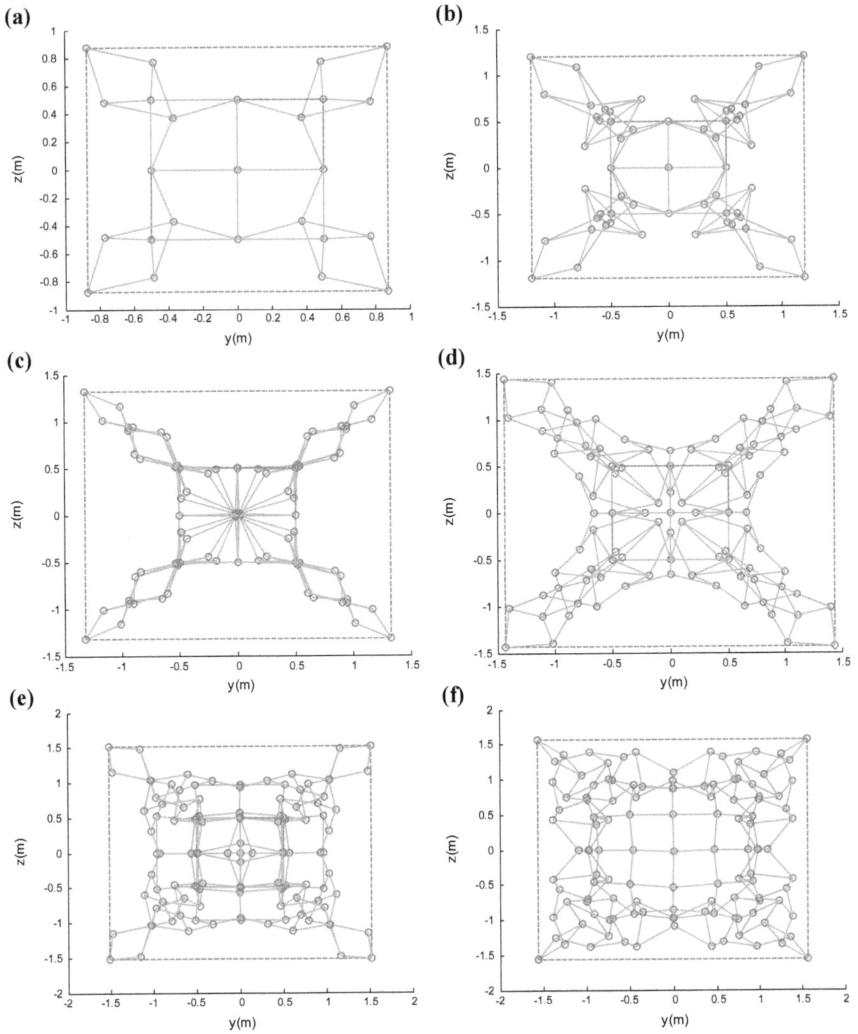

Fig. 3.3 Spread process of the square pattern

3.1.3 Star Pattern

The folding process of the star pattern has been drawn in Fig. 3.4. The full line in the figure is defined as outward folding, and the dash line is the inward folding the same with square pattern. And four maneuverable units have been marked on the way with a blue regular hexagon. As shown in (a) of Fig. 3.4, the quadrilateral net completes the fold in accordance with the "Peak-valley" crease and becomes the appearance in (b) which is in a three-dimensional configuration similar to the "star". And then, according to the "Peak-valley" crease, the four branches of the "star" are folded completely in (c). The four MUs are all on the top of the final shape. The net is finally stored in a plane in square case.

It should be noted that after the completion of this folding method, the net package becomes a three-dimensional configuration with an opening bottom and an empty interior. It has been clearly stated that the hollow configuration with around closed cannot be used as a folded wrap configuration for net because the interior hollow sections cannot be supported. However, the semi-hollow configuration of the "star" folded can easily be supported by a raised structure. In addition, as with the square pattern, the entire folding process is clear and simple, and the folding creases can be arbitrarily increased according to the size of the final launcher to realize the desired size. When the creases of the first step are sufficient, the thickness of the package after folding of the net is negligible. The so-called hollow problem does not confuse the actual launcher design.

Same as square pattern, the net is stored in a 3D space in star case as shown in Fig. 3.4, and the spread process of the star pattern is shown in Fig. 3.5.

The folded net wrap is still a fully symmetrical regular quadrilateral, with four maneuverable units at four corners. The final package size $0.5\,\mathrm{m} \times 0.5\,\mathrm{m}$ is the same as the square pattern, the number of the process creases is $4 \times 4 + 4 \times 4 = 32$ and the total flip area in grids is $\sum_{i=1}^{100} 180°\zeta(i)$.

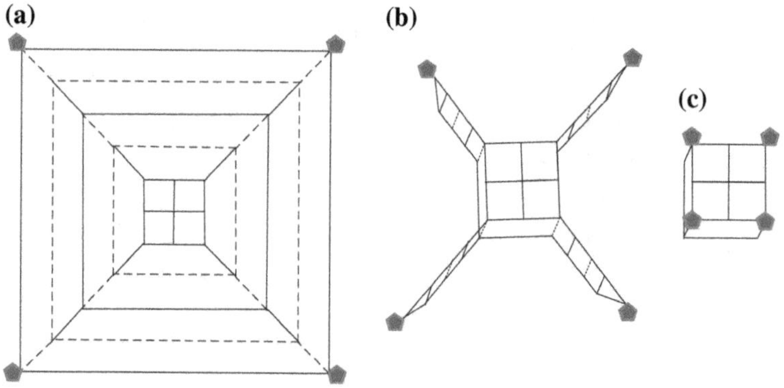

Fig. 3.4 Star pattern

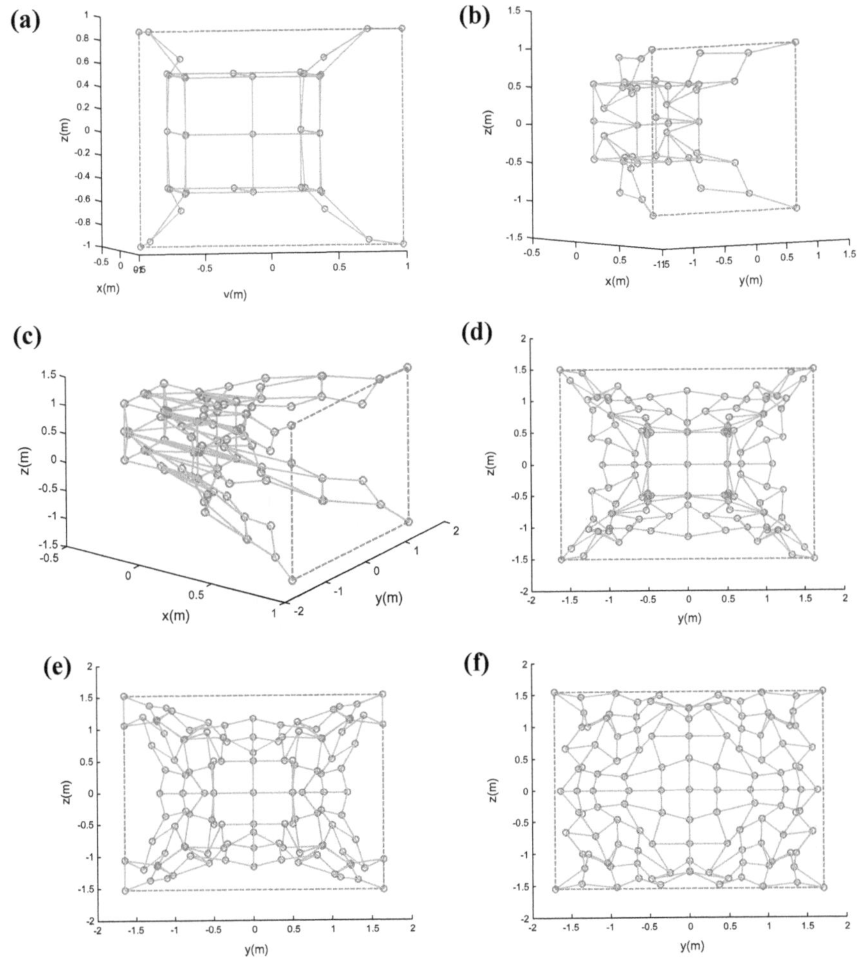

Fig. 3.5 Spread process of the star pattern

3.1.4 Cross-Shaped Pattern

The folding step of the cross-shaped pattern is shown in Fig. 3.6. As with the previous folding rule, the full line in the figure is defined as outward folding, and the dash line is the inward folding. Four maneuverable units have been marked on the way with a blue regular hexagon. First, as shown in (a), the quadrilateral net completes the folding process in accordance with the “Peak-valley” crease and becomes the appearance in (b), which is a three-dimensional configuration resembling a “cross”. And then, according to the “Peak-valley” crease, the four branches of the “cross” are folded completely in (c).

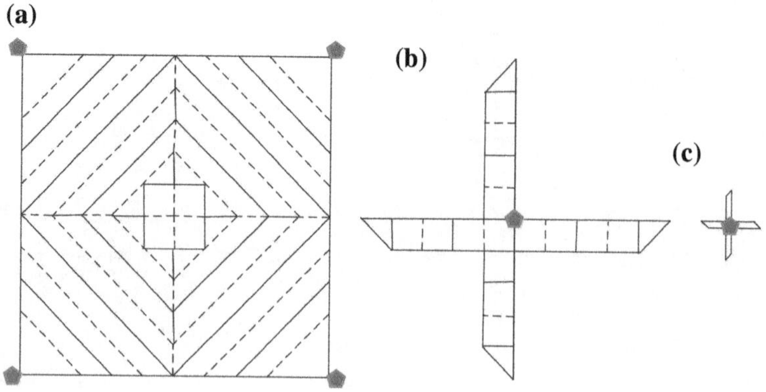

Fig. 3.6 Cross-shaped pattern

It is worth noting that in Fig. 3.6b, c, there is only one hexagonal symbol that represents the maneuverable unit. When the folded thickness of the net is ignored, the four maneuverable units are coincident in the cross-shaped pattern. The reason why the cross-shaped folding method is selected as candidate folding is that the four maneuverable units are "dispersed" at the instant of launching with both square and star patterns, and the maneuverable units are "centralized" in the cross-shaped pattern. The spread process of the cross-shaped pattern is shown in Fig. 3.7.

To evaluate the performance, three criteria are defined first:

(1) Maximum opening area—the largest area of a quadrangle composed of four MUs;
(2) Maximum depth—the depth of a net when it reaches the maximum opening area;
(3) Maximum volume—the actual volume of a net when it reaches maximum opening area.

A comparison based on three criteria of these three folding patterns is given in Table 3.1. It is clear that the maximum area of square pattern is bigger than the star pattern, while other two criteria are worse. Unquestioningly, the maximum opening area is the first importance, because it goes straight for whether a target capturing could succeed or not. However, when the maximum areas are almost the same (like our case), the maximum depth and volume of net are important. A deeper depth (or a bigger volume) indicates less impact between mesh knot and the target satellite. According to a comprehensive survey, star pattern is the final choice.

Through the simulation analysis and comparison of the above three-candidate folding methods, it can be derived that the arrangement relationship of these three-candidate folding methods considering the maximum opening area index is: cross-shaped pattern > square pattern > star pattern; considering the maximum depth indicator, the arrangement relationship is: star pattern > square pattern > cross-shaped pattern; the arrangement relationship with the indicator of

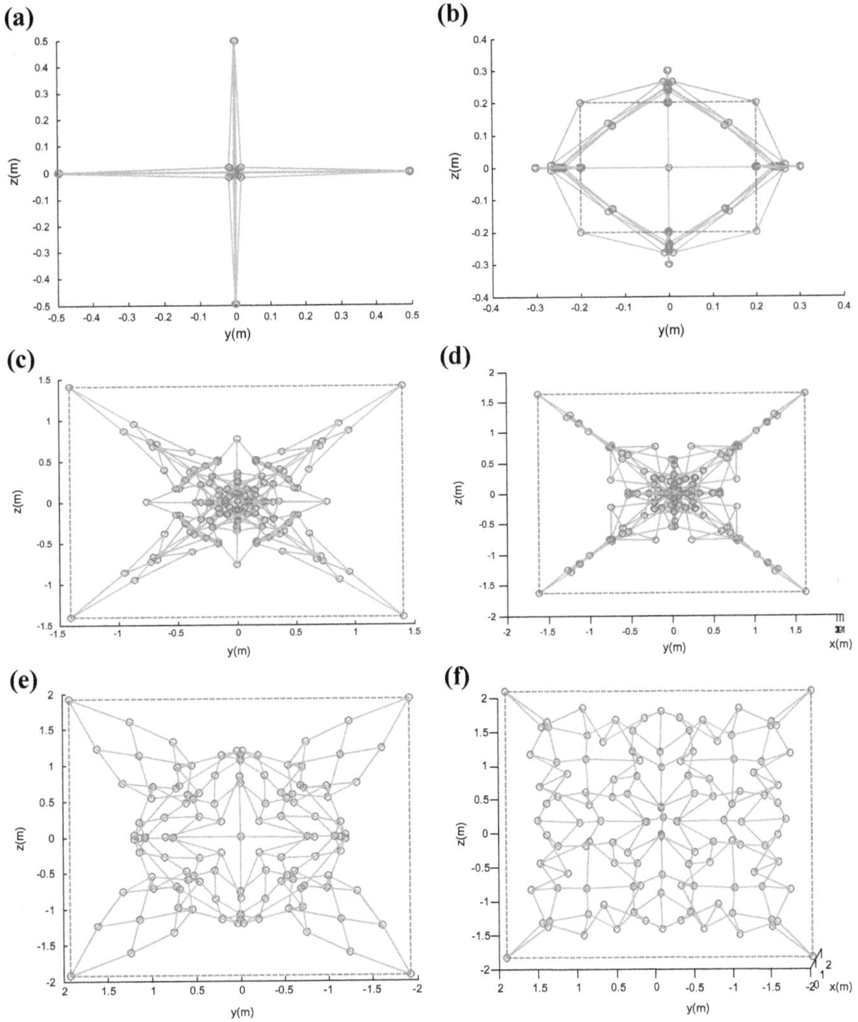

Fig. 3.7 Spread process of the cross-shaped pattern

Table 3.1 Parameters of a comparison of three patterns

Items	Square pattern	Star pattern	Cross-shaped pattern
Maximum opening area (m^2)	10.6010	9.8734	14.3801
Maximum depth (m)	1.1101	1.1490	0.7150
Maximum volume (m^3)	6.0284	7.1048	5.8815

maximum effective volume is: star pattern > square pattern > cross-shaped pattern. It is obvious that the larger the opening area of the net, the smaller the depth of the deployment net. The maximum depth of the star pattern is the largest owing to that its initial folded configuration is a three-dimensional configuration and the center of the net lags behind the launching plane of the maneuverable units. The maximum opening area of the cross-shaped pattern is the largest because the maneuverable unit is located in the center of the folded net and wrapped around by the net, thus the net will be rapidly deployed by the ejection of the maneuvering unit.

Judging from the parameters of a comparison of the three patterns, the differences between the three-candidate folding methods are not very obvious, and they all meet the folding requirements of the net. Therefore, an appropriate folding method can be selected according to the specific design of the launch tube, the type, and size of the target satellite to be captured.

3.2 Natural Flight of the System Without Control

The folded net is stored in releasing device assembled on platform satellite before released. This shooting plane is treated as a reference plane. We define a shooting coordinate system to describe the initial parameters of the TSNR. The origin is at the center of mass of TSNR. Since the net is exactly symmetric and the folded TSNR is also symmetric, this folding centroid is just the centroid when it totally deployed. In this shooting coordinate, x-axis is vertical to the storing plane, while y-axis is in the releasing plane and vertical to one of the sidelines of the TSNR, and z-axis is under the right-hand triad. The x-direction is toward to the target space debris. The basic parameters used in this manuscript are listed in the Table 3.2.

In this research, the TSNR is assumed to apply on GEO. In GEO, most of the satellites are communication satellites, meteorological satellites, missile warning satellites, and repeater satellite. These satellites are expensive and important to military and commercial strategies, which are valuable for retrieval. Our focus is indeed on these satellites, those who are expensive and useful passive satellites,

Table 3.2 Parameters of the system

Parameters	Magnitude
Mass of each MU (kg)	10
Mass of the net (kg)	0.5
Length of the sideline of net (m)	5
Length of the sideline of each mesh (m)	0.5
Material of the net	Kevlar
Diameter of the net's thread (mm)	1
Young's modulus of the net's thread (MPa)	13000
Damping factor of the net's thread	0.1
Orbit altitude (km)	36000

Fig. 3.8 The motions of the TSNR's sidelines

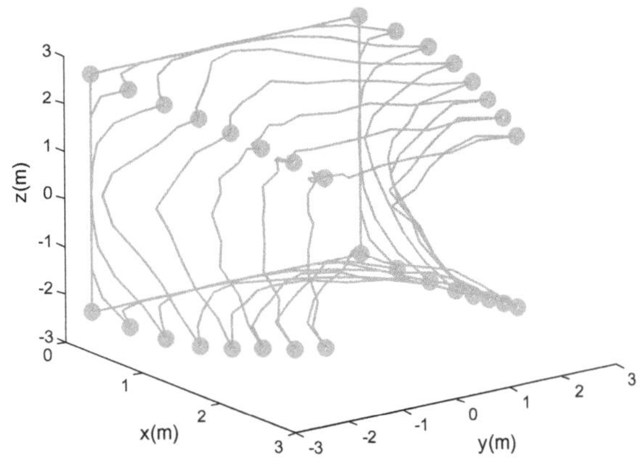

Table 3.3 Releasing performance

Time (s)	Net's opening area (m^2)	Flying distance (m)
1	6.2141	0.4880
2	6.0634	0.9451
5	4.4310	1.9551
9	2.1954	3.1286
15	0.1041	4.7232
16.52	0	5.1118

which are valuable to retrieve for other use. So our solution is focused solely on GEO.

Before the discussion of the releasing specifics of TSNR, we would like to begin with a basic case. The TSNR is assumed to be stored completely unfolded, and the initial velocities of the four MUs are exactly vertical to the shooting plane. Based on this simple assumption, we could know the basic motions of the TSNR with initial velocity $v_{x,0} = 0.5\,\text{m/s}$. A simulation result of the net's sideline after releasing is plotted in Fig. 3.8, and the exact data is shown in Table 3.3.

From Fig. 3.8 and Table 3.3, it is obvious that the MUs are moving toward the gathering direction, because of the elasticity of each tether segment. Although initial velocity is only working on *x*-direction, the shrinks on *y*- and *z*-direction are inevitable.

As described in [3], the net is simplified to be a matrix with masses, which is connected via elastic massless links. Each tether segment of mesh is abstracted as a spring and a damper. In the study of mechanical vibration, *n* connected springs have *n* degrees of freedom. After the launch, the net is released and deployed through the elasticity, while the damper also dissipates the energy. During the whole deployment, this "stretching-shrink (wind)" phenomenon always exists, although we have selected Kevlar, who is a better trade-off of lightweight, strong, and flexible. That is to say, the elasticity brings in additional energy consumption, so the initial releasing conditions should beis rewritten as $-EA|\xi| - a\dot{r}_{ij} = \alpha_0/2 + \sum_{j=1}^{\infty}\left(\alpha_j \cos j\omega t + \beta_j \sin j\omega t\right)$, where α_0, α_j and β_j can be defined as $\alpha_j = \frac{2}{T}\int_{-T/2}^{T/2} F(t)\cos j\omega t \mathrm{d}t$ $(j = 0, 1, 2, \cdots)$, and $\beta_j = \frac{2}{T}\int_{-T/2}^{T/2} F(t)\sin j\omega t \mathrm{d}t$ $(j = 1, 2, \cdots)$. In this situation, the net will shrink and come into a chaos.

3.3 Shooting Conditions of the Maneuverable Units

3.3.1 Shooting Angle

We design the shooting angle as the included angle between storing plane and the shooting velocity. Different releasing directions (30°, 45°, and 60°) will be simulated and analysed. We have introduced the definition of the maximum opening area of the net previously, and we will give another definition for the releasing of the net. The time of totally deployment (t_d) is defined as the deploying time when the significant opening area of net is the maximum. So corresponding to the previous initial shooting velocities (30°, 45°, and 60°), the simulation results are $t_d = 38.23$ s, $t_d = 18.89$ s, and $t_d = 40.17$ s, respectively.

Besides the deployed time, another consideration is the motion trajectories. Theoretically, after the time of total deployment, the net will be in chaos gradually, which is not what we expected. This is because the MUs are retarded after the shot, while the net's mass points are accelerated. Another important conclusion is that the shape of the net is not symmetrical, albeit the force distribution seems symmetrical during free flying.

The deploying time is just the turning point of the deployment. Before this point, the TSNR is free flying and the net is gradually deployed. At this point, the velocities of the MUs reach zeros, and the net expands in the maximum area. So after this point, the net begins to shrink. Therefore, the traditional unmaneuverable space net should enclose the target before it shrinks. It is obvious that the time of TSNR's chaos for $\alpha = 45°$ case is earlier than two other cases, since its totally deployed time is also earlier than other two. Finally, we choose $\alpha = 60°$ as the shooting angle. In this initial situation, both the maximum opening area and the volume of net are satisfied. Besides, longer free-flying time indicates that the controller can be employed later, which is just demanded in fuel consumption.

3.3.2 Shooting Velocity

We have discussed the folding pattern and shooting angle of the TSNR for releasing. In this section, the shooting velocity will be analyzed. For traditional unmaneuverable space net, the shooting velocity can be figured out through the optimal control. The dynamics of the system can be rewritten as

$$\dot{\mathbf{x}}(t) = \mathbf{f}[\mathbf{x}(t), t] + \mathbf{B}[\mathbf{x}(t), t]\mathbf{u}(t) \tag{3.7}$$

where $\mathbf{x}(t) \in R^n$ is the states of the TSNR, $\mathbf{u}(t) \in R^m$ is the control input ($n = 121$, $m = 4$ in this paper), and $\mathbf{B}[\mathbf{x}(t), t]$ is the $n \times m$ matrix. The terminal constraint for deployment can be understood as the states of net at capture moment, which is $\xi[\mathbf{x}(t_f), t_f] = 0$. So, the shooting velocity is just the optimal controller $\mathbf{u}^*(t)$ who drives the system from the initial state

$$\mathbf{x}(t_0) = \mathbf{x}_0 \tag{3.8}$$

to the final state $\mathbf{x}(t_f)$, and makes the objective function

$$J[\mathbf{u}(t)] = \int_{t_0}^{t_f} \mathrm{dt} \tag{3.9}$$

get the minimum.

However, this method is not always feasible, since the constraint $\xi[\mathbf{x}(t_f), t_f] = 0$ can only restrict the size of the opening area, but not the shape of net. So in this paper, we propose a maneuverable net. Before the controller design, we first analyze the shooting velocity to find an appropriate initial condition for our case.

With star pattern, and $\alpha = 60°$ as the shooting angle, simulation results for different shooting velocities are listed in Table 3.4. We use $|\mathbf{v}_{MUi}| = 0.1$ m/s as the shooting velocity, since the maximum opening area and volume of the net at the deploying time.

Although the control strategy will be involved at the end of the free flying and the states of net can be changed by the MUs, an appropriate releasing condition is necessary. A proper pattern scheme and initial condition could provide a better deployment, and lower fuel consumption. Based on the previous analysis, the conclusions are summarized as follows:

(1) Although we choose the star pattern in this paper, the square pattern is also a correct candidate, which is easier for assemble and launch mechanics design.

Table 3.4 A comparison with different shooting velocities

Items	$\lvert\mathbf{v}_{MUi}\rvert = 0.1$ m/s	$\lvert\mathbf{v}_{MUi}\rvert = 0.5$ m/s	$\lvert\mathbf{v}_{MUi}\rvert = 1$ m/s
Maximum opening area (m^2)	9.8470	8.9433	9.1729
Maximum Depth (m)	0.8971	1.0701	1.0690
Maximum volume (m^3)	6.9037	6.7496	6.6708

(2) Smaller shooting angle makes faster transversal deployment of a net, but slower longitudinal deployment. This is useful for a target which is close and with a large cross-sectional area.
(3) Similarly, larger shooting angle makes slower transversal deployment of a net, but faster longitudinal deployment, which is better for a far target.
(4) Higher shooting velocity leads to longer flying distance before thrusters involved. However, it also leads to an earlier chaos after the turning point (deploying time).

The releasing scheme is not fixed. All of the relevant parameters should be considered, such as shape of the target, relative distance, and the characteristics of the shooting appliance.

References

1. Demaine ED, O'Rourke J (2008) Geometric folding algorithms: linkages, origami, polyhedra. Cambridge University Press
2. Tibert G, Gardsback M (2002) Space webs, final report. Report, Royal institute of Technology, Stockholm, Sweden, AdrianaID: 054109
3. Benvenuto R, Salvi S, Lavagna M (2015) Dynamics analysis and GNC design of flexible systems for space debris active removal. Acta Astronaut 110:247–265

Chapter 4
Centralized Deployment Control

In configuration-keeping phase, the MSTN is an underactuation system in which only four MUs are controllable. All the motions of uncontrollable mass points of the net are treated as disturbance on the MUs. And the classic Adaptive Super-Twisting Sliding Mode Controller scheme is employed to solve the vibration problem.

In configuration-maneuvering phase, The MSTN can be concluded as an underactuated flexible system, which is composed of hundreds of state variables, but only four controllable Units. The oscillations of the knitted tethers caused by the flexibility and elasticity are sensitive to the control forces. The disturbance from the main tether, dynamics uncertainties and oscillations from the flexible net are complex uncertainties and a new control scheme is proposed.

4.1 Configuration-Keeping Control After Releasing

4.1.1 Problem Statement

In Chap. 2, the dynamic modeling and analysis of the MSTN are completed. The kinetic equations of arbitrary braided nodes and MUs of the system are shown in Eq. (4.2). In order to facilitate the controller design, the dynamic formula is rewritten as follows:

$$\begin{cases} \dot{\mathbf{x}}_I = f(\mathbf{x}_I, \mathbf{x}_{II}, t) \\ \dot{\mathbf{x}}_{II} = f(\mathbf{x}_I, \mathbf{x}_{II}, t) + g(\mathbf{x}_{II}, t)u(t) \\ \mathbf{y} = \mathbf{x}_{II}(t) \end{cases} \tag{4.1}$$

where $\mathbf{x}_I \in \mathbb{R}^{n-m}$ $(n > m)$ s all the tether weaving nodes except for the four MUs in the MSTN; $\mathbf{x}_{II} \in \mathbb{R}^m$ represents 4 MUs; $\mathbf{y}$ is the system output, which also is the

P. Huang and F. Zhang, *Theory and Applications of Multi-Tethers in Space*,
Springer Tracts in Mechanical Engineering,
https://doi.org/10.1007/978-981-15-0387-0_4

status of 4 MUs; $u(t)$ represents the controller input; $g(\mathbf{x}, t)$ represents the control function.

$$\dot{\mathbf{p}}_{ij} = \sum \boldsymbol{f}_{ij} + \boldsymbol{G} + \boldsymbol{F}_e + \mathbf{d} \tag{4.2}$$

The constraints of the MUs can be expressed as

$$\begin{aligned} \boldsymbol{\Psi} = \{ & \mathbf{x}_{II} | \mathbf{x}_{II} = [\mathbf{x}_{n-3} \quad \mathbf{x}_{n-2} \quad \mathbf{x}_{n-1} \quad \mathbf{x}_n]^T \in \boldsymbol{R}^m, \\ & \|\mathbf{x}_i - \mathbf{x}_{i+1}\|_2 \le l + \varepsilon, i = n-3, n-2, n-1, \\ & and\ \|\mathbf{x}_{n-3} - \mathbf{x}_n\|_2 \le l + \varepsilon \} \end{aligned} \tag{4.3}$$

where $l + \varepsilon$ is the maximum length of each two adjacent MUs, l is the natural rope length between two adjacent MUs and ε is the maximum elastic stretch of the connecting tether.

Rewrite Eq. (4.1) as follows:

$$\begin{cases} \dot{\mathbf{x}} = \boldsymbol{F}(\mathbf{x}, t) + \boldsymbol{G}(\mathbf{x}, t)\mathbf{u}(t) \\ \mathbf{y} = \mathbf{x}_{II}(t) \end{cases} \tag{4.4}$$

Therefore, each MU exists as a separate subsystem, and all vibrations of the MSTN dynamics are treated as external disturbances, and its concrete expression will be given in the later rewritten system dynamic equation.

Select the sliding mode control sliding mode parameters as

$$\mathbf{s}_i = \mathbf{s}_i(x_{i1}, x_{i2}) = \sum_{j=1}^{2} c^{j-1} x_{ij} \tag{4.5}$$

where $i = 1, 2, 3, 4$ represents four MUs; c represents a sliding surface parameter; x_{i1} and x_{i2} represent the position status and speed status, respectively. Since the control algorithms of the MUs are the same, all subsequent subscripts i are omitted in the subsequent derivation of the control algorithm for the sake of brevity.

The derivative of the sliding mode parameter over time can be expressed as

$$\begin{aligned} \dot{\mathbf{s}} &= \underbrace{\frac{\partial \mathbf{s}}{\partial t} + \left(\frac{\partial \mathbf{s}}{\partial \mathbf{x}}\right)^{\mathrm{T}} \boldsymbol{f}(\mathbf{x})}_{a(\mathbf{x},t)} + \underbrace{\left(\frac{\partial \mathbf{s}}{\partial \mathbf{x}}\right)^{\mathrm{T}} \boldsymbol{g}(\mathbf{x})\mathbf{u}(t)}_{b(\mathbf{x},t)} \\ &= a(\mathbf{x}, t) + u(\mathbf{x}, t) \end{aligned} \tag{4.6}$$

As already mentioned above, the key to this control problem is the complicated disturbance term, which is the system vibration caused by the flexibility of the flying net. In this chapter, if not specifically mentioned, the so-called interference represents the sum of the vibrations caused by the flexibility of the tether. So add the interference term to Eq. (4.6):

$$\dot{\mathbf{s}} = a(\mathbf{x}, t) + u(\mathbf{x}, t) + d(\mathbf{x}, t) \tag{4.7}$$

In order to constrain this distracter, the following assumptions are made:

Assumption 1 Disturbance in Eq. (4.7) $d(\mathbf{x}, t)$ satisfies $\|d(\mathbf{x}, t)\| \leq \sum_{i=1}^{\tau} c_i \|\mathbf{x}\|^i$, where τ and c_i are known terms.

According to references [1, 2] and Assumption 1, $d(\mathbf{x}, t)$ can be classified as vanishing disturbances, and this disturbance is a function of the state. The upper limit can be rewritten as

$$\begin{aligned} \|d(\mathbf{x}, t)\| &\leq \sum_{i=1}^{\tau} c_i \|\mathbf{x}\|^i \\ &\leq c_1 \|\mathbf{x}\| + c_2(\|\mathbf{x}\| + \|\mathbf{x}\|^{\tau_1}) + \cdots + c_{\tau-1}(\|\mathbf{x}\| + \|\mathbf{x}\|^{\tau_1}) + c_\tau \|\mathbf{x}\|^{\tau_1} \\ &\leq c_{a1} \|\mathbf{x}\| + c_{a2} \|\mathbf{x}\|^{\tau_1} \end{aligned} \tag{4.8}$$

where $c_{a1} \triangleq \sum_{i=1}^{\tau-1} c_i$.

Defining the system's state tracking error $\mathbf{e} \in \mathbf{R}^m$ is

$$\mathbf{e} = \mathbf{x}_{II} - \mathbf{x}_{II,c} \tag{4.9}$$

So Eq. (4.7) can be further written as

$$\begin{aligned} \dot{\mathbf{s}} &= a(\mathbf{e}, t) + u(\mathbf{e}, t) + d(\mathbf{e}, t) \\ &= \tilde{a}(\mathbf{e}, t) + u(\mathbf{e}, t) \end{aligned} \tag{4.10}$$

Assumption 2 According to Assumption 1, $d(\mathbf{x}, t)$ can be expressed as $d(\mathbf{x}, t) \leq g[\phi_1(s)]$. According to the Eq. (4.8), we can get

$$g[\phi_1(s)] = c_{a1} \|\mathbf{x}\| + c_{a2} \|\mathbf{x}\|^{\tau_1} \tag{4.11}$$

where $\phi_1(s)$ is a function of the sliding mode variable, which will be specified in the algorithm design.

Assumption 3 Assume that all states in the MSTN are measurable.

Therefore, the research content of this chapter is that a set of controllers need to be designed to ensure that the input–output dynamics equation (4.10) of the MSTN is disturbed by the upper bound $d(\mathbf{x}, t) \leq g[\phi_1(s)]$ (this upper bound is unknown but does exist). Under the effect, it will steadily fly to the target satellite to be captured.

4.1.2 Controller Design

According to the description of the control problem in the previous section, the specific design steps of an improved second-order adaptive sliding mode variable structure controller are given in this section. The controller inherits the high robustness of the traditional sliding mode controller. The structure is simple and easy to implement the characteristics of engineering practice, but also can reduce state chatter, to ensure the stable flight of the flying net robot. In this section, we first introduce the traditional second-order sliding mode variable structure control algorithm, and then give an improved second-order algorithm.

For a given system state equation:

$$\begin{cases} \dot{x}_1 = x_2 \\ \dot{x}_2 = u + f(x_1, x_2, t) \end{cases} \tag{4.12}$$

where u denotes the control input, $f(x_1, x_2, t)$ denotes external disturbances acting on the system, and it is assumed that the interference is bounded, $|f(x_1, x_2, t)| \leq L > 0$. The initial state of the system is $x_1(0) = x_{10}$, $x_2(0) = x_{20}$. The research content is to design a controller $u = u(x_1, x_2)$ in the presence of unknown interference $f(x_1, x_2, t)$, so that the state of the controlled object reaches 0, $\lim_{t \to \infty} x_1, x_2 = 0$.

The sliding mode variables of the design system are

$$\sigma = \sigma(x_1, x_2) = cx_1 + x_2, c > 0 \tag{4.13}$$

Then the system (4.12) dynamic equation is

$$\dot{\sigma} = cx_2 + f(x_1, x_2, t) + u, \sigma(0) = \sigma_0 \tag{4.14}$$

The Lyapunov equation about σ is

$$V = \frac{1}{2}\sigma^2 \tag{4.15}$$

In order to ensure the stability of the system (4.14) at the equilibrium point, the following two conditions need to be met: For any $\sigma \neq 0$, there is $\dot{V} < 0$; $\lim_{|\sigma| \to \infty} V = \infty$.

Since the second condition can always be guaranteed according to Eq. (4.15), only the first condition needs to be proved. In order to ensure global asymptotic stability, condition 1 is further deduced to obtain

$$\dot{V} \leq -\alpha V^{1/2}, \alpha > 0 \tag{4.16}$$

separating the state variables from the above equation and integrating the time $0 \le \tau \le t$ can yield:

$$V^{1/2}(t) \le -\frac{1}{2}\alpha t + V^{1/2}(0) \tag{4.17}$$

$V(t)$ reaches zero within a limited time:

$$t_r \le \frac{2V^{1/2}(0)}{\alpha} \tag{4.18}$$

Therefore, the controller u can control the sliding mode variable to zero within a limited time, and the derivative of the Lyapunov equation to time is

$$\dot{V} = \sigma\dot{\sigma} = \sigma(cx_2 + f(x_1, x_2, t) + u) \tag{4.19}$$

Assuming $u = -cx_2 + v$ and substituting Eq. (4.19), we can get

$$\dot{V} = \sigma(f(x_1, x_2, t) + v) = \sigma f(x_1, x_2, t) + \sigma v \le |\sigma|L + \sigma v \tag{4.20}$$

Order $v = -\rho \text{sgn}(\sigma)$, which

$$\text{sgn}(x) = \begin{cases} 1 & \text{if } x > 0 \\ -1 & \text{if } x < 0 \end{cases} \tag{4.21}$$

as well as

$$\text{sgn}(0) \in [-1, 1] \tag{4.22}$$

For any $\rho > 0$ and add it into the Eq. (4.20), we can get

$$\dot{V} \le |\sigma|L - |\sigma|\rho = -|\sigma|(\rho - L) \tag{4.23}$$

according to the Lyapunov equation (4.15) and the global asymptotic stability condition (4.16), we can get

$$\dot{V} \le -\alpha V^{1/2} = -\frac{\alpha}{\sqrt{2}}|\sigma|, \alpha > 0 \tag{4.24}$$

Combining Eqs. (4.23) and (4.24), we can get:

$$\dot{V} \le -|\sigma|(\rho - L) = -\frac{\alpha}{\sqrt{2}}|\sigma| \tag{4.25}$$

the controller gain ρ is

$$\rho = L + \frac{\alpha}{\sqrt{2}} \tag{4.26}$$

The first-order sliding mode algorithm is

$$u = -cx_2 - \rho \text{sgn}(\sigma) \tag{4.27}$$

It can be seen that the first-order sliding mode control algorithm (4.27) contains discontinuities $\text{sgn}(\sigma)$, which is the root cause of chattering. The purpose of the controller is to control the sliding mode variable σ to 0, but the control u occurs only in the first derivative of the sliding mode variable, so the best way is to use the derivative of the controller as a new controller.

A common system:

$$\dot{x} = a(t,x) + b(t,x)u \tag{4.28}$$

where $x \in \mathbb{R}^n$. The system output is

$$\sigma = \sigma(t,x) \tag{4.29}$$

a, b, σ all have the required derivative orders. For simplicity, consider the simplest case $\sigma, u \in \mathbb{R}$.

The full derivative for time about σ is

$$\dot{\sigma} = \sigma'_t + \sigma'_x \alpha + \sigma'_x bu \tag{4.30}$$

Assuming $\sigma'_x b \equiv 0$, the second-order full derivative of σ is

$$\ddot{\sigma} = \sigma''_{tt} + 2\sigma''_{tx}\alpha + \sigma'_x a'_t + \left[\sigma''_{xx}(a+bu)\right]a + \sigma'_x\left[a'_x(a+bu)\right] \tag{4.31}$$

So

$$\begin{cases} \ddot{\sigma} = h(t,x) + g(t,x)u \\ h(t,x) = \sigma''_{tt} + 2\sigma''_{tx}\alpha + \sigma'_x a'_t + \left(\sigma''_{xx}a\right)a + \sigma'_x\left(a'_x a\right) \\ g(t,x) = \left(\sigma''_{xx}b\right)a + \sigma'_x\left(a'_x b\right) \end{cases} \tag{4.32}$$

It can be obtained from Eqs. (4.30), (4.31) and (4.32) that the relative order of the system is 1 when $\sigma'_x b \neq 0$; and the relative order of the system is 2 when $\sigma'_x b \equiv 0$ and $\left(\sigma''_{xx}b\right)a + \sigma'_x\left(a'_x b\right) \neq 0$.

Assuming that the relative order of the system does exist, the control function u that constitutes σ is a discontinuous feedback. When the relative order of the system is 1, σ is continuous but $\dot{\sigma}$ is discontinuous; when the relative order of the system is 2, $\dot{\sigma}$ is continuous and $\ddot{\sigma}$ is discontinuous. Therefore, the traditional sliding mode control is relative to the first-order, and the second-order sliding mode needs to be realized by the second order of the relatively discontinuous controller.

The dynamic system is still composed of (4.28) and (4.29), where $x \in \mathbb{R}^n$ is the system state, $u \in \mathbb{R}$ is the control input, σ is the only measurement parameter, functions a, b, σ are smooth and has the required derivative order. The purpose of the controller design is to allow $\sigma \equiv 0$ for a limited time.

Assume that the measurable output $\sigma = \sigma(t,x)$ has the second-order for time, $\sigma'_x b \equiv 0$ and $(\sigma''_{xx} b)a + \sigma'_x(a'_x b) \neq 0$. According to Eq. (4.28), find the second derivative of Eq. (4.29), which can be obtained from Eq. (4.32):

$$\ddot{\sigma} = h(t,x) + g(t,x)u \tag{4.33}$$

where $h = \ddot{\sigma}|_{u=0}$ and $g = \frac{\partial}{\partial u}\ddot{\sigma} \neq 0$ are smooth functions.

Assuming:

$$0 < K_m \leq g \leq K_M, |h| \leq C \tag{4.34}$$

Obviously, no feedback controller $u = \varphi(\sigma, \dot{\sigma})$ can solve the above control problem. In fact, satisfy $\sigma \equiv 0$ means satisfying $\ddot{\sigma} \equiv 0$, which shows $\varphi(0,0) = -h(t,x)/g(t,x)$ and $\sigma = \dot{\sigma} = 0$. However, the existence of interference in the system makes $\sigma = \dot{\sigma} = 0$ difficult to implement, because the controller has no effect at all in this simple linear system: $\ddot{\sigma} = c + ku$, $K_m \leq k \leq K_M$, $|c| \leq C$, $\varphi(0,0) \neq -c/k$.

Assume that Eq. (4.34) can be guaranteed globally. Equations (4.32) and (4.34) mean that the dynamic equation is a differential equation consisting of inclusion relations:

$$\ddot{\sigma} \in [-C, C] + [K_m, K_M]u \tag{4.35}$$

The controller of Eq. (4.35) may consider using a second-order sliding mode algorithm to drive the states σ, $\dot{\sigma}$ to zero. The inclusion relationship in Eq. (4.35) indicates that the equation can't "remember" the initial state of the system (4.28).

So, the control problem is to find a feedback controller:

$$u = \varphi(\sigma, \dot{\sigma}) \tag{4.36}$$

which makes the state trajectories in the system (4.35) converge to the cross section $\sigma = \dot{\sigma} = 0$ within a limited time.

The second-order sliding mode control algorithm has achieved a lot of pioneering research. The more common second-order algorithms include Twisting Controller (Fig. 4.1a), Suboptimal Controller (Fig. 4.1b), and Quasi-Continues Controller (Fig. 4.1c).

However, the above three second-order sliding mode algorithms all require real-time measuring $\dot{\sigma}$ or at least $\text{sgn}(\dot{\sigma})$, which is to ensure that $\sigma = \dot{\sigma} = 0$, σ, and $\dot{\sigma}$ must be measured in real time. Although this condition is reasonable, it cannot be

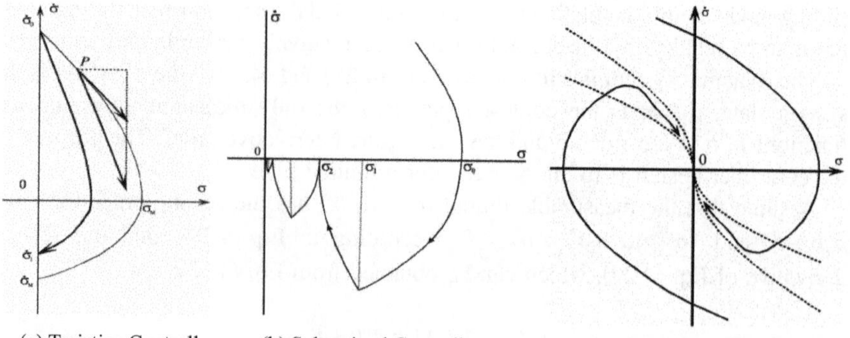

(a) Twisting Controller (b) Suboptimal Controller (c) Quasi-Continues Controller

Fig. 4.1 The second-order sliding mode trajectories

Fig. 4.2 The super-twist second-order sliding mode

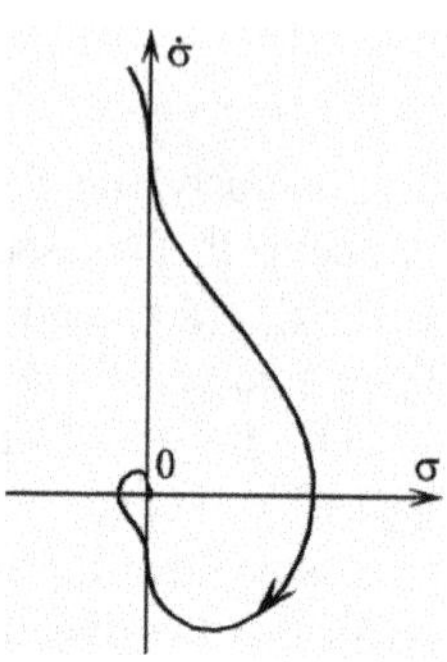

guaranteed in many cases. The super-twist second-order sliding mode control algorithm (Fig. 4.2) only needs to know σ of a sliding mode, but it has the effect of the second-order sliding mode algorithm.

Inspired by Moreno [3], Nagesh [4], and Wang [5], the proposed improved super-twist adaptive sliding mode algorithm is as follows:

$$\begin{cases} u = -\alpha_1\left(v_1|s|^{\frac{1}{2}}\mathrm{sgn}(s) + v_2 s\right) + w \\ \dot{w} = -\alpha_2\left(\frac{1}{2}v_1^2\mathrm{sgn}(s) + \frac{3}{2}v_1v_2|s|^{\frac{1}{2}}\mathrm{sgn}(s) + v_2^2 s\right) \\ \dot{\alpha}_1 = \begin{cases} \frac{2\kappa+\varepsilon}{1+4\varepsilon^2}\phi_1^2(s)\frac{\partial}{\partial s}\phi_1(s) & |s| > 0 \\ 0 & |s| = 0 \end{cases} \\ \alpha_2 = \chi + 2\varepsilon\alpha_1 - \frac{2\varepsilon\chi}{\theta_1} - \frac{2\varepsilon}{\theta_2} \end{cases} \tag{4.37}$$

where u is the control input of the system's dynamic equation (4.10) about the sliding mode parameter; the constants $\theta_1 > 0$, $\theta_2 > 0$ are all positive numbers; the structure parameters $v_1 > 0$, $v_2 \geq 0$, $\chi > 0$ are all positive numbers; α_1 and α_2 are adaptive scalar gain; $\phi_1(s)$ can be specifically defined as $\phi_1(s) = v_1|s|^{\frac{1}{2}}\mathrm{sgn}(s) + v_2 s$;

ε is an arbitrarily small normal number. Similar to the definition of $\phi_1(s)$, for the sake of a more concise and elegant mathematical expression, define $\phi_2(s) = \frac{1}{2}v_1^2 \operatorname{sgn}(s) + \frac{3}{2}v_1 v_2 |s|^{\frac{1}{2}} \operatorname{sgn}(s) + v_2^2 s$.

4.1.3 Stability Analysis of the Closed-Loop System

Assuming there is indeed an unknown constant $\varpi > 0$, $d(\mathbf{x}, t)$ satisfies Assumptions 1 and 2. For any system initial state $\mathbf{x}(0)$ and $\boldsymbol{\sigma}(0)$, under the control algorithm (4.37), the system's sliding mode parameters (4.5) can reach the sliding mode surface $\boldsymbol{\sigma} = \dot{\boldsymbol{\sigma}} = \mathbf{0}$ and remain in the sliding mode surface within a limited time. The proof is as follows:

Bring the control algorithm (4.37) into the system's dynamic equation (4.10), and according to the definitions of $\phi_1(s)$ and $\phi_2(s)$, a simplified dynamic equation can be obtained as

$$\begin{aligned} \dot{s} &= -\alpha_1 \phi_1(s) + w \\ \dot{w} &= -\alpha_2 \phi_2(s) \end{aligned} \tag{4.38}$$

According to the definition of $\phi_1(s)$ and $\phi_2(s)$, we can get

$$\frac{\partial}{\partial s}\phi_1(s) = \frac{1}{2}v_1|s|^{-\frac{1}{2}} + v_2 \tag{4.39}$$

The parameters v_1 and v_2 are already defined in Eq. (4.37), so

$$\frac{\partial}{\partial s}\phi_1(s) \geq 0 \tag{4.40}$$

$$\phi_2(s) = \phi_1(s)\frac{\partial}{\partial s}\phi_1(s) \tag{4.41}$$

Defining $\boldsymbol{\xi} = [\phi_1(s) \quad 0]^T$, the derivative of ξ is

$$\dot{\boldsymbol{\xi}} = \begin{bmatrix} \dot{\phi}_1(s) \\ 0 \end{bmatrix} = \frac{\partial}{\partial s}\phi_1(s)\left\{\begin{bmatrix} -\alpha_1 & 1 \\ 0 & 0 \end{bmatrix}\begin{bmatrix} \phi_1(s) \\ 0 \end{bmatrix} + \begin{bmatrix} w \\ 0 \end{bmatrix}\right\} = \frac{\partial}{\partial s}\phi_1(s)(\mathbf{A}\xi + \boldsymbol{\rho}) \tag{4.42}$$

where $\mathbf{A} = \begin{bmatrix} -\alpha_1 & 1 \\ 0 & 0 \end{bmatrix}$, $\boldsymbol{\rho} = \begin{bmatrix} \rho_1 \\ 0 \end{bmatrix} = \begin{bmatrix} w \\ 0 \end{bmatrix}$.

According to Assumptions 1, 2, and Eq. (4.3), $\boldsymbol{\rho}$ in Eq. (4.42) can be derived $\|\boldsymbol{\rho}\| = \|\rho_1\| = \|w\| \leq g[\phi_1(s)]$.

Define a positive-definite symmetric matrix $\mathbf{P} = \begin{bmatrix} \chi & -2\varepsilon \\ -2\varepsilon & 1 \end{bmatrix}$, where χ is a normal number (scalar). To satisfy the Lyapunov equation, construct an equation as follows:

$$\mathbf{A}^T\mathbf{P} + \mathbf{PA} + \delta\mathbf{I} + g^2\mathbf{C}^T\mathbf{C} + P\mathbf{\Theta}^{-1}\mathbf{P} = \mathbf{Q} \tag{4.43}$$

where $\mathbf{C} = [1 \quad 0]^T$, $\mathbf{\Theta} = \begin{bmatrix} 1 & 0 \\ 0 & 0 \end{bmatrix}$.

So:

$$Q = \begin{bmatrix} -2\alpha_1\chi + 4\varepsilon\alpha_2 + \delta + g^2 + \chi + 4\varepsilon^2 & \chi + 2\varepsilon\alpha_1 - \alpha_2 - 2\varepsilon\chi \\ \chi + 2\varepsilon\alpha_1 - \alpha_2 - 2\varepsilon\chi & -4\varepsilon + 4\varepsilon^2 + \delta \end{bmatrix} \tag{4.44}$$

Constructing a Lyapunov equation $V_1 = \boldsymbol{\xi}^T\mathbf{P}\boldsymbol{\xi}$, its derivative for time is

$$\begin{aligned}
\dot{V}_1 &= \dot{\boldsymbol{\xi}}^T\mathbf{P}\boldsymbol{\xi} + \boldsymbol{\xi}^T\mathbf{P}\dot{\boldsymbol{\xi}} \\
&= \left[\frac{\partial}{\partial s}\phi_1(s)(\mathbf{A}\boldsymbol{\xi} + \boldsymbol{\rho})\right]^T \mathbf{P}\boldsymbol{\xi} + \boldsymbol{\xi}^T\mathbf{P}\left[\frac{\partial}{\partial s}\phi_1(s)(\mathbf{A}\boldsymbol{\xi} + \boldsymbol{\rho})\right] \\
&= \frac{\partial}{\partial s}\phi_1(s)\left[(\mathbf{A}\boldsymbol{\xi} + \boldsymbol{\rho})^T\mathbf{P}\boldsymbol{\xi} + \boldsymbol{\xi}^T\mathbf{P}(\mathbf{A}\boldsymbol{\xi} + \boldsymbol{\rho})\right] \\
&= \frac{\partial}{\partial s}\phi_1(s)\left[\boldsymbol{\xi}^T\mathbf{A}^T\mathbf{P}\boldsymbol{\xi} + \boldsymbol{\rho}^T\mathbf{P}\boldsymbol{\xi} + \boldsymbol{\xi}^T\mathbf{PA}\boldsymbol{\xi} + \boldsymbol{\xi}^T\mathbf{P}\boldsymbol{\rho}\right] \\
&= \frac{\partial}{\partial s}\phi_1(s)\left[\boldsymbol{\xi}^T(\mathbf{A}^T\mathbf{P} + \mathbf{PA})\boldsymbol{\xi} + \boldsymbol{\rho}^T\mathbf{P}\boldsymbol{\xi} + \boldsymbol{\xi}^T\mathbf{P}\boldsymbol{\rho}\right] \\
&= \frac{\partial}{\partial s}\phi_1(s)\left\{\begin{bmatrix} \boldsymbol{\xi} \\ \boldsymbol{\rho} \end{bmatrix}^T \begin{bmatrix} \mathbf{A}^T\mathbf{P} + \mathbf{PA} & \mathbf{P} \\ \mathbf{P} & 0 \end{bmatrix} \begin{bmatrix} \boldsymbol{\xi} \\ \boldsymbol{\rho} \end{bmatrix}\right\}
\end{aligned} \tag{4.45}$$

According to the Lyapunov asymptotic stability theorem, from the Eqs. (4.43) and (4.45), the asymptotic stability of the system needs $Q \le 0$, which means

$$\begin{cases} \alpha_1 \le \frac{1}{2(\chi - 4\varepsilon^2)}\left[\delta + g^2 + \chi^2 + 4\varepsilon\chi - 8\varepsilon^2\chi + 4\varepsilon^2\right] \\ \alpha_2 = \chi + 2\varepsilon\alpha_1 - 2\varepsilon\chi \\ \delta < 4\varepsilon - 4\varepsilon^2 \end{cases} \tag{4.46}$$

According to the definitions of g, C and Θ, the Eq. (4.45) can be further derived as

$$\dot{V}_1 \le \frac{\partial}{\partial s}\phi_1(s)\left\{\begin{bmatrix} \boldsymbol{\xi} \\ \boldsymbol{\rho} \end{bmatrix}^T \left(\begin{bmatrix} \mathbf{A}^T\mathbf{P} + \mathbf{PA} & \mathbf{P} \\ \mathbf{P} & 0 \end{bmatrix} + \begin{bmatrix} g^2\mathbf{C}^T\mathbf{C} & 0 \\ 0 & -\mathbf{\Theta} \end{bmatrix}\right) \begin{bmatrix} \boldsymbol{\xi} \\ \boldsymbol{\rho} \end{bmatrix}\right\} \tag{4.47}$$

Because $\begin{bmatrix} \boldsymbol{\xi} \\ \boldsymbol{\rho} \end{bmatrix}^T \begin{bmatrix} g^2\mathbf{C}^T\mathbf{C} & 0 \\ 0 & -\boldsymbol{\Theta} \end{bmatrix} \begin{bmatrix} \boldsymbol{\xi} \\ \boldsymbol{\rho} \end{bmatrix} = \boldsymbol{\xi}^T g^2 \mathbf{C}^T\mathbf{C}\boldsymbol{\xi} - \boldsymbol{\rho}^T\boldsymbol{\Theta}\boldsymbol{\rho} = \left(\phi_1^2(s) + w^2\right)g^2 - w^2 \geq 0$ (derived from Assumption 2 and Eq. (4.42)), the derivative of the Lyapunov equation can be further written as

$$
\begin{aligned}
\dot{V}_1 &\leq \frac{\partial}{\partial s}\phi_1(s)\left\{\begin{bmatrix} \boldsymbol{\xi} \\ \boldsymbol{\rho} \end{bmatrix}^T \begin{bmatrix} \mathbf{A}^T\mathbf{P} + \mathbf{P}\mathbf{A} + g^2\mathbf{C}^T\mathbf{C} & \mathbf{P} \\ \mathbf{P} & -\boldsymbol{\Theta} \end{bmatrix} \begin{bmatrix} \boldsymbol{\xi} \\ \boldsymbol{\rho} \end{bmatrix}\right\} \\
&= \frac{\partial}{\partial s}\phi_1(s)\left\{\begin{bmatrix} \boldsymbol{\xi} \\ \boldsymbol{\rho} \end{bmatrix}^T \begin{bmatrix} \mathbf{Q} & \mathbf{P} \\ \mathbf{P} & -\boldsymbol{\Theta} \end{bmatrix} \begin{bmatrix} \boldsymbol{\xi} \\ \boldsymbol{\rho} \end{bmatrix}\right\} \\
&\leq -\frac{\partial}{\partial s}\phi_1(s)\varpi\|\xi\|^2
\end{aligned}
\tag{4.48}
$$

The complete Lyapunov equation is defined as $V(\boldsymbol{\xi}, \alpha_1, \alpha_2) = V_1(\boldsymbol{\xi}) + \frac{1}{2\gamma_1}\left(\alpha_1 - \alpha_1^*\right)^2 + \frac{1}{2\gamma_2}\left(\alpha_2 - \alpha_2^*\right)^2$, where α_1^* and α_2^* are normal numbers, respectively, and are the upper limits of $\alpha_1(t)$ and $\alpha_2(t)$. So

$$
\begin{aligned}
\dot{V}(\boldsymbol{\xi}, \alpha_1, \alpha_2) &= -rV_1^{\frac{1}{2}}(\boldsymbol{\xi}) + \frac{1}{\gamma_1}\left(\alpha_1 - \alpha_1^*\right)\dot{\alpha}_1 + \frac{1}{\gamma_2}\left(\alpha_2 - \alpha_2^*\right)\dot{\alpha}_2 \\
&= -rV_1^{\frac{1}{2}}(\boldsymbol{\xi}) - \frac{\omega_1}{\sqrt{2r_1}}\left|\alpha_1 - \alpha_1^*\right| - \frac{\omega_2}{\sqrt{2r_2}}\left|\alpha_2 - \alpha_2^*\right| \\
&\quad + \frac{1}{\gamma_1}\left(\alpha_1 - \alpha_1^*\right)\dot{\alpha}_1 + \frac{1}{\gamma_2}\left(\alpha_2 - \alpha_2^*\right)\dot{\alpha}_2 + \frac{\omega_1}{\sqrt{2r_1}}\left|\alpha - \alpha^*\right| + \frac{\omega_2}{\sqrt{2r_2}}\left|\alpha_2 - \alpha_2^*\right| \\
&\leq -\min(r, \omega_1, \omega_2, \omega_3)\left[V_1(\boldsymbol{\xi}) + \frac{1}{2r_1}\left(\alpha_1 - \alpha_1^*\right)^2 + \frac{1}{2r_2}\left(\alpha_2 - \alpha_2^*\right)^2\right]^{\frac{1}{2}} \\
&\quad + \frac{1}{\gamma_1}\left(\alpha_1 - \alpha_1^*\right)\dot{\alpha}_1 + \frac{1}{\gamma_2}\left(\alpha_2 - \alpha_2^*\right)\dot{\alpha}_2 + \frac{\omega_1}{\sqrt{2r_1}}\left|\alpha_1 - \alpha_1^*\right| + \frac{\omega_2}{\sqrt{2r_2}}\left|\alpha_2 - \alpha_2^*\right|
\end{aligned}
\tag{4.49}
$$

Since for any $\forall t \geq 0$ there are $\alpha(t) - \alpha^* < 0$ and $\beta(t) - \beta^* < 0$, so:

$$
\dot{V}(\boldsymbol{\xi}, \alpha_1, \alpha_2) \leq -\min(r, \omega_1, \omega_2)V^{\frac{1}{2}} + \Pi \tag{4.50}
$$

where $\Pi = -\left|\alpha_1 - \alpha_1^*\right|\left(\frac{1}{\gamma_1}\dot{\alpha}_1 - \frac{\omega_1}{\sqrt{2r_1}}\right) - \left|\alpha_2 - \alpha_2^*\right|\left(\frac{1}{\gamma_2}\dot{\alpha}_2 - \frac{\omega_2}{\sqrt{2r_2}}\right)$. Only need to make $\Pi = 0$ and carefully select the parameters ω_1, γ_1, ω_2 and γ_2 to get the recognition rate in Eq. (4.37), and $\dot{V}(\boldsymbol{\xi}, \alpha_1, \alpha_2, \vartheta) \leq -\min(r, \omega_1, \omega_2)V^{\frac{1}{2}}$.

At this point, the stability proof of the entire control algorithm is completed. For any initial state $\mathbf{x}(0)$ and $\mathbf{s}(0)$, and given suitable adaptive gains α_1 and α_2, using the super-twist adaptive sliding mode control algorithm in Eq. (4.37) and the corresponding adaptive law can drive the system state to a given slip mold surface $s = \dot{s} = 0$.

We use scalar forms for the proof of the control scheme previously. The net of MSTN is assumed to be a simple net with 11 × 11 nodes, each state of the node can be distinguished via subscript. $\mathbf{x}_{Ka}$, $\mathbf{x}_{Kb}$, $\mathbf{x}_{Kc}$, and $\mathbf{x}_{Kd}$ denote four corner MUs in MSTN. Recalling the dynamics model in Chap. 2, it can be rewritten exactly as below, so the scalar control scheme can work on each of the dynamics equations:

$$\begin{cases} \ddot{x}_{Aa} = f\begin{pmatrix} x_{Ba}, x_{Bb}, x_{Bc}, x_{Bd}, y_{Ba}, y_{Bb}, y_{Bc}, y_{Bd}, z_{Ba}, z_{Bb}, z_{Bc}, z_{Bd}, \\ \dot{x}_{Ba}, \dot{x}_{Bb}, \dot{x}_{Bc}, \dot{x}_{Bd}, \dot{y}_{Ba}, \dot{y}_{Bb}, \dot{y}_{Bc}, \dot{y}_{Bd}, \dot{z}_{Ba}, \dot{z}_{Bb}, \dot{z}_{Bc}, \dot{z}_{Bd}, t \end{pmatrix} \\ \ddot{y}_{Aa} = f\begin{pmatrix} x_{Ba}, x_{Bb}, x_{Bc}, x_{Bd}, y_{Ba}, y_{Bb}, y_{Bc}, y_{Bd}, z_{Ba}, z_{Bb}, z_{Bc}, z_{Bd}, \\ \dot{x}_{Ba}, \dot{x}_{Bb}, \dot{x}_{Bc}, \dot{x}_{Bd}, \dot{y}_{Ba}, \dot{y}_{Bb}, \dot{y}_{Bc}, \dot{y}_{Bd}, \dot{z}_{Ba}, \dot{z}_{Bb}, \dot{z}_{Bc}, \dot{z}_{Bd}, t \end{pmatrix} \\ \ddot{z}_{Aa} = f\begin{pmatrix} x_{Ba}, x_{Bb}, x_{Bc}, x_{Bd}, y_{Ba}, y_{Bb}, y_{Bc}, y_{Bd}, z_{Ba}, z_{Bb}, z_{Bc}, z_{Bd}, \\ \dot{x}_{Ba}, \dot{x}_{Bb}, \dot{x}_{Bc}, \dot{x}_{Bd}, \dot{y}_{Ba}, \dot{y}_{Bb}, \dot{y}_{Bc}, \dot{y}_{Bd}, \dot{z}_{Ba}, \dot{z}_{Bb}, \dot{z}_{Bc}, \dot{z}_{Bd}, t \end{pmatrix} \\ \vdots \\ \ddot{x}_{Ka} = f(x_{Ja}, x_{Jb}, y_{Ja}, y_{Jb}, z_{Ja}, z_{Jb}, \dot{x}_{Ja}, \dot{x}_{Jb}, \dot{y}_{Ja}, \dot{y}_{Jb}, \dot{z}_{Ja}, \dot{z}_{Jb}, t) + u_{Ka\_x} \\ \ddot{y}_{Ka} = f(x_{Ja}, x_{Jb}, y_{Ja}, y_{Jb}, z_{Ja}, z_{Jb}, \dot{x}_{Ja}, \dot{x}_{Jb}, \dot{y}_{Ja}, \dot{y}_{Jb}, \dot{z}_{Ja}, \dot{z}_{Jb}, t) + u_{Ka\_y} \\ \ddot{z}_{Ka} = f(x_{Ja}, x_{Jb}, y_{Ja}, y_{Jb}, z_{Ja}, z_{Jb}, \dot{x}_{Ja}, \dot{x}_{Jb}, \dot{y}_{Ja}, \dot{y}_{Jb}, \dot{z}_{Ja}, \dot{z}_{Jb}, t) + u_{Ka\_z} \\ \vdots \\ \ddot{x}_{Kd} = f(x_{Jg}, x_{Jh}, y_{Jg}, y_{Jh}, z_{Jg}, z_{Jh}, \dot{x}_{Jg}, \dot{x}_{Jh}, \dot{y}_{Jg}, \dot{y}_{Jh}, \dot{z}_{Jg}, \dot{z}_{Jh}, t) + u_{Kd\_x} \\ \ddot{y}_{Kd} = f(x_{Jg}, x_{Jh}, y_{Jg}, y_{Jh}, z_{Jg}, z_{Jh}, \dot{x}_{Jg}, \dot{x}_{Jh}, \dot{y}_{Jg}, \dot{y}_{Jh}, \dot{z}_{Jg}, \dot{z}_{Jh}, t) + u_{Kd\_y} \\ \ddot{z}_{Kd} = f(x_{Jg}, x_{Jh}, y_{Jg}, y_{Jh}, z_{Jg}, z_{Jh}, \dot{x}_{Jg}, \dot{x}_{Jh}, \dot{y}_{Jg}, \dot{y}_{Jh}, \dot{z}_{Jg}, \dot{z}_{Jh}, t) + u_{Kd\_z} \end{cases} \tag{4.51}$$

The control block diagram of super-twisting adaptive sliding mode control algorithm is as Fig. 4.3.

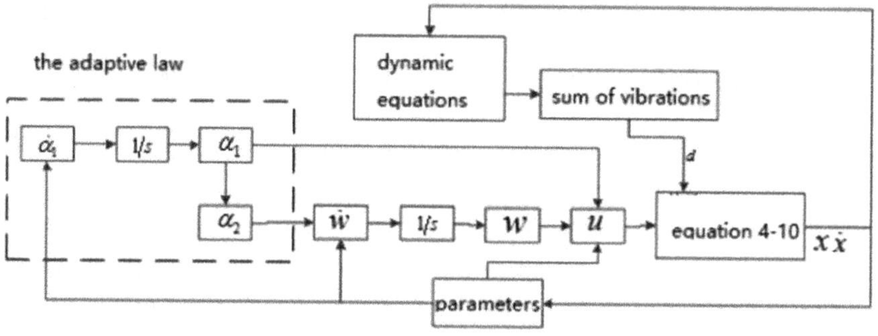

Fig. 4.3 The control block diagram

4.1.4 Numerical Simulation

According to the research on the releasing characteristics of the MSTN in Chap. 3, the initial launch conditions of the MUs are: shooting angle $\alpha = 60^\circ$, shooting velocity $|\mathbf{v}_{MUi}| = 0.1$ m/s, and the folding mode of the flying net in the launch tube is a star type folding. The specific spatial simulation environment is: the orbital altitude of the geosynchronous orbit is 36,000 km, the orbital angular velocity is $\omega = 9.2430 \times 10^{-5}$ rad/s, and the orbital inclination is 0. The relevant simulation parameters of the MSTN are shown in Table 4.1.

The MUs of the MSTN is free to fly after being ejected by platform satellites. At this stage, the net will be passively deployed. At the final point of free flight, the speed of the MUs in both directions on the deployment plane drops to zero, indicating that the free-flying ends; at the moment, the actuator of the MU opens and enters the control flying phase. In the catapult release and approach flight of the MSTN, the coordinate system of the relative target satellite is defined as shown in Fig. 4.4. In this chapter, the MUs open the actuator at time t = 38.09 s, and the positions of the four MUs at the free flight end point are $(1.4378, 1.4390, 1.4390)$, $(1.4378, 1.4390, -1.4390)$, $(1.4378, -1.4390, 1.4390)$, and $(1.4378, -1.4390, -1.4390)$; corresponding to this initial state, the expected states of the four MUs are $(1.4378 + 0.1(t - 38.09), 1.7, 1.7)$, $(1.4378 + 0.1(t - 38.09), 1.7, -1.7)$, $(1.4378 + 0.1(t - 38.09), -1.7, 1.7)$ and $(1.4378 + 0.1(t - 38.09), -1.7, -1.7)$. The desired speed for the four MUs in the direction x is 0.1 m/s, and the desired speed in the direction y and z is zero.

Based on the above-described simulation environment, numerical simulation is used to the system's dynamic equation (Eq. (4.20)) based on the control algorithm (Eq. (4.37)). In Fig. 4.3, in order to make full use of the robustness of the sliding mode controller and make the controller easy to design, divide the dynamic equations of the system into the dynamic equations of the MUs, and the kinematic equations of the net movement and the kinematic part of the net is used as interference input. The dashed box in the figure shows the adaptive law of the sliding mode parameters in the super-twisting adaptive sliding mode algorithm. w as the

Table 4.1 Parameters of the system

Parameters	Magnitude
Mass of each MU (kg)	20
Mass of the net (kg)	2
Length of the sideline of net (m)	5
Length of the sideline of each mesh (m)	0.5
Material of the net	Kevlar
Diameter of the net's thread (mm)	1
Young's modulus of the net's thread (MPa)	124000
Damping factor of the net's thread	0.01

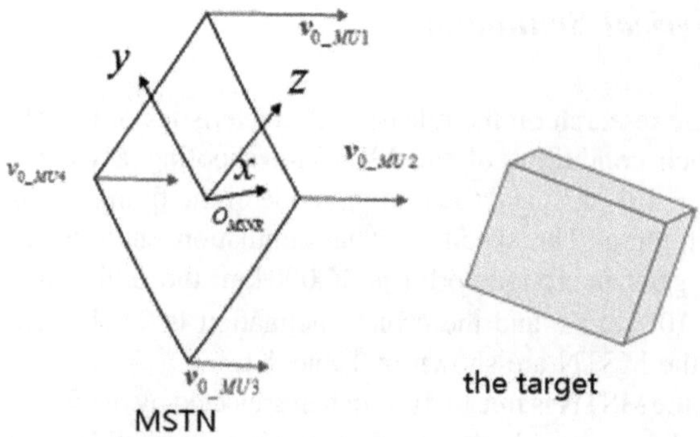

Fig. 4.4 The releasing coordinate of MSTN

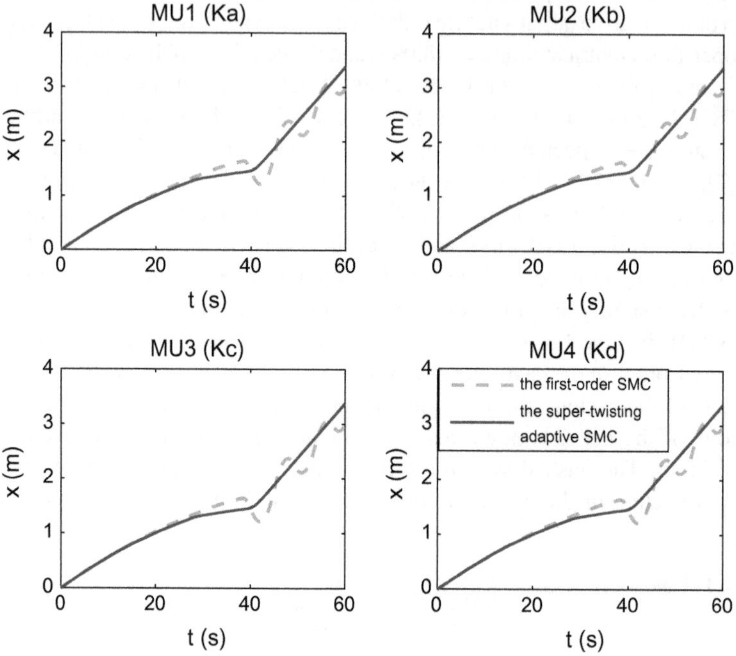

Fig. 4.5 MUs' position on x-direction

second-order sliding mode part constitutes the entire control algorithm u. The specific sliding mode parameters are shown in Eq. (4.5).

Figures 4.5, 4.6, and 4.7 are four MUs' positions in releasing and initial approaching phases in the first-order sliding mode controller and the improved super-twisting adaptive sliding mode controller proposed in this chapter, respectively. The dotted lines represent the first-order sliding mode algorithm, and the

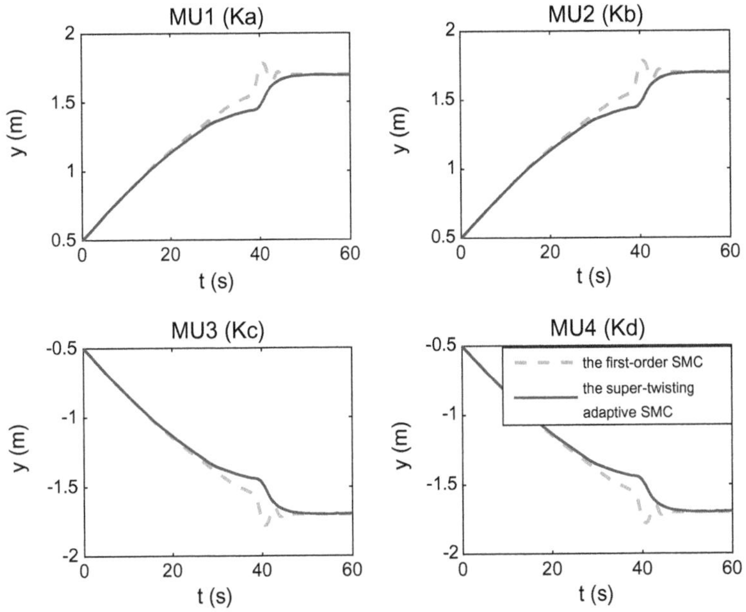

Fig. 4.6 MUs' position on y-direction

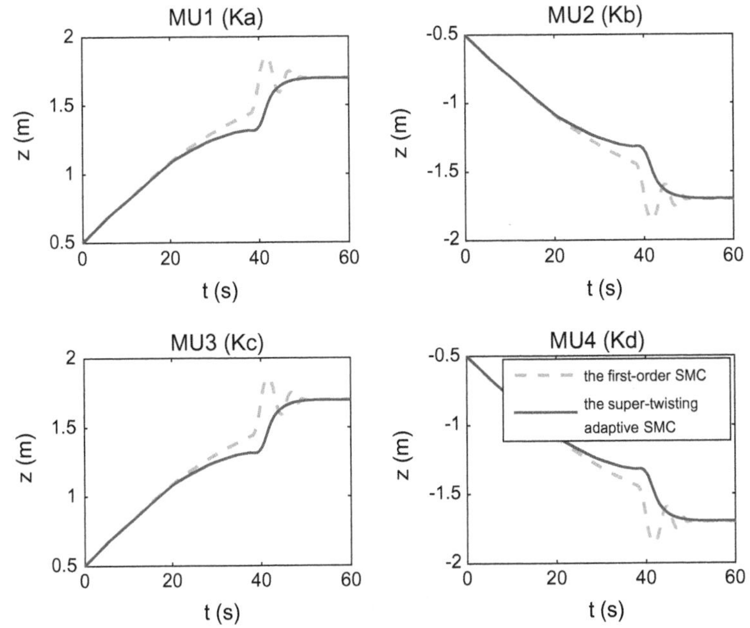

Fig. 4.7 MUs' position on z-direction

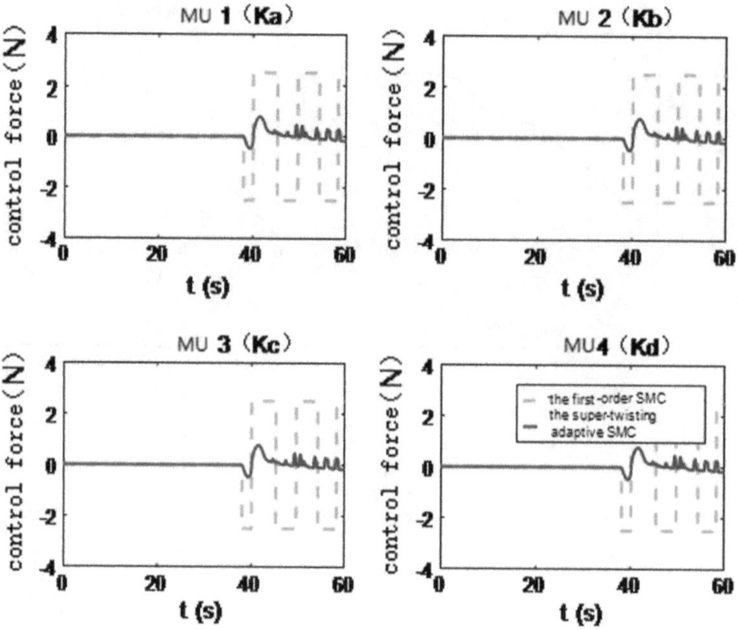

Fig. 4.8 MUs' control force on x-direction

solid lines indicate the algorithm proposed in this chapter. From the three simulation results, it can be seen that the improved control algorithms proposed in this chapter are performing well regardless of the position direction. However, the directions y and z of the first-order sliding mode are superior to the direction x. This is because the desired states in both directions y and z are constants, and the desired state in the direction x is a smooth curve, which makes the MUs' state has been "chasing" the desired state, and due to the non-adaptable controller output, it will inevitably "catch up" and continue to chase the desired state in reverse. The algorithm proposed in this chapter can ensure that the system state reaches and stabilizes in the desired state in any direction of motion.

Figures 4.8, 4.9, and 4.10 are the MUs' control force on the expanded plane of the net under different algorithms. It can be seen that although in Figs. 4.6 and 4.7, the net can be quickly deployed under the control of the first-order sliding mode algorithm, and is stabilized on the desired sliding surface after unfolding. But in content with Figs. 4.9 and 4.10, it can be known that this stable state is premised on the sacrifice of high-frequency operation of the controller. This is a typical first-order sliding mode algorithm. As a comparison, we can see that the control algorithm proposed in this chapter can greatly reduce the controller's output amplitude and frequency. Compared with the previous simulation result of MUs' state, the simulation results of this part of the control input better illustrate the superiority of the improved algorithm.

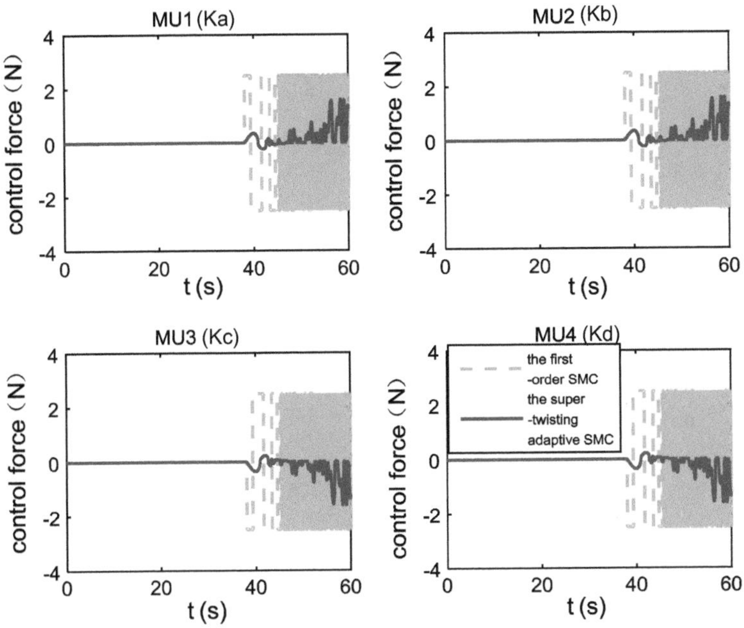

Fig. 4.9 MUs' control force on y-direction

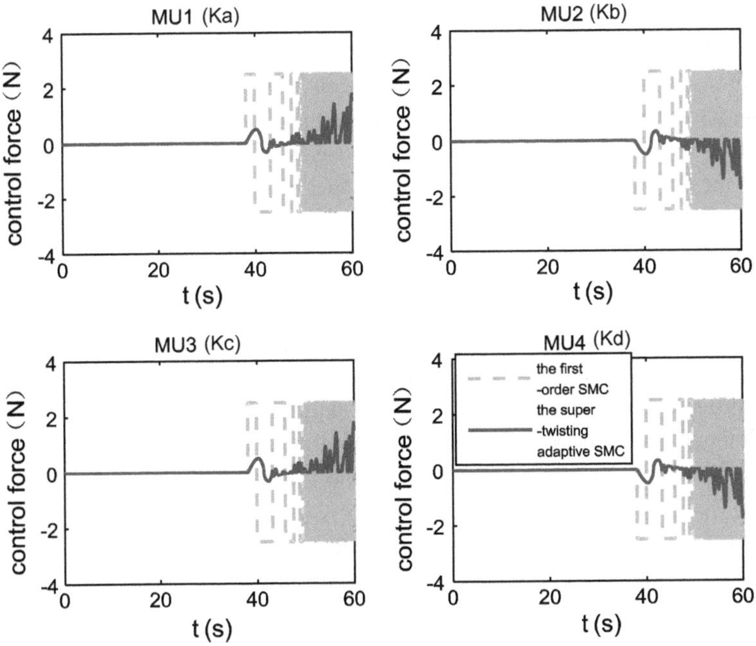

Fig. 4.10 MUs' control force on z-direction

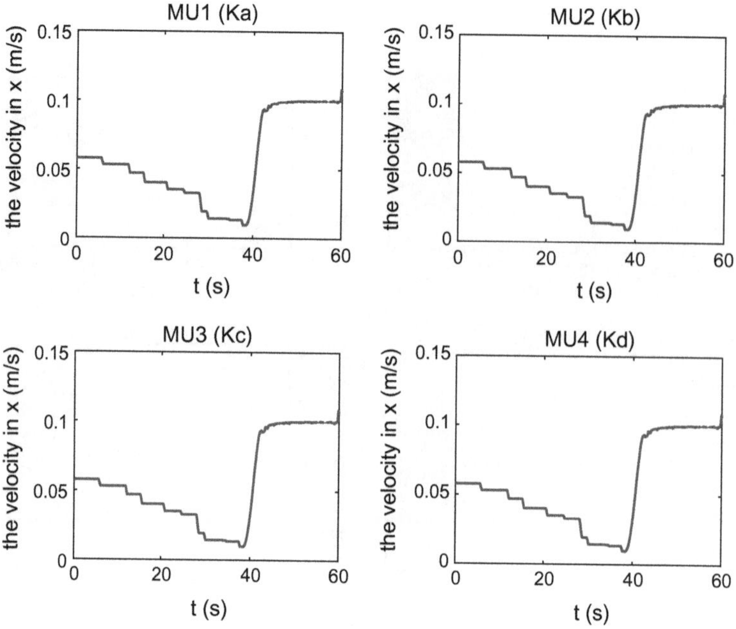

Fig. 4.11 MUs' velocity on x-direction

Figures 4.11, 4.12, and 4.13 show the simulation results of the velocity change of the MUs in three directions. From the definition of the sliding mode parameter $\mathbf{s}_i = \mathbf{s}_i(x_{i1}, x_{i2}) = \sum_{j=1}^{2} c^{j-1} x_{ij}$, it can be seen that the sliding mode surface is composed of the state position information and the velocity information. Under the control of the controller, the condition of the system to achieve this sliding surface is that the first derivative of the system state (error) and state (error) is zero. From the simulation results, it can be seen that the MUs can reach the desired state in the direction x, and velocity can reach to zero in the directions y and z, and there is slight vibration around zero. This shows that the MUs can almost maintain a steady state with zero speed after reaching the desired state. In addition, it can be seen that before t = 38.09 s, the speed of the net in the directions y and z have reached zero, indicating that the net has been expanded to the limit during the free-flying phase. If the thruster is not introduced again, the net will shrink into chaos.

Although it can be known from the definition of the sliding mode parameter $\mathbf{s}_i = \mathbf{s}_i(e_{i1}, e_{i2}) = \sum_{j=1}^{2} c^{j-1} e_{ij}$, as long as the state and its first derivative are all zero, it means that the sliding mode surface is reached, but when different sliding mode surface parameters are selected, there will be different properties of the sliding mode surface. The sliding surface parameters in this chapter are selected as $c = 0.4$. The system starts to introduce the controller at t = 38.09 s and executes the

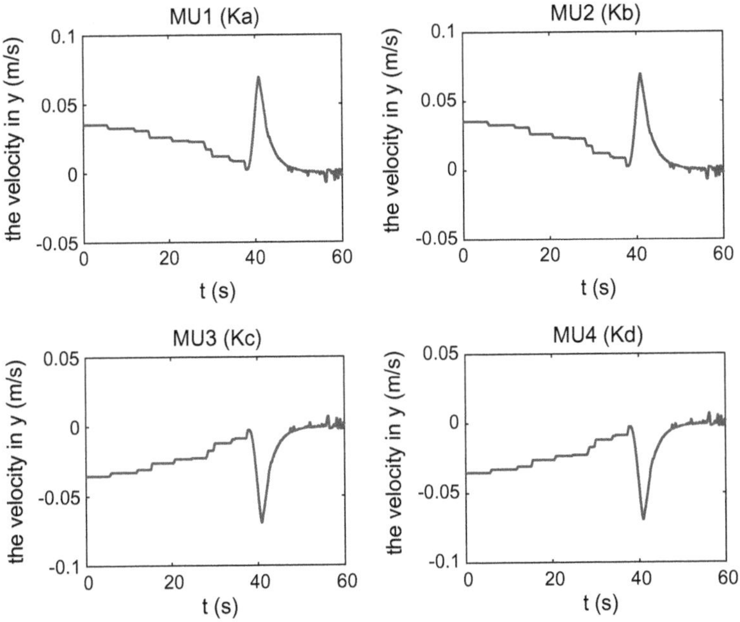

Fig. 4.12 MUs' velocity on y-direction

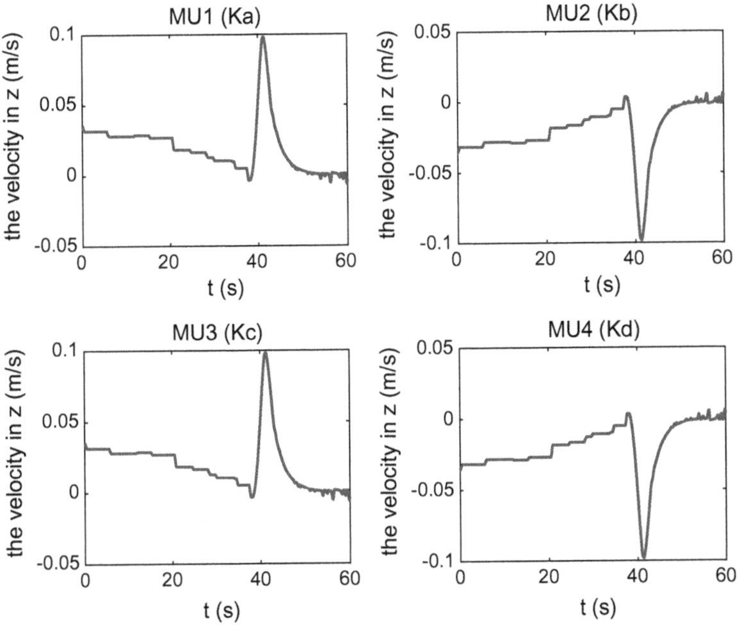

Fig. 4.13 MUs' velocity on z-direction

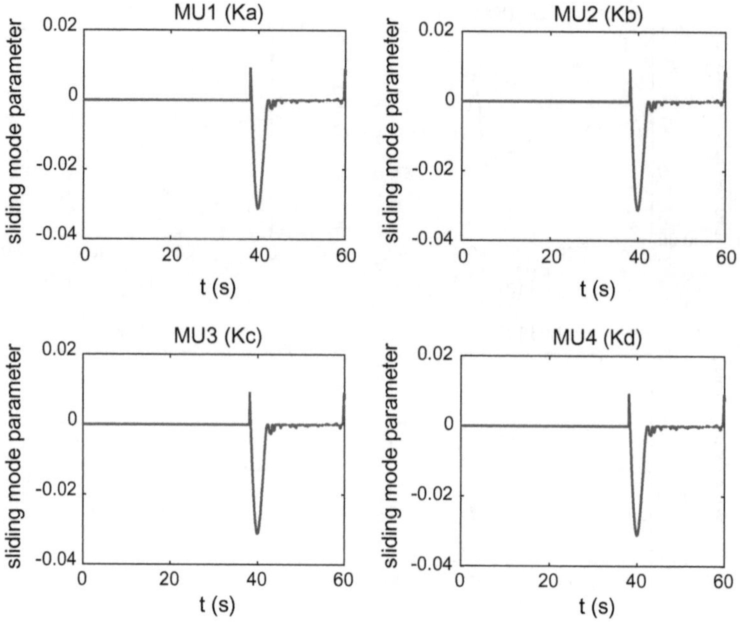

Fig. 4.14 Sliding mode parameter s_x

control command by the actuator of the MUs. At t = 38.09 s, the relative closest point is selected from the phase plan consisting of the initial error state $(e_0, \dot{e}_0)$ of the system. At the same time, according to different systems, the weight of e and $\dot{e}$ are different. Figures 4.14, 4.15, and 4.16 show the sliding mode parameters s_x, s_y, and s_z of the four MUs in three directions. From the simulation results, we can see that under the control of the improved adaptive super-sliding mode algorithm, we can quickly reach the designed sliding surface and can almost remain on the sliding surface.

Figure 4.17a–c show the full-system configuration of the MSTN in t = 38.09 s and t = 60 s under the algorithm and the first-order sliding mode algorithm proposed in this chapter. All the red circles in the figure indicate the tether weaving nodes of the net, and the blue solid lines indicate the braided tethers of the net. The quadrilateral between the adjacent solid blue lines in the figure is the weaving grid of the net. The red circles at the four corners represent four MUs. The quadrilateral formed by the green dashed line is the effective net opening formed by the MUs and is determined by the size of the subject satellite's satellite. In the actual catching task, the final effective net opening of the MSTN approaching phase is one of the keys to whether the task can be completed. From the comparison between (a) and (b), it can be seen that, although the velocity of the MUs in the directions of y and z is already zero at this time, the net obviously does not fully develop. Under the control of the algorithm proposed in this chapter, the net continues to fly toward the target satellite in the direction x and continues to expand in the directions of y and

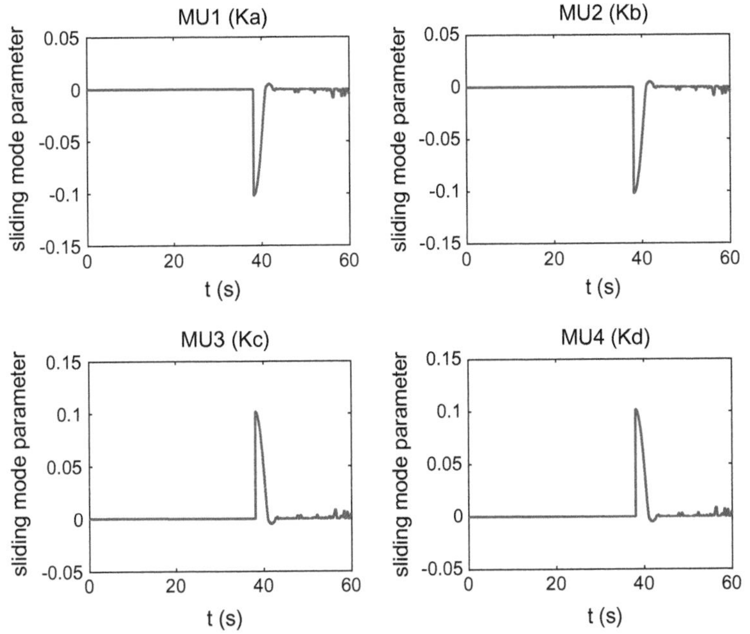

Fig. 4.15 Sliding mode parameter s_y

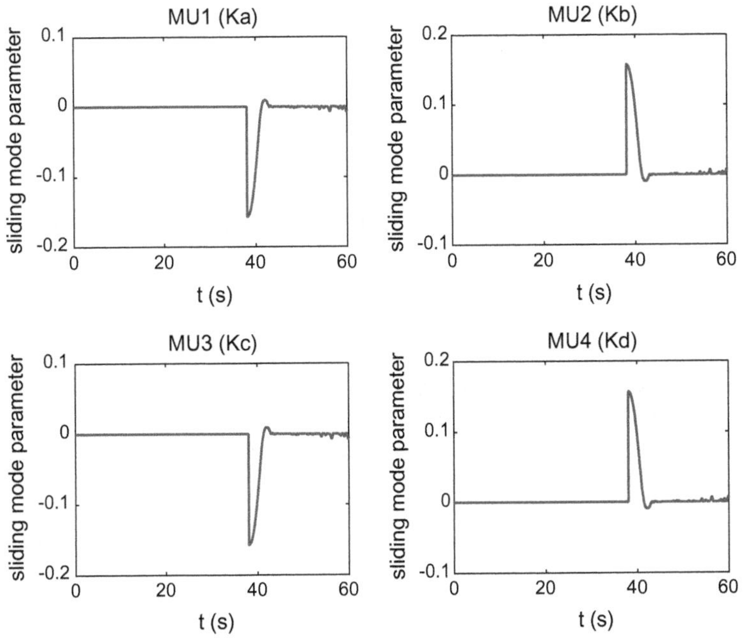

Fig. 4.16 Sliding mode parameter s_z

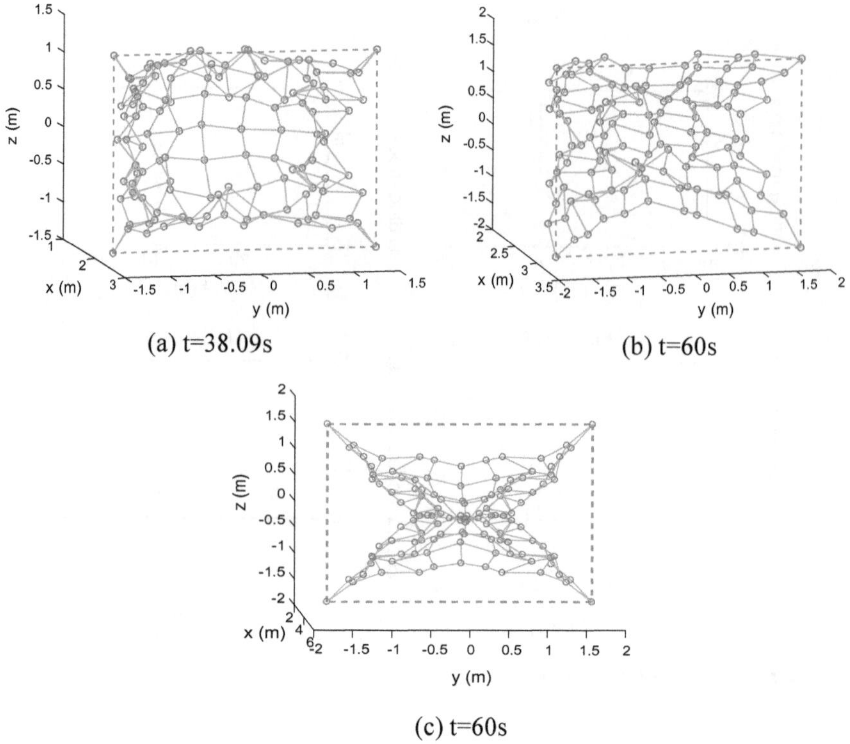

Fig. 4.17 Configuration of MSTN

z until it reaches the desired state. As shown in (b), the net has fully deployed at time t = 60 s. From the comparison between (b) and (c), it can be seen that with the first-order sliding mode algorithm, the MUs can achieve the desired value, but the net doesn't deploy at time t = 38.09 s. It can be seen that although the effective areas of (b) and (c) are the same, the effective volume of the configuration in (b) is approximately $19.8\,\mathrm{m}^3$, while in (b), the effective volume is only about $4.1\,\mathrm{m}^3$. Therefore, the control algorithm proposed in this chapter can not only ensure that the MUs reaches the desired state, but also enables the uncontrollable net part to maintain a good configuration. It is worth noting that although the design of the net is 5 m on the side and the expected state is only 3.4 m (1.7×2), the net has almost reached its deployment limit. This is because if the net is completely unfolded into a plane 5 m $\times$ 5 m, the braided tether is completely tensioned and is not conducive to control. In order to ensure the stability of the net and the shallow U-shaped configuration, the linear link distance between two adjacent MUs should not exceed 3.5 m.

4.2 Configuration-Keeping Control Under the Unknown Disturbance

4.2.1 Problem Statement

Since the MSTN has only four MUs that can execute control commands, this is a typical underdrive system. In order to facilitate the controller design, the dynamic equation is rewritten as follows:

$$\begin{cases} \ddot{\mathbf{x}}_{Ka} = f(\mathbf{x}_{Ka}, \mathbf{x}_{Ja}, \mathbf{x}_{Jb}, \dot{\mathbf{x}}_{Ka}, \dot{\mathbf{x}}_{Ja}, \dot{\mathbf{x}}_{Jb}, t) + \mathbf{u}_{Ka} + \tilde{\mathbf{d}}_{Ka} \\ \ddot{\mathbf{x}}_{Kb} = f(\mathbf{x}_{Kb}, \mathbf{x}_{Jc}, \mathbf{x}_{Jd}, \dot{\mathbf{x}}_{Kb}, \dot{\mathbf{x}}_{Jc}, \dot{\mathbf{x}}_{Jd}, t) + \mathbf{u}_{Kb} + \tilde{\mathbf{d}}_{Kb} \\ \ddot{\mathbf{x}}_{Kc} = f(\mathbf{x}_{Kc}, \mathbf{x}_{Je}, \mathbf{x}_{Jf}, \dot{\mathbf{x}}_{Kc}, \dot{\mathbf{x}}_{Je}, \dot{\mathbf{x}}_{Jf}, t) + \mathbf{u}_{Kc} + \tilde{\mathbf{d}}_{Kc} \\ \ddot{\mathbf{x}}_{Kd} = f(\mathbf{x}_{Kd}, \mathbf{x}_{Jg}, \mathbf{x}_{Jh}, \dot{\mathbf{x}}_{Kd}, \dot{\mathbf{x}}_{Jg}, \dot{\mathbf{x}}_{Jh}, t) + \mathbf{u}_{Kd} + \tilde{\mathbf{d}}_{Kd} \end{cases} \tag{4.52}$$

where $\tilde{\mathbf{d}}_i$ is the resultant external force acting in the dynamic equation except the control force on the ith MU. Take $\tilde{\mathbf{d}}_{Ka}$, for example, all external forces acting on the MU Ka can be written as $\tilde{\mathbf{d}}_{Ka} = \tilde{\mathbf{f}}_{Ja} + \tilde{\mathbf{f}}_{Jb} + \mathbf{G}_{Ka} = \ddot{\mathbf{x}}_{Ja} m_{Ja} + \ddot{\mathbf{x}}_{Jb} m_{Jb} - \mu^I \mathbf{R}_{Ka} / {}^I R_{Ka}^3$. So Eq. (4.52) can be further used for each MU as an example written as

$$\begin{cases} \ddot{\mathbf{x}}_i = f(\mathbf{x}) + g(\mathbf{x}) \mathbf{u}_i + \tilde{\mathbf{d}}_i \\ \mathbf{y}_i = \mathbf{x}_i(t) \end{cases} \tag{4.53}$$

where: $\mathbf{x} \subset \mathbb{R}^n \in X$ is the controlled system state variable, X is a tight set; $\mathbf{x}_i \in \mathbb{R}^3$ is the controllable state variable in the system; $\mathbf{u}_i \subset \mathbb{R}^3 \in U$ is the control algorithm function of each controllable state, U is a tight set; $f(\mathbf{x}) \in \mathbb{R}^3$ is a function that can be differentiated and only partially known by the system state.

Due to model errors and spatial environmental disturbances, the dynamic equation (4.53) was further rewritten as

$$\begin{aligned} \ddot{\mathbf{x}}_i &= [\bar{f}(\mathbf{x}) + \Delta f(\mathbf{x})] + [\bar{g}(\mathbf{x}) + \Delta g(\mathbf{x})] \mathbf{u}_i + \tilde{\mathbf{d}}_i + \mathbf{d}_o \\ &= \bar{f}(\mathbf{x}) + \bar{g}(\mathbf{x}) \mathbf{u}_i + \bar{\bar{\mathbf{d}}}_i \end{aligned} \tag{4.54}$$

where the superscript "—" represents the part known in the dynamic equation; "Δ" represents the unknown dynamic model error; $\mathbf{d}_o$ is the external disturbance in the space unknown environment; $\bar{\bar{\mathbf{d}}}_i$ is the sum of the calculation error of the model, the vibration superposition of the uncontrollable part and the unknown disturbances in the space environment, it is specifically expressed as $\bar{\bar{\mathbf{d}}}_i = \Delta f(\mathbf{x}) + \Delta g(\mathbf{x}) \mathbf{u}_i + \tilde{\mathbf{d}}_i + \mathbf{d}_o$.

The assumptions used in the derivation of the control algorithm in this section are:

Assumption 4 Assuming that a suitable sliding mode parameter $\sigma_i = \sigma_i(\mathbf{x}, t) \in \mathbb{R}$ has been selected (i denotes the ith MU), and the sliding mode surface $\sigma = \sigma(\mathbf{x}, t) = 0$ determined by the sliding mode parameter can reach under control.

The relative order of the dynamic equation and the sliding mode parameter of the MSTN with respect to the sliding mode parameter ($u_i \to \sigma_i$) is 1. Therefore, the system's new input–output dynamic equation can be expressed as

$$\begin{aligned}\dot{\sigma}_i &= \frac{\partial \sigma_i}{\partial t} = \frac{\partial \sigma_i}{\partial \mathbf{x}_i}\frac{\partial \mathbf{x}_i}{\partial t} + \frac{\partial \sigma_i}{\partial t} \\ &= \underbrace{\frac{\partial \sigma_i}{\partial t} + \frac{\partial \sigma_i}{\partial \mathbf{x}_i}\bar{f}(\mathbf{x}, t)}_{\varphi_i(\mathbf{x}_i, t)} + \underbrace{\frac{\partial \sigma_i}{\partial \mathbf{x}_i}\bar{g}(\mathbf{x}, t)\, u_i}_{b_i(\mathbf{x}_i, t)} + \underbrace{\frac{\partial \sigma_i}{\partial \mathbf{x}_i}\bar{\bar{\mathbf{d}}}_i(\mathbf{x}, t)}_{\gamma_i(\mathbf{x}_i, t)} \\ &= \bar{w}_i(\mathbf{x}_i, t) + \gamma_i(\mathbf{x}_i, t)\end{aligned} \tag{4.55}$$

where $\bar{w}_i(\mathbf{x}_i, t) = \varphi_i(\mathbf{x}_i, t) + b_i(\mathbf{x}_i, t)u_i$; $\gamma_i(\mathbf{x}_i, t)$ is the complex control interference. Defined $\bar{w}_i(\mathbf{x}_i, t)$ as the controller's nominal input.

Assumption 5 Complex control disturbances $\gamma_i(\mathbf{x}_i, t)$ can be further expressed by sliding mode parameters as $|\gamma_i(\mathbf{x}_i, t)| \leq \xi|\sigma_i|^{\frac{1}{2}}$, which means $\gamma_i(\mathbf{x}_i, t)$ has an unknown but real upper bound ξ.

Assumption 6 Assume that all states in the system are measurable.

Therefore, the research content of this section is that design a set of controllers of the MSTN to ensure that the sliding mode variable input–output dynamic equation (4.55) can fly to the target satellite under complicated asymmetrical interference with the upper bound $|\gamma_i(\mathbf{x}_i, t)| \leq \xi|\sigma_i|^{\frac{1}{2}}$.

4.2.2 Preliminary

The fuzzy algorithm is an intelligent control algorithm based on human prior knowledge. The basic concepts of some fuzzy algorithms used here include

Clear set: For a normal set or "clear" set $x_i \in X_i$ $(i = 1, 2, \ldots, n)$ and $y_i \in Y_i$ $(i = 1, 2, \ldots, n)$, the scales are, respectively, distinct sets of and.

Feature function: Assume X_i is a set on the domain S, if $x_i \in X_i$, then the feature function of the set. X_i is $\mu_{X_i}(x_i) = 1$.

Membership degree function and Fuzzy set: Assume X_i is a set on the domain S, the membership function of the set X_i assigns each $x_i \in X_i$ value or degree of membership, $\mu : X \to [0, 1]$. Then the set becomes a fuzzy set.

From the above definitions, it can be seen that the membership function takes the value domain [0, 1], and the feature function only takes the boundary value on the interval. Therefore, the clear set is a special case of fuzzy sets [6]. There are many kinds of membership function choices, several commonly used include triangular membership function, trapezoidal membership function, Gaussian membership function. The membership function is a nonnegative function ($0 \leq \mu(x) \leq 1$), and the choice of membership function depends on prior knowledge of humans.

Language rules: They are a set of conditional-conclusive rules that are obeyed from input to output in a fuzzy algorithm. For example:

$$\textbf{\textit{IF}} \quad \textit{premise} \quad \textbf{\textit{THEN}} \quad \textit{consequent} \tag{4.56}$$

4.2.3 Controller Design

The fuzzy adaptive sliding mode control algorithm proposed in this section is

$$\begin{cases} \bar{w}_i = -k_{i1}(t)\left[\sigma_i/\|\sigma_i\|_2^{\frac{1}{2}} + k_{i3}\sigma_i\right] + v_i \\ \dot{v}_i = -k_{i2}(t)\left[\frac{1}{2}\sigma_i/\|\sigma_i\|_2 + \frac{3}{2}k_{i3}\sigma_i/\|\sigma_i\|_2^{\frac{1}{2}} + k_{i3}^2\sigma_i\right] \end{cases} \tag{4.57}$$

where $k_1(t) = k_1(\sigma_i, \dot{\sigma}_i, t)$, and $k_2(t) = k_2(\sigma_i, \dot{\sigma}_i, t)$ are adaptive parameters, k_3 is a normal-number control gain and $\|\cdot\|_2$ represents the Euclidean norm.

According to the Assumption 5, there is always a position constant $\xi > 0$, so that the unknown uncertainty $\gamma(\mathbf{x}, t)$ including the model calculation error and the superposition of the vibration of the uncontrolled part under the interference of the main tether meets the Assumption 5. For any initial state $\mathbf{x}(0)$ and corresponding initial sliding mode parameter $\boldsymbol{\sigma}(0)$, driven by the control algorithm (4.57), the controlled system (4.55) can reach the sliding mode surface $\boldsymbol{\sigma} = \dot{\boldsymbol{\sigma}} = \mathbf{0}$ within a limited time, and the adaptive law of the adaptive parameter in (4.57) is

$$\begin{aligned} \dot{k}_1(t) &= \begin{cases} \omega_1\sqrt{\rho_1/2} & \sigma_i \neq 0 \\ 0 & \sigma_i = 0 \end{cases} \\ k_2(t) &= 2\varepsilon k_1(t) + \left[\hat{\boldsymbol{\theta}}_\gamma \delta(\mathbf{x})\right]/(\eta'\eta_1) + \chi + 4\varepsilon^2 \end{aligned} \tag{4.58}$$

where ω_1, ρ_1, and χ are arbitrary given normal numbers, ε is any given real numbers.

Put Eq. (4.57) into the dynamic input–output equation (4.55) based on sliding mode parameters, we can get

$$\begin{cases} \dot{\sigma}_i = -k_{i1}(t)\left[\sigma_i/\|\sigma_i\|_2^{\frac{1}{2}} + k_{i3}\sigma_i\right] + v_i + \gamma_i \\ \dot{v}_i = -k_{i2}(t)\left[\frac{1}{2}\sigma_i/\|\sigma_i\|_2 + \frac{3}{2}k_{i3}\sigma_i/\|\sigma_i\|_2^{\frac{1}{2}} + k_{i3}^2\sigma_i\right] \end{cases} \tag{4.59}$$

Due to the complexity of the interference γ_i, it cannot be calculated at all or through measurement. So adaptive fuzzy algorithm is used to approximate unknown uncertainties γ_i.

Define the fuzzy domain $A_i^{l_j}$, where $l_j = 1, 2, \ldots, p_j$, $j = 1, \ldots, m$ (m is the number of control inputs), and $l_1 \ldots l_n$ is the arrangement of elements $l_1, \ldots, l_n$. The fuzzy rules of the multiplicator $\prod_{j=1}^{n} p_j$ are acquired to approximate the unknown interference γ_i. Multiplier fuzzy rules can be specifically expressed as

IF x_1 is $A_1^{l_1}$ and … and x_n is $A_1^{l_n}$,

THEN $\hat{\gamma}_i$ is $E^{l_1,\ldots,l_n}$.

The unknown uncertainty γ_i obtained by the fuzzy logic operation is

$$\hat{\gamma}_i\left(\mathbf{x}\middle|\hat{\boldsymbol{\theta}}_\gamma\right) = \frac{\sum_{l_1=1}^{p_1}\cdots\sum_{l_n=1}^{p_n} \bar{y}_\gamma^{l_1,\ldots,l_n}\left(\prod_{i=1}^{n}\mu_{A_i^{l_j}}(x_i)\right)}{\sum_{l_1=1}^{p_1}\cdots\sum_{l_n=1}^{p_n}\left(\prod_{i=1}^{n}\mu_{A_i^{l_j}}(x_i)\right)} \tag{4.60}$$

where $\mu_{A_i^{l_j}}(x_i)$ is the membership function of the state x_i; the free parameters $\bar{y}_\gamma^{l_1,\ldots,l_n}$ make up the domain $\hat{\boldsymbol{\theta}}_\gamma \in \mathbb{R}^{\prod_{i=1}^{n} p_i}$, and the vectors of $\prod_{i=1}^{n} p_i$ dimension about $\mathbf{x}$ are represented as $\delta(\mathbf{x})$ in the fuzzy algorithm. So Eq. (4.60) can be further written as

$$\hat{\gamma}_i\left(\mathbf{x}\middle|\boldsymbol{\theta}_\gamma\right) = \hat{\boldsymbol{\theta}}_\gamma^T \delta(\mathbf{x}) \tag{4.61}$$

where:

$$\delta_{l_1,\ldots,l_n}(x) = \frac{\prod_{i=1}^{n}\mu_{A_i^{l_j}}(x_i)}{\sum_{l_1=1}^{p_1}\cdots\sum_{l_n=1}^{p_n}\left(\prod_{i=1}^{n}\mu_{A_i^{l_j}}(x_i)\right)} \tag{4.62}$$

Therefore, the approximation of unknown uncertain γ_i based on the fuzzy algorithm is $\hat{\gamma}_i\left(\mathbf{x}\middle|\hat{\boldsymbol{\theta}}_\gamma\right) = \hat{\boldsymbol{\theta}}_\gamma^T \delta(\mathbf{x})$, where the optimal identification vector $\hat{\boldsymbol{\theta}}_\gamma^T$ can be marked as $\boldsymbol{\theta}_\gamma^{*T}$, and its concrete expression is

$$\boldsymbol{\theta}_\gamma^* = \arg\min_{\boldsymbol{\theta}_\gamma \in \Omega_\gamma}\left\{\sup_{\mathbf{x}\in\mathbb{R}^n}\left|\hat{\gamma}_i\left(\mathbf{x}\middle|\hat{\boldsymbol{\theta}}_\gamma\right) - \gamma_i\right|\right\} \tag{4.63}$$

where Ω_γ represents the collection of $\boldsymbol{\theta}_\gamma$ (domain).

So γ_i can be expressed as

$$\gamma_i(\mathbf{x}|\boldsymbol{\theta}_\gamma) = \boldsymbol{\theta}_\gamma^{*T}\delta(\mathbf{x}) + \varepsilon_\gamma \tag{4.64}$$

where ε_γ denotes the approximation uncertainty of the unknown fuzzy uncertainty algorithm.

Based on Eqs. (4.61) and (4.64), the control algorithm (4.57) proposed in this section can be further written as

$$\begin{cases} \dot{\sigma}_i = -k_1(t)\left[\sigma_i/\|\sigma_i\|_2^{\frac{1}{2}} + k_3\sigma\right] + v_i + \boldsymbol{\theta}_\gamma^{*T}\boldsymbol{\delta}(\mathbf{x}) + \varepsilon_\gamma \\ \dot{v}_i = -k_2(t)\left[\frac{1}{2}\sigma_i/\|\sigma_i\|_2 + \frac{3}{2}k_3\sigma_i/\|\sigma_i\|_2^{\frac{1}{2}} + k_3^2\sigma_i\right] \end{cases} \tag{4.65}$$

Define a new variable $\boldsymbol{\eta}^T = \begin{pmatrix} \sigma_i/\|\sigma_i\|_2^{\frac{1}{2}} + k_3\sigma_i & v_i \end{pmatrix}$, specifically defined $\eta_1 = \sigma_i/\|\sigma_i\|_2^{\frac{1}{2}} + k_3\sigma_i$, $\dot{\eta}_2 = \frac{1}{2}\sigma_i/\|\sigma_i\|_2 + \frac{3}{2}k_3\sigma_i/\|\sigma_i\|_2^{\frac{1}{2}} + k_3^2\sigma_i$. So we can get a new conclusion about $\boldsymbol{\eta}$:

$$\dot{\boldsymbol{\eta}} = \eta_1'\begin{pmatrix} -k_1(t) & 1 \\ -k_2(t) & 0 \end{pmatrix}\begin{pmatrix} \eta_1 \\ \eta_2 \end{pmatrix} + \begin{pmatrix} \gamma_i(\mathbf{x}|\boldsymbol{\theta}_\gamma) \\ 0 \end{pmatrix} \tag{4.66}$$

where $\eta_1' = \frac{\partial\eta_1}{\partial\sigma_i}$.

Proposing a new definition $\frac{\gamma_i(\mathbf{x}|\boldsymbol{\theta}_\gamma)}{\eta_1'}\cdot\frac{1}{a(t)} = \eta_1$, the Eq. (4.66) can be further derived as

$$\dot{\boldsymbol{\eta}} = \eta_1'\begin{pmatrix} -k_1(t) + a(t) & 1 \\ -k_2(t) & 0 \end{pmatrix}\begin{pmatrix} \eta_1 \\ \eta_2 \end{pmatrix} = \eta_1'\mathbf{A}\boldsymbol{\eta} \tag{4.67}$$

A candidate equation for determining the Lyapunov equation of the controlled closed-loop system is

$$V(\boldsymbol{\eta}, k_1, k_2) = V_0(\boldsymbol{\eta}) + \frac{1}{2\rho_1}\left(k_1(t) - k_1^*\right)^2 + \frac{1}{2\rho_2}\left(k_2(t) - k_2^*\right)^2 + V_1\left(\tilde{\theta}_\gamma\right) \tag{4.68}$$

where $V_0(\boldsymbol{\eta}) = \frac{1}{2}\boldsymbol{\eta}^T\mathbf{P}\boldsymbol{\eta}$, $V_1\left(\tilde{\theta}_\gamma\right) = \frac{1}{2\varpi}\tilde{\boldsymbol{\theta}}_\gamma^T\tilde{\boldsymbol{\theta}}_\gamma$; k_1^*, k_2^*, ρ_1, ρ_2 and ϖ are both normal numbers; defining a positive definite symmetric matrix $\mathbf{P} = \begin{pmatrix} \chi + 4\varepsilon^2 & -2\varepsilon \\ -2\varepsilon & 1 \end{pmatrix}$; $\tilde{\boldsymbol{\theta}}_\gamma$ is defined by the optimal fuzzy parameter $\boldsymbol{\theta}_\gamma^*$ and the identified fuzzy parameter $\hat{\boldsymbol{\theta}}_\gamma$,$\tilde{\boldsymbol{\theta}}_\gamma = \boldsymbol{\theta}_\gamma^* - \hat{\boldsymbol{\theta}}_\gamma$. So γ_i can be further expressed as

$$\begin{aligned}\tilde{\gamma}_i(\mathbf{x}|\boldsymbol{\theta}_\gamma) &= \gamma_i(\mathbf{x}|\boldsymbol{\theta}_\gamma) - \hat{\gamma}_i(\mathbf{x}|\boldsymbol{\theta}_\gamma) \\ &= \boldsymbol{\theta}_\gamma^{*T}\delta(\mathbf{x}) + \varepsilon_\gamma - \hat{\boldsymbol{\theta}}_\gamma^T\delta(\mathbf{x}) \\ &= \tilde{\boldsymbol{\theta}}_\gamma^T\delta(\mathbf{x}) + \varepsilon_\gamma\end{aligned} \tag{4.69}$$

The derivative of the candidate equation $V_0(\boldsymbol{\eta})$ in the Lyapunov equation in (4.68) to time is

$$\begin{aligned}\dot{V}_0(\boldsymbol{\eta}) &= \dot{\boldsymbol{\eta}}^T\mathbf{P}\boldsymbol{\eta} + \boldsymbol{\eta}^T\mathbf{P}\dot{\boldsymbol{\eta}} \\ &= \eta_1'\boldsymbol{\eta}^T\left(\mathbf{A}^T\mathbf{P} + \mathbf{PA}\right)\boldsymbol{\eta} \\ &= -\eta_1'\boldsymbol{\eta}^T\mathbf{Q}\boldsymbol{\eta}\end{aligned} \tag{4.70}$$

where $\mathbf{Q} = -\left(\mathbf{A}^T\mathbf{P} + \mathbf{PA}\right)$ [3].

According to the Eqs. (4.67) and (4.70), $\mathbf{Q} = \begin{pmatrix} q_{11} & q_{12} \\ q_{21} & q_{22}\end{pmatrix}$ can be defined as

$$\begin{cases} q_{11} = k_1(t)(\chi + 4\varepsilon^2) + 4\varepsilon(-k_2(t) + a(t)) \\ q_{12} = \frac{1}{2}[k_2(t) - 2\varepsilon k_1(t) - a(t) - (\chi + 4\varepsilon^2)] \\ q_{21} = \frac{1}{2}[k_2(t) - 2\varepsilon k_1(t) - a(t) - (\chi + 4\varepsilon^2)] \\ q_{22} = 2\varepsilon \end{cases} \tag{4.71}$$

It can be seen that in the matrix $\mathbf{Q}$, $q_{12} = q_{21}$, satisfying the equation $\mathbf{Q} = -\left(\mathbf{A}^T\mathbf{P} + \mathbf{PA}\right)$ consisting of a positive definite symmetric matrix $\mathbf{P}$.

Defining $k_2(t) = 2\varepsilon k_1(t) + a(t) + (\chi + 4\varepsilon^2)$, the matrix $\mathbf{Q}$ can be derived further as

$$\mathbf{Q} = \begin{pmatrix} 2\chi a(t) + 4\varepsilon(\chi + 4\varepsilon^2) - 2\chi k_1(t) & 0 \\ 0 & 2\varepsilon \end{pmatrix} = \begin{pmatrix} q_{11} & q_{12} \\ q_{21} & q_{22}\end{pmatrix} \tag{4.72}$$

The unknown uncertainty $a(t)$ in $k_2(t)$ cannot be measured or calculated due to its complexity, and can only be approximated by an adaptive fuzzy algorithm $\hat{a}(t) = \hat{\gamma}_i\left(\mathbf{x}\middle|\hat{\boldsymbol{\theta}}_\gamma\right)/(\eta_1'\eta_1)$. Therefore, Eq. (4.72) can be further written as

$$\mathbf{Q} = \begin{pmatrix} 2\chi\hat{a}(t) + 4\varepsilon(\chi + 4\varepsilon^2) - 2\chi k_1(t) & 0 \\ 0 & 2\varepsilon \end{pmatrix} = \begin{pmatrix} q_{11} & 0 \\ 0 & q_{22}\end{pmatrix} \tag{4.73}$$

Define $q_{11} = 2\chi\hat{a}(t) + \Phi$ and $\Phi = 4\varepsilon(\chi + 4\varepsilon^2) - 2\chi k_1(t)$. Therefore, the derivative of the Lyapunov candidate equation $V_0(\boldsymbol{\eta})$ (Eq. (4.70)) can be further deduced as

$$\begin{aligned}\dot{V}_0(\boldsymbol{\eta}) &= -\eta_1'\left(q_{11}\eta_1^2 + q_{22}\eta_2^2\right)\\ &= -2\chi\hat{\gamma}_i\eta_1 - \eta_1'\Phi\eta_1^2 - \eta_1' q_{22}\eta_2^2\\ &= -2\chi\hat{\gamma}_i\eta_1 + 2\chi\gamma_i\eta_1 - 2\chi\gamma_i\eta_1 - \eta_1'\Phi\eta_1^2 - \eta_1' q_{22}\eta_2^2\\ &= 2\chi\tilde{\gamma}_i\eta_1 + \varepsilon_\gamma - 2\chi\gamma_i\eta_1 - \eta_1'\Phi\eta_1^2 - \eta_1' q_{22}\eta_2^2\end{aligned} \tag{4.74}$$

where $\tilde{\gamma}_i = \gamma_i - \hat{\gamma}_i + \varepsilon_\gamma$.

At the same time, the derivative of $V_1\left(\tilde{\boldsymbol{\theta}}_\gamma\right) = \frac{1}{2\varpi}\tilde{\boldsymbol{\theta}}_\gamma^T\tilde{\boldsymbol{\theta}}_\gamma$ can be deduced as

$$\dot{V}_1\left(\tilde{\boldsymbol{\theta}}_\gamma\right) = \frac{1}{\varpi}\tilde{\boldsymbol{\theta}}_\gamma^T\dot{\tilde{\boldsymbol{\theta}}}_\gamma \tag{4.75}$$

According to the Eq. (4.69) and $\dot{\tilde{\boldsymbol{\theta}}}_\gamma = \dot{\boldsymbol{\theta}}_\gamma^* - \dot{\hat{\boldsymbol{\theta}}}_\gamma = -\dot{\hat{\boldsymbol{\theta}}}_\gamma$, Eq. (4.75) can be further written as

$$\dot{V}_1\left(\tilde{\boldsymbol{\theta}}_\gamma\right) = \frac{1}{\varpi}\tilde{\boldsymbol{\theta}}_\gamma^T\dot{\hat{\boldsymbol{\theta}}}_\gamma \tag{4.76}$$

So

$$\begin{aligned}&\dot{V}_0(\boldsymbol{\eta}) + \dot{V}_1\left(\tilde{\boldsymbol{\theta}}_\gamma\right)\\ &= 2\chi\tilde{\gamma}_i\eta_1 + \frac{1}{\varpi}\tilde{\boldsymbol{\theta}}_\gamma^T\dot{\hat{\boldsymbol{\theta}}}_\gamma - 2\chi\gamma_i\eta_1 - \eta_1'\Phi\eta_1^2 - \eta_1' q_{22}\eta_2^2 + \varepsilon_\gamma\\ &= \tilde{\boldsymbol{\theta}}_\gamma^T\left[2\chi\delta(\mathbf{x})\eta_1 - \frac{1}{\varpi}\dot{\hat{\boldsymbol{\theta}}}_\gamma\right] - 2\chi\gamma_i\eta_1 - \eta_1'\Phi\eta_1^2 - \eta_1' q_{22}\eta_2^2 + \varepsilon_\gamma\end{aligned} \tag{4.77}$$

The self-learning law for an adaptive fuzzy algorithm is

$$\dot{\hat{\boldsymbol{\theta}}}_\gamma = 2\chi\varpi\eta_1\delta(\mathbf{x}) \tag{4.78}$$

4.2.4 Stability Analysis of the Closed-Loop System

The closed-loop input–output system based on the control algorithm in this section can be asymptotically stable for a limited time, driven by the self-learning law (4.78).

The specific proof is as follows:

According to the fuzzy self-learning law in Eq. (4.78), Eq. (4.77) can be further deduced:

$$\begin{aligned}
&\dot{V}_0(\boldsymbol{\eta}) + \dot{V}_1\left(\tilde{\boldsymbol{\theta}}_\gamma\right) \\
&= -2\chi\gamma_i\eta_1 - \eta_1'\Phi\eta_1^2 - \eta_1' q_{22}\eta_2^2 + \varepsilon_\gamma \\
&= -2\chi a(t)\eta_1'\eta_1\eta_1 - \eta_1'\Phi\eta_1^2 - \eta_1' q_{22}\eta_2^2 + \varepsilon_\gamma \\
&= -\eta_1'[2\chi a(t) + \Phi]\eta_1^2 - \eta_1' q_{22}\eta_2^2 + \varepsilon_\gamma \\
&= -\eta_1'\bar{q}_{11}\eta_1^2 - \eta_1' q_{22}\eta_2^2 + \varepsilon_\gamma \\
&= -\eta_1'\boldsymbol{\eta}^T\bar{\mathbf{Q}}\boldsymbol{\eta} + \varepsilon_\gamma
\end{aligned} \tag{4.79}$$

where $\bar{\mathbf{Q}} = \begin{pmatrix} [2\chi\gamma_i(\mathbf{x}|\boldsymbol{\theta}_\gamma)]/(\eta_1'\eta_1) + 4\varepsilon(\chi + 4\varepsilon^2) - 2\chi k_1(t) & 0 \\ 0 & 2\varepsilon \end{pmatrix}$.

Assuming that the approximation error ε_γ of the adaptive fuzzy approximation algorithm is sufficiently small, an inequality $\frac{1}{4}\varepsilon \geq \varepsilon_\gamma + D$ can be defined where D is an arbitrarily small normal number.

In order to ensure that the candidate equation $V_0(\boldsymbol{\eta}) + V_1\left(\tilde{\boldsymbol{\theta}}_\gamma\right)$ in the Lyapunov equation satisfies the Lyapunov theorem and the system is asymptotically stable in a finite time. Define $\bar{\bar{\mathbf{Q}}} = \bar{\mathbf{Q}} - \frac{1}{4}\varepsilon\mathbf{I}$ and the definition of the expression $\bar{\bar{\mathbf{Q}}}$ is

$$\bar{\bar{\mathbf{Q}}} = \begin{pmatrix} 2\chi a(t) + 4\varepsilon(\chi + 4\varepsilon^2) - 2\chi k_1(t) - \frac{1}{4}\varepsilon & 0 \\ 0 & \frac{7}{8}\varepsilon \end{pmatrix} = \begin{pmatrix} \bar{\bar{q}}_{11} & 0 \\ 0 & \bar{\bar{q}}_{22} \end{pmatrix} \tag{4.80}$$

According to the Schur complement theorem, a sufficient condition to guarantee the positive definiteness of the matrix $\bar{\bar{\mathbf{Q}}}$ and its minimum eigenvalue $\lambda_{\min}\left(\bar{\bar{\mathbf{Q}}}\right) > 0$ is

$$\begin{cases} \bar{\bar{q}}_{11} \geq 0 \, and \; \bar{\bar{q}}_{22} \geq 0 \\ \det(\bar{\bar{\mathbf{Q}}}) \end{cases} \tag{4.81}$$

From the sliding mode algorithm adaptive law $k_2(t) = 2\varepsilon k_1(t)$ + $\hat{\gamma}_i(\mathbf{x}\big|\hat{\theta}_\gamma)/(\eta_1'\eta_1) + (\chi + 4\varepsilon^2)$, we can get all the constraints:

$$\begin{cases} \varepsilon > 0 \\ k_1(t) < \hat{a}(t) + \frac{2\varepsilon(\chi + 4\varepsilon^2)}{\chi} - \frac{\varepsilon}{8\chi} \\ k_2(t) = 2\varepsilon k_1(t) + \hat{\gamma}_i(\mathbf{x}|\boldsymbol{\theta}_\gamma)/\eta_1'\eta_1 + \chi + 4\varepsilon^2 \end{cases} \tag{4.82}$$

$$\begin{pmatrix} 2[k_1(t)(\chi + 4\varepsilon^2) + 2\varepsilon(-k_2(t) + a(t))] - \frac{\varepsilon}{4} & \aleph \\ k_2(t) - 2\varepsilon k_1(t) - a(t) - (\chi + 4\varepsilon^2) & \frac{15}{4}\varepsilon \end{pmatrix} \tag{4.83}$$

According to the formulas (4.70) and (4.82), the numerical value of $V_0(\boldsymbol{\eta}) + V_1\left(\tilde{\boldsymbol{\theta}}_\gamma\right)$ can be further constrained to

$$\begin{aligned}\dot{V}_0(\boldsymbol{\eta}) + \dot{V}_1(\tilde{\boldsymbol{\theta}}_\gamma) &= -\eta_1'\boldsymbol{\eta}^T\bar{\bar{\mathbf{Q}}}\boldsymbol{\eta} \leq -\frac{\varepsilon}{4}\eta_1'\boldsymbol{\eta}^T\boldsymbol{\eta} \\ &= -\frac{\varepsilon}{4}\left(\frac{1}{2}|\sigma|^{-\frac{1}{2}} + k_3\right)\boldsymbol{\eta}^T\boldsymbol{\eta}\end{aligned} \tag{4.84}$$

According to the definition of positive definite quadratic equation $\bar{V}(\boldsymbol{\eta}, \tilde{\boldsymbol{\theta}}_\gamma) = V_0(\boldsymbol{\eta}) + V_1(\tilde{\boldsymbol{\theta}}_\gamma) = \frac{1}{2}\boldsymbol{\eta}^T\mathbf{P}\boldsymbol{\eta} + \frac{1}{2\varpi}\tilde{\boldsymbol{\theta}}_\gamma^T\tilde{\boldsymbol{\theta}}_\gamma$, we can get

$$\left[\lambda_{\min}(\mathbf{P})\|\boldsymbol{\eta}\|_2^2 + \frac{1}{\varpi}\|\tilde{\boldsymbol{\theta}}_\gamma\|_2^2\right] \leq 2\bar{V}(\boldsymbol{\eta}, \tilde{\boldsymbol{\theta}}_\gamma) \leq \left[\lambda_{\max}(\mathbf{P})\|\boldsymbol{\eta}\|_2^2 + \frac{1}{\varpi}\|\tilde{\boldsymbol{\theta}}_\gamma\|_2^2\right] \tag{4.85}$$

where: $\|\boldsymbol{\eta}\|_2$ is the Euclidean norm of $\boldsymbol{\eta}$:

$$\|\boldsymbol{\eta}\|_2^2 = \eta_1^2 + \eta_2^2 = |\sigma| + 2k_3|\sigma|^{\frac{3}{2}} + k_3^2\sigma^2 + \hat{v}^2 \tag{4.86}$$

Designing a new function $\frac{1}{\varpi}\|\tilde{\boldsymbol{\theta}}_\gamma\|_{2\Upsilon(t)}^{2\ 1} = \|\boldsymbol{\eta}\|_2^2$, we can get the following inequality:

$$|\eta_1| \leq \|\boldsymbol{\eta}\|_2 \leq \frac{2\bar{V}^{\frac{1}{2}}(\boldsymbol{\eta}, \tilde{\boldsymbol{\theta}}_\gamma)}{[\lambda_{\min}(\mathbf{P}) + \Upsilon(t)]^{\frac{1}{2}}} \tag{4.87}$$

$$\|\boldsymbol{\eta}\|_2 \geq \frac{2\bar{V}^{\frac{1}{2}}(\boldsymbol{\eta}, \tilde{\boldsymbol{\theta}}_\gamma)}{[\lambda_{\max}(\mathbf{P}) + \Upsilon(t)]^{\frac{1}{2}}} \tag{4.88}$$

And according to the Eqs. (4.86) and (4.87), we can further obtain inequality:

$$|\sigma|^{-\frac{1}{2}} \geq \frac{[\lambda_{\min}(\mathbf{P}) + \Upsilon(t)]^{\frac{1}{2}}}{2\bar{V}^{\frac{1}{2}}(\boldsymbol{\eta}, \tilde{\boldsymbol{\theta}}_\gamma)} \tag{4.89}$$

So, $\dot{\bar{V}}(\boldsymbol{\eta}, \tilde{\boldsymbol{\theta}}_\gamma)$ can be eventually derived as

$$\begin{aligned}\dot{\bar{V}}(\boldsymbol{\eta}, \tilde{\boldsymbol{\theta}}_\gamma) &= -\frac{\varepsilon}{8}|\sigma|^{-\frac{1}{2}}\|\boldsymbol{\eta}\|_2^2 - \frac{1}{4}\varepsilon k_3\|\boldsymbol{\eta}\|_2^2 \\ &\leq -\frac{\varepsilon}{8}|\sigma|^{-\frac{1}{2}}\frac{2\bar{V}(\boldsymbol{\eta}, \tilde{\boldsymbol{\theta}}_\gamma)}{[\lambda_{\max}(\mathbf{P}) + \Upsilon(t)]^{\frac{1}{2}}} - \frac{1}{4}\varepsilon\frac{2\bar{V}(\boldsymbol{\eta}, \tilde{\boldsymbol{\theta}}_\gamma)}{[\lambda_{\max}(\mathbf{P}) + \Upsilon(t)]^{\frac{1}{2}}} \\ &\leq -\frac{\varepsilon}{8}\frac{[\lambda_{\min}(\mathbf{P}) + \Upsilon(t)]^{\frac{1}{2}}}{[\lambda_{\max}(\mathbf{P}) + \Upsilon(t)]^{\frac{1}{2}}}\bar{V}^{\frac{1}{2}}(\boldsymbol{\eta}, \tilde{\boldsymbol{\theta}}_\gamma) - \frac{1}{2}\frac{\varepsilon k_3\bar{V}(\boldsymbol{\eta}, \tilde{\boldsymbol{\theta}}_\gamma)}{[\lambda_{\max}(\mathbf{P}) + \Upsilon(t)]^{\frac{1}{2}}}\end{aligned} \tag{4.90}$$

Equation (4.90) can be simplified and rewritten as

$$\dot{\bar{V}}(\boldsymbol{\eta}, \tilde{\boldsymbol{\theta}}_\gamma) \le -\gamma_1 \bar{V}^{-\frac{1}{2}}(\boldsymbol{\eta}, \tilde{\boldsymbol{\theta}}_\gamma) - \gamma_2 V_0(\boldsymbol{\eta}) \tag{4.91}$$

where $\gamma_1 = \left\{\varepsilon[\lambda_{\min}(\mathbf{P}) + \Upsilon(t)]^{\frac{1}{2}}\right\} / \left\{8[\lambda_{\max}(\mathbf{P}) + \Upsilon(t)]^{\frac{1}{2}}\right\}$, $\gamma_2 = \varepsilon k_3 / \left\{2[\lambda_{\max}(\mathbf{P}) + \Upsilon(t)]^{\frac{1}{2}}\right\}$.

Thus, with the control algorithm proposed in this section, for any initial state $\mathbf{x}(0)$ and corresponding initial sliding mode parameter $\boldsymbol{\sigma}(0)$, the sliding mode surface $\sigma = \dot{\sigma} = 0$ composed of sliding mode parameters can be reached within a limited time.

Integrate Eq. (4.91), we can get

$$\bar{V}(\boldsymbol{\eta}, \tilde{\boldsymbol{\theta}}_\gamma) = e^{-\gamma_2 t}\left[\bar{V}^{\frac{1}{2}}(\mathbf{0}) + \frac{\gamma_1}{\gamma_2}\left(1 - e^{\frac{\gamma_1}{\gamma_2}t}\right)\right]^2 \tag{4.92}$$

Therefore, the candidate equation of the Lyapunov equation $\bar{V}(\boldsymbol{\eta}, \tilde{\boldsymbol{\theta}}_\gamma) = \frac{1}{2}\boldsymbol{\eta}^T\mathbf{P}\boldsymbol{\eta} + \frac{1}{2\varpi}\tilde{\boldsymbol{\theta}}_\gamma^T\tilde{\boldsymbol{\theta}}_\gamma$ can converge to zero within a finite time, and this finite time can be specifically expressed as [4]:

$$T_f = \frac{2}{\gamma_2}\ln\left(\frac{\gamma_1}{\gamma_2}\bar{V}^{\frac{1}{2}}(\mathbf{0}) + 1\right) \tag{4.93}$$

So far it can be concluded that under the control of the control algorithm (4.57) and the corresponding adaptive laws $k_1(t)$ and $k_2(t)$, the variable $\boldsymbol{\eta}^T = \left(\left(|\sigma|^{\frac{1}{2}}\sigma\right)/\|\sigma\|_2 + k_3\sigma \quad \hat{v}\right)$ can converge to zero within a limited time. In this case, there is always two constants k_1^* and k_2^* so that in any initial state the sliding surface $\sigma = \dot{\sigma} = 0$ is reachable.

The final derivative of the Lyapunov equation $V(\boldsymbol{\eta}, \tilde{\boldsymbol{\theta}}_\gamma, k_1, k_2)$ is

$$\begin{aligned}
&\dot{V}(\boldsymbol{\eta}, \tilde{\boldsymbol{\theta}}_\gamma, k_1, k_2) \\
&\le -\gamma_1\bar{V}^{\frac{1}{2}}(\boldsymbol{\eta}, \tilde{\boldsymbol{\theta}}_\gamma) - \gamma_2\bar{V}(\boldsymbol{\eta}, \tilde{\boldsymbol{\theta}}_\gamma) + \frac{1}{\rho_1}(k_1(t) - k_1^*) + \frac{1}{\rho_2}(k_2(t) - k_2^*) \\
&\le -\bar{\gamma}\bar{V}^{\frac{1}{2}}(\boldsymbol{\eta}, \tilde{\boldsymbol{\theta}}_\gamma) - \frac{\omega_1}{\sqrt{2\rho_1}}|k_1 - k_1^*| - \frac{\omega_2}{\sqrt{2\rho_2}}|k_2 - k_2^*| + \frac{1}{\rho_1}(k_1 - k_1^*) \\
&\quad + \frac{1}{\rho_2}(k_2 - k_2^*) + \frac{\omega_1}{\sqrt{2\rho_1}}|k_1 - k_1^*| + \frac{\omega_2}{\sqrt{2\rho_2}}|k_2 - k_2^*| \\
&\le -\min(r_1, \omega_1, \omega_2)\left[\bar{V}(\boldsymbol{\eta}, \tilde{\boldsymbol{\theta}}_\gamma) + \frac{1}{2\rho_1}(k_1 - k_1^*)^2 + \frac{1}{2\rho_2}(k_2 - k_2^*)^2\right]^{\frac{1}{2}} \\
&\quad + \frac{1}{\rho_1}(k_1 - k_1^*)\dot{k}_1 + \frac{1}{\rho_2}(k_2 - k_2^*)\dot{k}_2 + \frac{\omega_1}{\sqrt{2\rho_1}}|k_1 - k_1^*| + \frac{\omega_2}{\sqrt{2\rho_2}}|k_2 - k_2^*|
\end{aligned} \tag{4.94}$$

When the proposed adaptive law (4.58) is used, the adaptive parameters $k_1(t)$ and $k_2(t)$ are bounded. So for $\forall t \geq 0$, there are always k_1^* and k_2^* making $k_1(t) - k_1^* < 0$ and $k_2(t) - k_2^* < 0$. Then Eq. (4.94) can be further deduced as

$$\dot{V}\left(\mathbf{\eta}, \tilde{\mathbf{\theta}}_\gamma, \alpha, \beta, \vartheta\right) \leq -\min(\gamma, \omega_1, \omega_2) V^{\frac{1}{2}} + \Pi \tag{4.95}$$

where $\Pi = -\left|k_1 - k_1^*\right|\left(\frac{1}{\rho_1}\dot{k}_1 - \frac{\omega_1}{\sqrt{2\rho_1}}\right) - \left|k_2 - k_2^*\right|\left(\frac{1}{\rho_2}\dot{k}_2 - \frac{\omega_2}{\sqrt{2\rho_2}}\right)$.

In order to ensure the closed-loop system's asymptotic stability over a finite time, it needs to be guaranteed $\Pi = 0$. So we can get the adaptive laws of $k_1(t)$ and $k_2(t)$:

$$\dot{k}_1(t) = \omega_1\sqrt{\frac{\rho_1}{2}}, \quad \dot{k}_2(t) = \omega_2\sqrt{\frac{\rho_2}{2}} \tag{4.96}$$

If the sliding mode parameter σ_i in the sliding mode control algorithm is specifically selected as $\sigma_i = (d/dt + \lambda)e_i$ and $e_i = x_{id} - x_i$ is the tracking error of the system state, it can be obtained under the control algorithm (4.57), the sliding mode adaptive law (4.58), the fuzzy approximation algorithm (4.61), and the fuzzy algorithm self-learning law (4.78), the tracking error of the system state can satisfy $e_i \to 0$ in limited time.

After constructing the new sliding mode parameter $\mathbf{\eta}^T(\sigma_i) = \left(\sigma_i/\|\sigma_i\|_2^{\frac{1}{2}} + k_3\sigma_i\right) v_i$, both the sliding mode parameter σ and its derivative $\dot{\sigma}$ can reach to zero within a limited time, which means the tracking error of the system state can converge to zero within a limited time.

4.2.5 *Numerical Simulation*

The simulation environment is similar to the previous section. In this section, the MUs open the actuator at time $t = 38.09$ s, and the positions of the four MUs at the end of free-flying phase are $(1.4378, 1.4390, 1.3162)$, $(1.4378, 1.4390, -1.3162)$, $(1.4378, -1.4390, 1.3162)$ and $(1.4378, -1.4390, -1.3162)$; corresponding to this initial state, the expected states of the four MUs are $(1.4378 + 0.1(t - 38.09), 1.7, 1.7)$, $(1.4378 + 0.1(t - 38.09), 1.7, -1.7)$, $(1.4378 + 0.1(t - 38.09), -1.7, 1.7)$, and $(1.4378 + 0.1(t - 38.09), -1.7, -1.7)$. The desired speed for the four MUs in the direction x is 0.1 m/s, and the desired speed in the directions y and z is zero.

Based on the above simulation environment, the dynamic equations of the system can be divided into the dynamic equations of the MUs and the kinematic part of the net. The kinematic part of the net is mainly used to calculate the interference and the unknown uncertainties formed by the superposition of vibrations acting on each MU. The fuzzy approximation algorithm with self-learning law approaches this unknown uncertainty, and then the approximation result is input to the adaptive law of the sliding mode algorithm. Sliding mode adaptive laws with interference terms can adjust the final controller output to complete the control of the MUs.

Since the algorithm proposed in this section is for unknown uncertainties, the unknown model uncertainty is first given in the numerical simulation. In addition, another major interference is the equivalent external force acting on each MU after the vibration of the entire net is superimposed. In order to distinguish the unknown uncertainties of the connected tethers from the net vibration forces, the model error and space environmental disturbance force are not added during the free-flying phase. After the controller is opened, these two disturbances are joined. Figures 4.18, 4.19, and 4.20 show the unknown uncertain forces of the four MUs in the directions x, y, and z. It can be seen that in the free flying phase, since only the vibration of the net is superimposed, the interference force of the system is small. But when the main tether is suddenly connected with the pulling force, the actuator itself will also exacerbate the net oscillation. Therefore, it is unknown that the indeterminate force is nearly three times that of the free flight segment. The simulation results directly show that the unknown uncertainties acting on each MU are not the same, which achieves the purpose of testing the control algorithm.

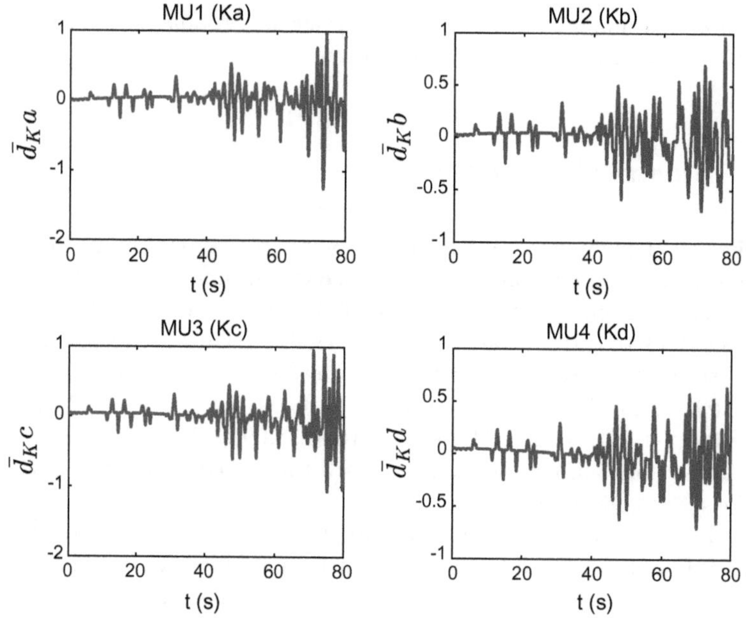

Fig. 4.18 The unknown uncertain on x-position

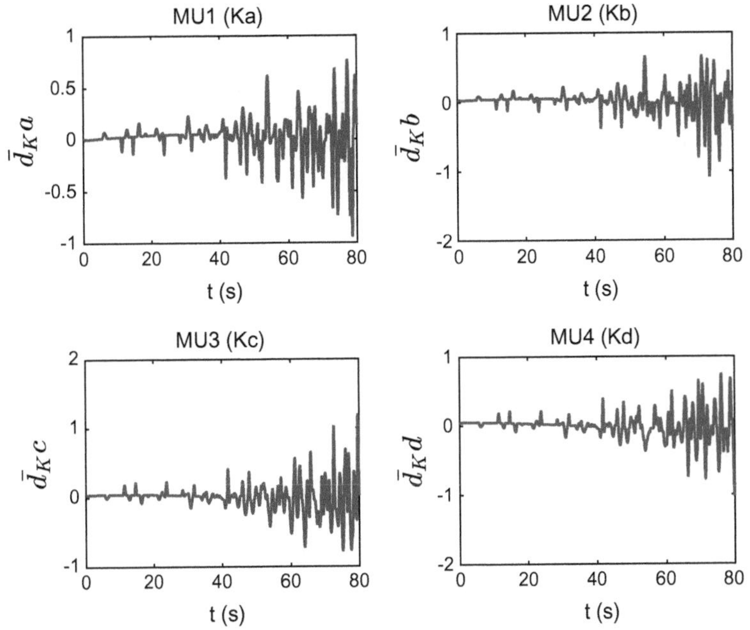

Fig. 4.19 The unknown uncertain on y-position

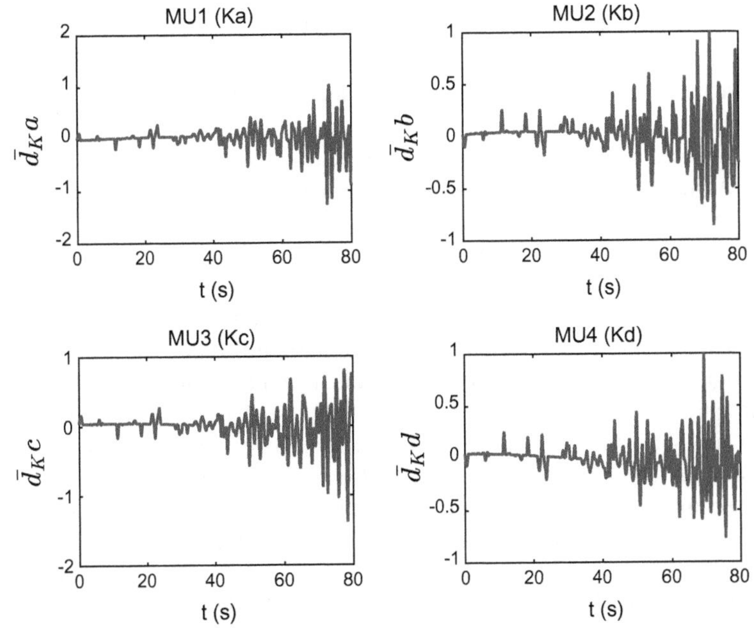

Fig. 4.20 The unknown uncertain on z-position

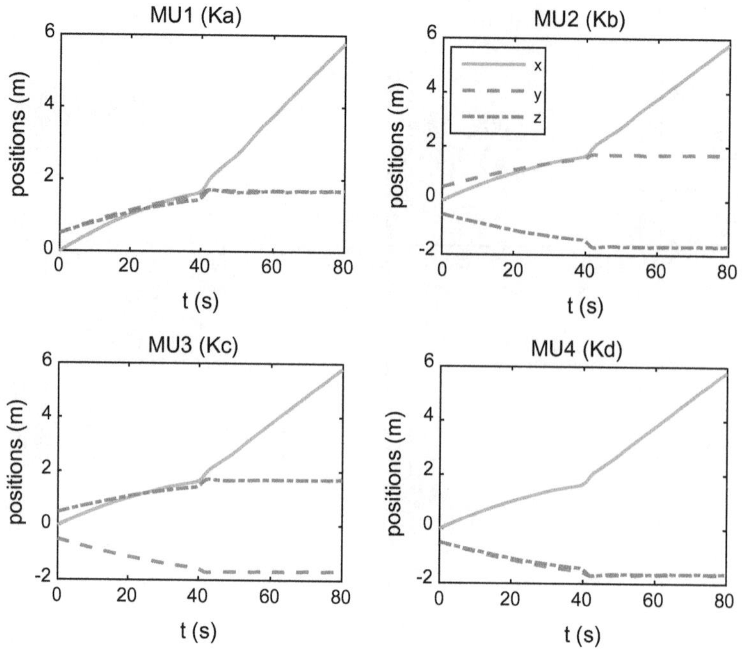

Fig. 4.21 Positions of MUs

Figures 4.21 and 4.22 show the positions and velocities of the MUs in the three directions x, y, and z. According to the simulation environment, there is still a gap between the MU state and the expected state when the controller is turned on. Under the above-mentioned continuous and complex interference, all position states of the MUs can be stabilized and rapidly approached to the desired state and remain stable. Although the speed state is slightly dithered after reaching the desired value, the vibration frequency and amplitude are acceptable.

Figures 4.23, 4.24, and 4.25 show the sliding mode parameters of the MUs in the three directions x, y, and z, respectively. According to the design of sliding mode parameters $\sigma_i = (d/dt + \lambda)e_i$, it can be obtained that the sliding mode parameters are the first-order combination of system position and velocity. In the simulation, $\lambda = 1$. From the simulation results, it can be seen that the status of each MU is quickly reached by the control algorithm and reaches the sliding surface $\sigma_i = 0$, and almost remains on the sliding surface. By comparison, it was found that the sliding mode parameters of the four MUs in the direction x remained the worst after reaching zero. This is because, in directions y and z, the desired position state and velocity state are constant, but in the direction x, the position expectation is a dynamic function. This leads to the need for the actuator to work in order to track the desired state during the control process. It can also be seen from Fig. 4.22 that the speed stability in the direction x is slightly worse than the other two directions. In general, the simulation results of the sliding mode parameters further confirm the

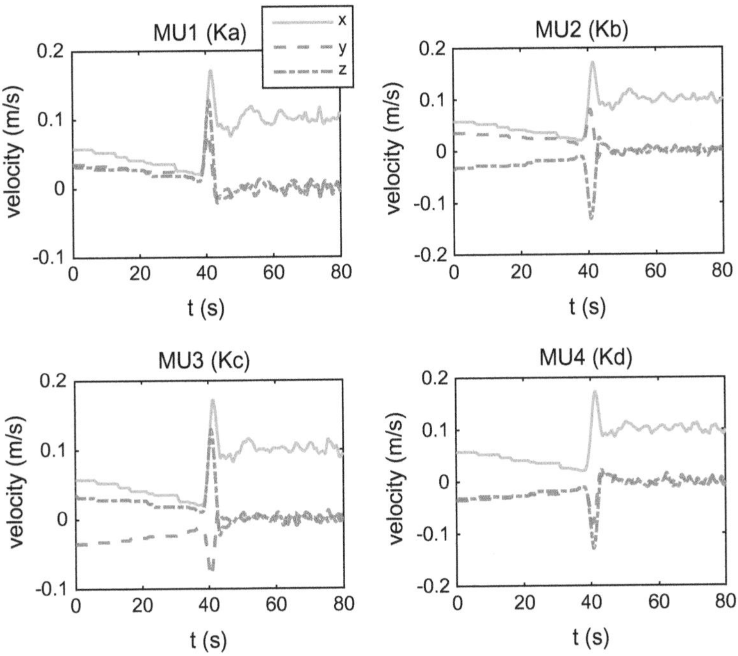

Fig. 4.22 Velocities of Mus

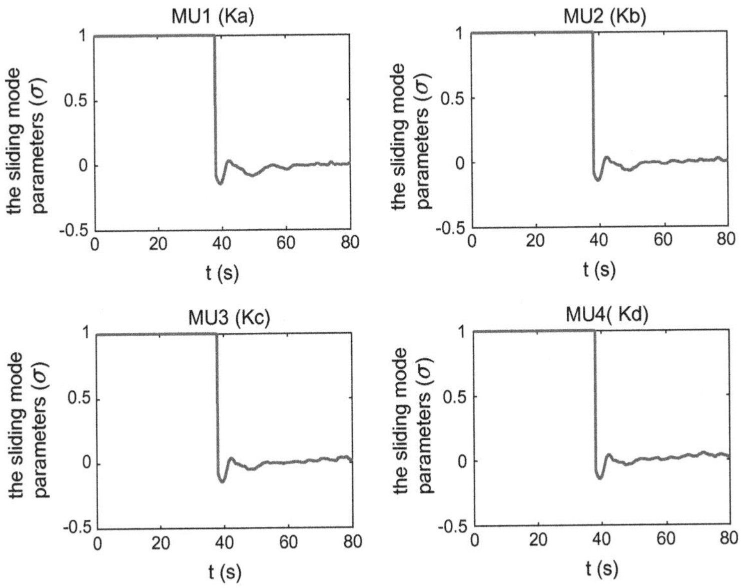

Fig. 4.23 The sliding mode parameters on position-x

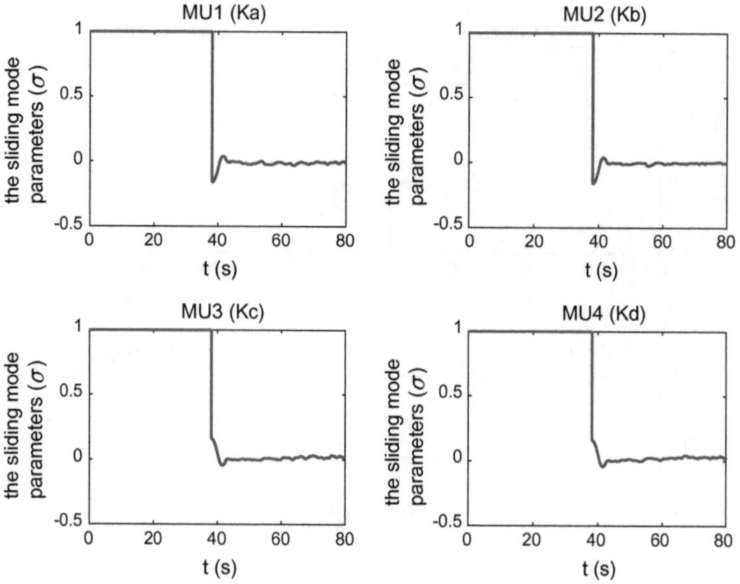

Fig. 4.24 The sliding mode parameters on position-y

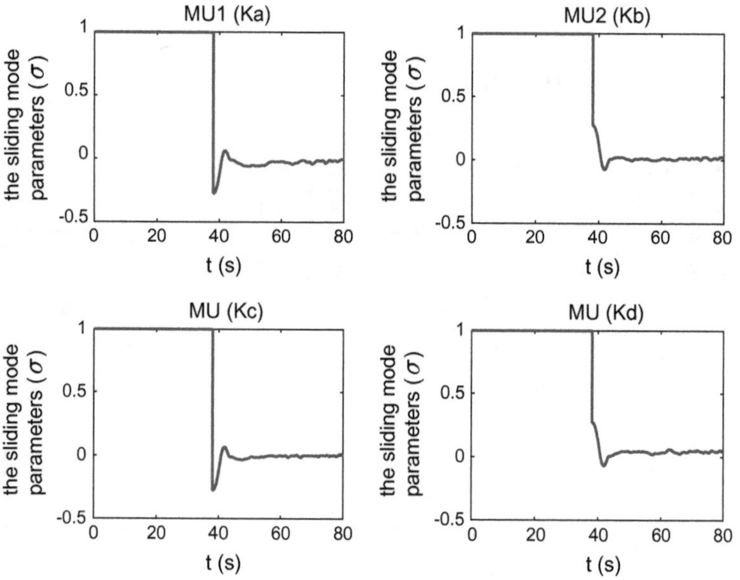

Fig. 4.25 The sliding mode parameters on position-z

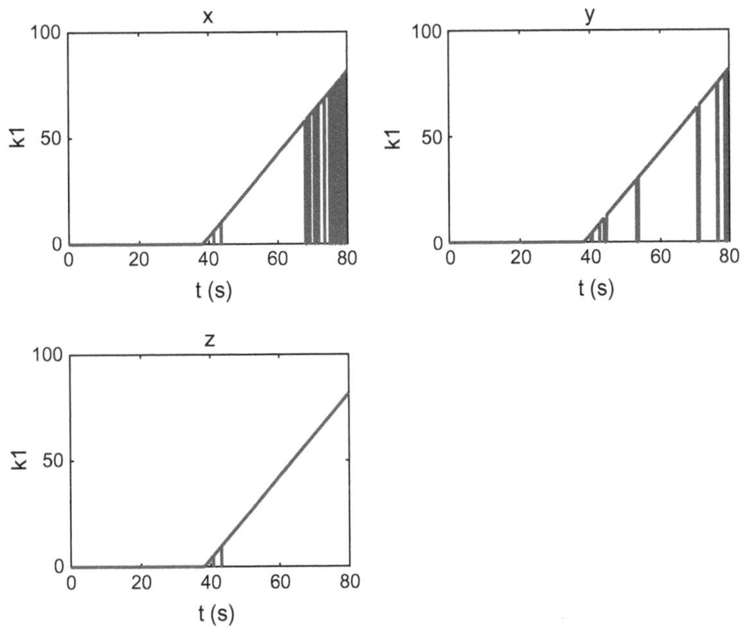

Fig. 4.26 K1 of MU1 (Ka)

simulation results of the MUs in Fig. 4.21, that is, tracking the desired state in real time and maintaining stability.

Figures 4.26, 4.27, 4.28, and 4.29 show the adaptive law in the algorithm proposed in this section. Figures 4.26 and 4.27 are adaptive laws $k_1(t)$ of MU1 and MU2 in three directions x, y, and z. Figures 4.28 and 4.29 are the adaptive laws $k_2(t)$ of the MU3 and MU4 in the three directions x, y and z. As the system state enters the sliding mode boundary layer, $\dot{k}_1(t) = 0$, the instantaneous integral value of the adaptive law $k_1(t)$ is zero. Therefore, in the simulation results shown in Figs. 4.26 and 4.27, there are more longitudinal solid lines. Compared to other results, the MU2 (Kb) has a longer dwelling surface dwelling time in the direction y. From the adaptive law $k_2(t) = 2\varepsilon k_1(t) + \left[\hat{\boldsymbol{\theta}}_\gamma \delta(\mathbf{x})\right] / (\eta' \eta_1) + \chi + 4\varepsilon^2$, we know that when the system state enters the boundary layer of the sliding surface and $|\sigma_i| < v$, the momentary integral value of $\dot{k}_1(t) = 0$ and $k_1(t)$ is zero, and $k_2(t)$ can be simplified as $k_2(t) = \left[\hat{\boldsymbol{\theta}}_\gamma \delta(\mathbf{x})\right] / (\eta' \eta_1) + \chi + 4\varepsilon^2$. It can be seen that the changes of $k_2(t)$ are mainly used to overcome unknown uncertainty $\hat{\gamma}$.

Figures 4.30, 4.31, 4.32 show the control input of the MUs in x, y and z. From the simulation results, we can see that in addition to the moment the controller is turned on, due to the relatively large error between the actual state and the desired state, the controller gives a high amplitude control force. During the other phases of the control process, the amplitude and frequency of control force output are much

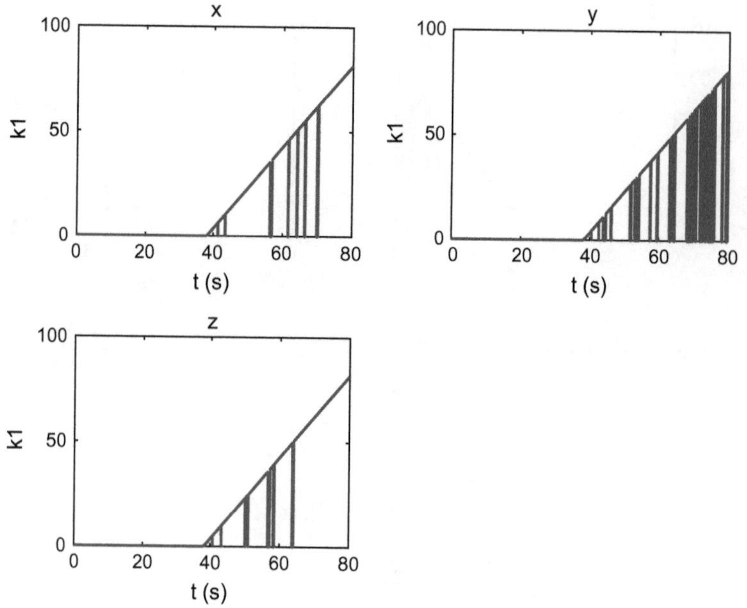

Fig. 4.27 K1 of MU2 (Kb)

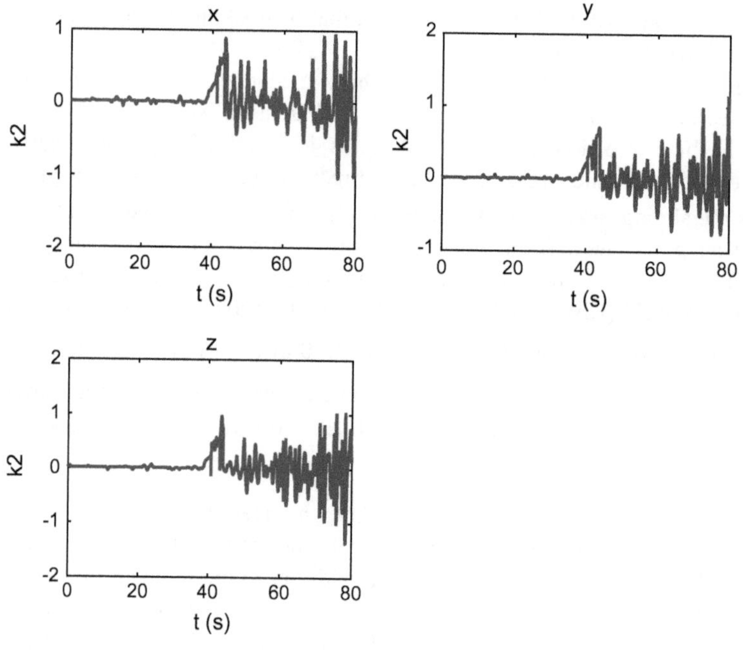

Fig. 4.28 K2 of MU3 (Kc)

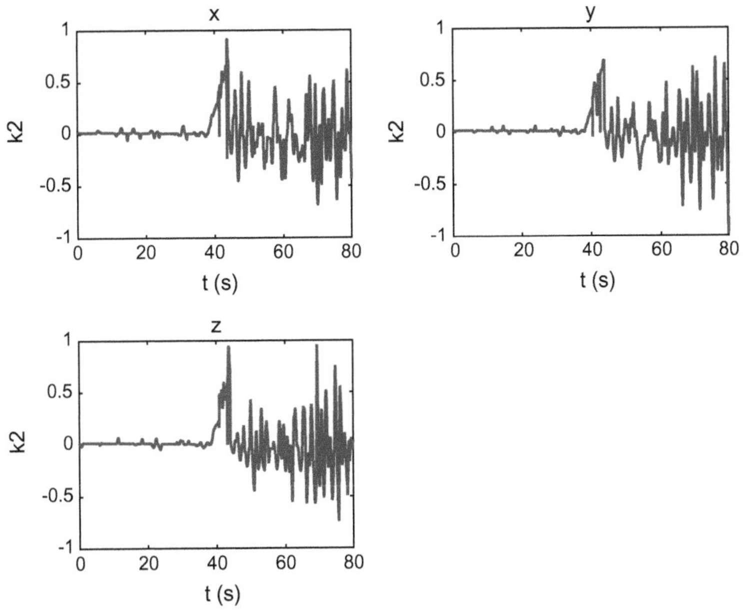

Fig. 4.29 K2 of MU4 (Kd)

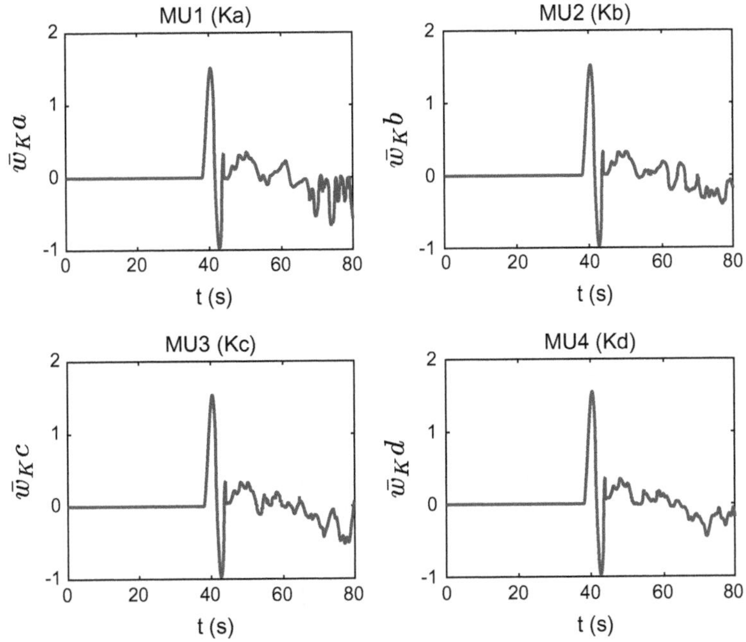

Fig. 4.30 Control force on position-x

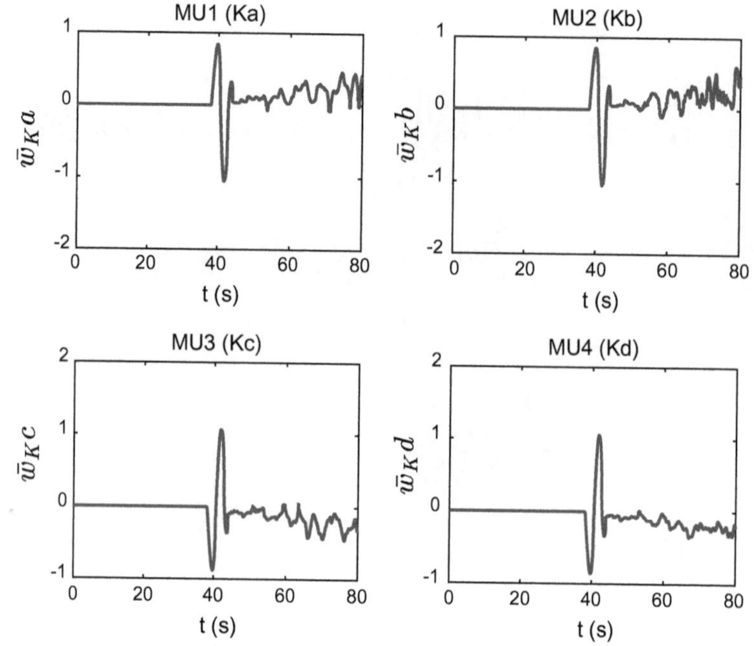

Fig. 4.31 Control force on position-y

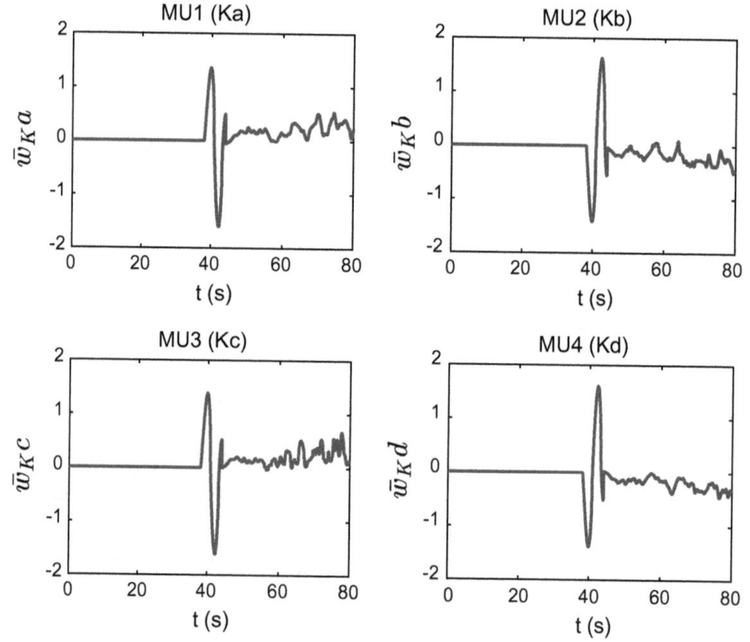

Fig. 4.32 Control force on position-z

smaller than the traditional first-order sliding mode control. Comparing the interference force and the control force, it can be found that the disturbance force, whether in amplitude or in frequency, is higher than the control force. That is, facing higher frequency higher amplitude interference force, through the effective controller design, using only a small amount of control force, to suppress the disturbance of the entire system of disturbance. The whole set of algorithms proposed in this section include control algorithm, sliding mode adaptive law, fuzzy approximation algorithm, and fuzzy algorithm self-learning law. In the face of unknown uncertainty, the desired trajectory can be tracked quickly and efficiently to ensure that the MSTN can move steadily toward the targets.

4.3 Coordinated Control for the Maneuverable Space Tethered Net

Compared with previous studies, the model in this section can reflect the variation of the net shape and the coupling between the four maneuvering robots. Meanwhile, the designed controller has the ability to ensure the net to move along the expected trajectory and keep the required shape. First, considering that the material of the space net has extremely high stiffness when in tension and the braided structure will become slack under axial pressure, we assume that Young's modulus is infinite in the former case and zero in the latter case. Thus, the effects of the net on the motion of maneuvering robots can be split into two aspects: one is the inertia and gravitational force caused by the distributed mass, which will be transmitted to the robots; the other is unilateral state constraints caused by the inextensible cord in the net. Second, we divide the net into four sections and use the T3 element in the shell theory to approximate the shape of one section, which makes the distributed mass of the net equivalent to the lumped mass at the net center and the four robots. Meanwhile, we employ the theory about the contact dynamics of the rigid robot [7] to describe the effect of the unilateral state constraints. Based on the extended Hamilton's principle, we obtain the dynamical equations accompanied by a linear complementarity problem which should be solved before getting the solution of dynamical equations. Third, given the fact that the two ends of an inextensible cord will have the same velocity when it suddenly becomes tent, we use a similar method as in [8] to obtain the velocity jump of points in the net. Lastly, referring to the contact control of rigid robots [9, 10], we designed a coordinated controller which is composed of the classical PD adjuster and the inverse dynamics module. However, since the shape of the net is not only decided by the four maneuvering robots, the conventional Jacobian matrix and Moore–Penrose pseudo-inverse are not applicable for the inverse dynamics of MSTN. Therefore, we propose a method to transfer the inverse dynamics solving problem to a double-level optimization problem.

4.3.1 Improved Dynamic Model

Maneuverable space tethered net (MSTN) is composed of the connecting tether, the flexible net and four rigid robots. The dynamics of the rigid-flexible combinative system is too complicated to build an accurate dynamic model with all details. In order to analyze the dynamics of MSTN, we first introduce the following five basic assumptions:

(i) The platform moves along a nearly circular Keplerian orbit. The mass of the platform is far heavier than the connecting tether, the flexible net, and the mass of the maneuvering robots. Its orbit and attitude are not affected by MSTN. Therefore, it can be simplified as a mass point.

(ii) Due to the passive deployment mechanism, there will always be a certain tension in the connecting tether during the whole approaching phase. In addition, the stiffness of the tether is very high while its density is very low. Hence, it can be approximated as a massless rod.

(iii) Compared with the flexible net, the volume of the maneuvering robot is so small that it can be neglected. Therefore, as in Fig. 4.33, the robot is simplified to a controllable mass point in building the dynamic model.

(iv) When the flexible net is fully unfolded, the mass distribution is uniform and the mesh of the net is very small compared with the whole size of the fully expanded net. In the approaching phase, due to the requirement to keep the net mouth fully opened, the flexible net will not deform greatly. Therefore, the net can be approximated as a uniform shell.

(v) The tether connecting point is in the center of the net. As shown in Fig. 4.33, the four cords which connect the net center and the maneuvering robots equally divide the whole net into four sections. In the approaching phase, all sections are approximately flattened by the maneuvering robots.

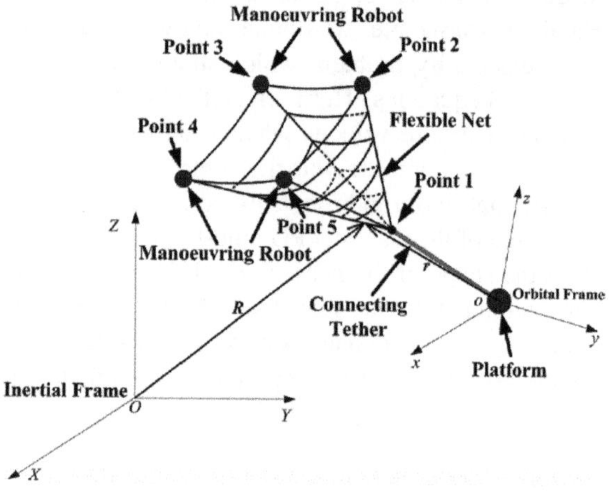

Fig. 4.33 Simplified description of MSTN

Since the length is far bigger than the diameter of the section, the cord used to braid the flexible net will become slack when it is under axial pressure. Therefore, the tension–strain relation can be written as

$$N = \begin{cases} EA(\varepsilon + \alpha\dot{\varepsilon}) & \varepsilon \geq 0 \\ 0 & \varepsilon < 0 \end{cases} \tag{4.98}$$

where ε is the strain of the tether, α is the damping coefficient, A is the sectional area of the cord, and E is Young's modulus of the cord. Generally speaking, the net is made of fibers with extremely high stiffness. The Young's modulus of these fibers can be more than 130 GPa. In addition, since the movement of the net in the approaching phase is rather smooth, the tension in the net should not be too big. Therefore, the strain of the tether is so small that it can be neglected. Hence, in order to further simplify the description of the net, we introduce the following approximation:

$$\lim_{\varepsilon \to 0^+} E = +\infty \tag{4.99}$$

This means that the net can be arbitrarily folded while it is impossible to be elongated. Therefore, the elastic potential of the net can be calculated as

$$\Pi = \int_0^{\varepsilon} N \mathrm{d}\varepsilon \to 0 \tag{4.100}$$

Hence, it can be neglected in dynamics modeling.

In order to facilitate the description of MSTN, we first number the tether connecting point and the four maneuvering robots as shown in Fig. 4.33. Then, we introduce the following two coordinate systems:

(i) The inertial frame *OXYZ*: the origin *O* is located in the center of the Earth; the *X*-axis is toward the ascending point of the orbit; the *Z-axis* has the same direction as the normal vector of the orbit plane; the *Y-axis* completes the frame following the righthand principle.

(ii) The orbital frame *oxyz*: the origin *o* is located in the mass center of the space platform; the *x*-axis has the same direction as the tangent vector of the orbit; the *z*-axis is directed from the origin *o* to the Earth center *O*; the *y*-axis has the same direction as the negative normal vector of the orbit plane.

According to Assumption (v), the flexible net is equally divided into four sections. The section consisting of point 1, point 2, and point 3 is taken out as an example and shown in Fig. 4.34. In order to get the best performance of capture, the

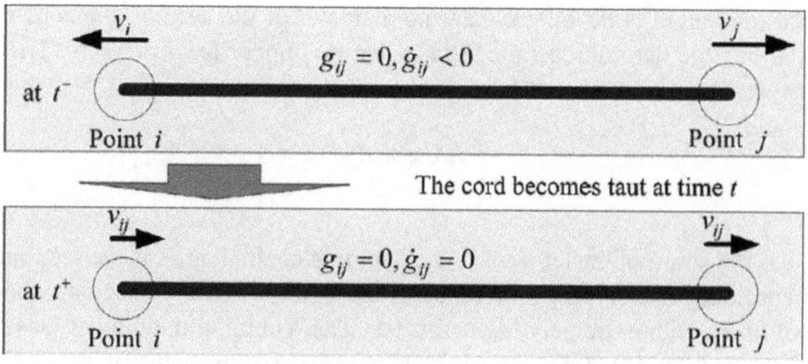

Fig. 4.34 The scenario of becoming taut

net mouth is expected to be fully open in the approaching phase. Meanwhile, due to the friction of the tether-deploying mechanism, there must be a certain dragging force in the tether connecting the platform and the net. Therefore, the section will keep nearly flat in the approaching phase, and it can be approximated as a triangle shell, namely ΔABC. Thus, the T3 element in the shell theory is employed here to describe the shape of this section. Hence, for an arbitrary point D in this section, its position vector in the inertial frame can be approximately calculated as

$$\mathbf{R} \approx s_1\mathbf{R}_1 + s_2\mathbf{R}_2 + s_3\mathbf{R}_3 \tag{4.101}$$

where s_1, s_2, and s_3 stand for the area coordinates of point D. They satisfy

$$s_1 = \frac{\bar{S}_{\Delta \mathrm{BCD}}}{\bar{S}_{\Delta \mathrm{ABC}}}, s_2 = \frac{\bar{S}_{\Delta \mathrm{CAD}}}{\bar{S}_{\Delta \mathrm{ABC}}}, s_3 = \frac{\bar{S}_{\Delta \mathrm{ABD}}}{\bar{S}_{\Delta \mathrm{ABC}}}$$

where $\bar{S}_{\Delta}$ stands for the area of the triangle without any deformation. Therefore, the Lagrange function of this section can be written as

$$\begin{aligned} L_{123} &= \iint_{\Delta \mathrm{ABC}} \frac{1}{2}\rho \dot{\boldsymbol{R}}^{\mathrm{T}}\dot{\boldsymbol{R}} \mathrm{d}\Sigma - \iint_{\Delta \mathrm{ABC}} \left(-\rho \frac{GM}{\|\boldsymbol{R}\|}\right) \mathrm{d}\Sigma \\ &= \frac{m_W}{4}\int_0^1 \mathrm{d}s_1 \int_0^{1-s_1} \left(\frac{1}{2}\dot{\boldsymbol{R}}^{\mathrm{T}}\dot{\boldsymbol{R}} + \rho\frac{GM}{\|\boldsymbol{R}\|}\right)\mathrm{d}s_2 \end{aligned} \tag{4.102}$$

where ρ is the areal density of the equivalent shell, m_W is the mass of the whole net, G is the universal gravitational constant, and M is the mass of the Earth. The variation of the Lagrange function can be written as

$$\delta \int_{t_1}^{t_2} L_{123} \mathrm{d}t = \int_{t_1}^{t_2} \left[\frac{m_W}{4} \int_0^1 \mathrm{d}s_1 \int_0^{1-s_1} \delta \boldsymbol{R}^{\mathrm{T}} \left(-\ddot{\boldsymbol{R}} - \rho \frac{GM}{\|\boldsymbol{R}\|^3} \boldsymbol{R} \right) \mathrm{d}s_2 \right] \mathrm{d}t \tag{4.103}$$

Compared with the orbital motion, the relative motion of MSTN is trivial. Since the position vector $\boldsymbol{R}$ involves the orbital motion of the platform, using it directly to describe the approaching phase will cause computation problems. Hence, we need to transfer Eq. (4.103) into the relative orbit coordinate system. According to the C–W equation, the integral term in Eq. (4.103) can be calculated as

$$\ddot{\boldsymbol{R}} + \rho \frac{GM}{\|\boldsymbol{R}\|^3} \boldsymbol{R} \approx \ddot{\boldsymbol{r}} + \boldsymbol{M}_{\dot{r}} \dot{\boldsymbol{r}} + \boldsymbol{M}_r \boldsymbol{r} \tag{4.104}$$

where

$$\boldsymbol{M}_{\dot{r}} = \begin{bmatrix} 0 & 0 & -2\omega \\ 0 & 0 & 0 \\ 2\omega & 0 & 0 \end{bmatrix}, \quad \boldsymbol{M}_r = \begin{bmatrix} 0 & 0 & 0 \\ 0 & \omega^2 & 0 \\ 0 & 0 & -3\omega^2 \end{bmatrix}$$

Here ω is the orbital angular velocity of the space platform. In addition, the variation of the position vector can be calculated as

$$\delta \boldsymbol{R} = \delta[\boldsymbol{R}_o(t) + \boldsymbol{r}] = \delta \boldsymbol{r} \tag{4.105}$$

where $\boldsymbol{R}_o$ is the position vector of the origin of the orbital coordinate system. Inserting Eqs. (4.104) and (4.105) into Eq. (4.103) yields

$$\begin{aligned} \int_{t_1}^{t_2} \delta L_{123} \mathrm{d}t &\approx \int_{t_1}^{t_2} \left[\frac{m_W}{4} \int_0^1 \mathrm{d}s_1 \int_0^{1-s_1} -\delta \boldsymbol{r}^{\mathrm{T}} (\ddot{\boldsymbol{r}} + \boldsymbol{M}_{\dot{r}} \dot{\boldsymbol{r}} + \boldsymbol{M}_r \boldsymbol{r}) \mathrm{d}s_2 \right] \mathrm{d}t \\ &\approx \int_{t_1}^{t_2} -\delta \boldsymbol{r}_N^{\mathrm{T}} \left[\frac{m_W}{4} \left(\boldsymbol{M}_1^{123} \ddot{\boldsymbol{r}}_N + \boldsymbol{M}_2^{123} \dot{\boldsymbol{r}}_N + \boldsymbol{M}_3^{123} \boldsymbol{r}_N \right) \right] \mathrm{d}t \end{aligned} \tag{4.106}$$

where $\boldsymbol{r}_N = \left[\boldsymbol{r}_1^{\mathrm{T}}, \boldsymbol{r}_2^{\mathrm{T}}, \boldsymbol{r}_3^{\mathrm{T}}, \boldsymbol{r}_4^{\mathrm{T}}, \boldsymbol{r}_5^{\mathrm{T}} \right]^{\mathrm{T}}$, $\boldsymbol{M}_1^{123} = \boldsymbol{M}^{123} \otimes \boldsymbol{I}_{3\times3}$, $\boldsymbol{M}_2^{123} = \boldsymbol{M}^{123} \otimes \boldsymbol{M}_{\dot{r}}$, $\boldsymbol{M}_3^{123} = \boldsymbol{M}^{123} \otimes \boldsymbol{M}_r$. $\otimes$ stands for the direct product of matrixes, and $\boldsymbol{M}^{123}$ is the characteristic matrix corresponding to the section consisting of point 1, point 2, and point 3. The characteristic matrix for different sections can be calculated as

$$\boldsymbol{M}_{5\times5}^{lmn}(i,j) = \begin{cases} 1/12 & i,j \in \{l,m,n\}, i = j \\ 1/24 & i,j \in \{l,m,n\}, i \neq j \\ 0 & \text{otherwise} \end{cases}$$

Similarly, the Lagrange functions of other three sections and their variations can also be obtained.

Since we have assumed that the cord connecting two points cannot be elongated, the distance between two points must conform to the unilateral constraint. For example, the distance between point 1 and point 2 must satisfy

$$\|\boldsymbol{r}_1 - \boldsymbol{r}_2\| \leq L_{12} \tag{4.107}$$

where L_{12} is the undeformed length of the cord connecting point 1 and point 2. Similar unilateral constraints can also be obtained between other points. In order to involve the effects of these unilateral constraints in the modeling, we introduce the interval function. It is defined as

$$g_{ij} \overset{def}{=} L_{ij} - \|\boldsymbol{r}_i - \boldsymbol{r}_j\| \geq 0 \tag{4.108}$$

where L_{ij} is the undeformed cord length between point i and point j. According to the theory of contact dynamics [7], the corresponding constraint forces must satisfy

$$\lambda_{ij} = \begin{cases} 0 & g_{ij} > 0 \\ 0 & g_{ij} = 0, \dot{g}_{ij} > 0 \\ 0 & g_{ij} = 0, \dot{g}_{ij} = 0, \ddot{g}_{ij} > 0 \\ \geq 0 & g_{ij} = 0, \dot{g}_{ij} = 0, \ddot{g}_{ij} = 0 \end{cases} \tag{4.109}$$

Generally speaking, when solving the dynamical equations with numerical integration algorithms, we only know the zeroth-order and the first-order terms, that is, g_{ij} and $\dot{g}_{ij}$. Therefore, whether the constraint is in the first two cases can be easily distinguished. If the constraint is in the first two cases, the constraint force can be decided as zero and points can move freely. Thus, the solving procedure is the same as for the conventional dynamical systems without the unilateral constraint. However, if the constraint is not in the first two cases, the second-order term $\ddot{g}_{ij}$ and the corresponding constraint force λ_{ij} need to be solved before getting the acceleration of points. The method will be discussed in detail when the dynamical equations of MSTN are obtained. Here, in order to facilitate the expression, we number all the eight unilateral constraints of MSTN as shown in Table 4.2. In the following pages, the subscript of λ and g will be written as the number of the corresponding constraint.

Table 4.2 Number of connections

Number	Connected points	Number	Connected points
1	Point 1 and point 2	5	Point 2 and point 3
2	Point 1 and point 3	6	Point 3 and point 4
3	Point 1 and point 4	7	Point 4 and point 5
4	Point 1 and point 5	8	Point 5 and point 2

The Lagrange function of MSTN can be calculated as

$$L = L_{123} + L_{134} + L_{145} + L_{152} + \sum_{i=2}^{5} m_M \left[\frac{1}{2} \boldsymbol{R}_i^{\mathrm{T}} \boldsymbol{R}_i + \frac{GM}{\|\boldsymbol{R}_i\|} \right] \tag{4.110}$$

where m_M is the mass of the maneuvering robot. The virtual work of unconventional forces acting on MSTN (including the constraint forces) can be calculated as

$$\delta' W = -\delta \boldsymbol{r}_1^{\mathrm{T}} \frac{\boldsymbol{r}_1}{\|\boldsymbol{r}_1\|} F_T + \sum_{i=2}^{5} \delta \boldsymbol{R}_i^{\mathrm{T}} \boldsymbol{F}_i + \delta \boldsymbol{g}_N \boldsymbol{\lambda}_N \tag{4.111}$$

where F_T is the dragging force in the connecting tether and $\boldsymbol{F}_i$ $(i = 2, \ldots, 5)$ is the control force acting on the maneuvering robots. According to the extended Hamilton's principle, MSTN needs to satisfy

$$\int_{t_0}^{t_1} (\delta L + \delta' W) \mathrm{d}t = 0 \tag{4.112}$$

Inserting Eqs. (4.110) and (4.111) into Eq. (4.112) and then simplifying it with the partial integration method yields

$$\begin{aligned} \int_{t_0}^{t_1} \Bigg\{ & -\delta \boldsymbol{r}_N^{\mathrm{T}} \left[\frac{m_W}{4} \left(\boldsymbol{M}_1^W \ddot{\boldsymbol{r}}_N + \boldsymbol{M}_2^W \dot{\boldsymbol{r}}_N + \boldsymbol{M}_3^W \boldsymbol{r}_N \right) \right] \\ & - \sum_{i=2}^{5} \delta \boldsymbol{r}_i^{\mathrm{T}} \left[m_M \left(\ddot{\boldsymbol{r}}_i + \boldsymbol{M}_{\dot{r}} \dot{\boldsymbol{r}}_i + \boldsymbol{M}_r \boldsymbol{r}_i \right) \right] \\ & + \delta \boldsymbol{r}_N^{\mathrm{T}} \boldsymbol{F} + \delta \boldsymbol{r}_N^{\mathrm{T}} \left(\frac{\partial \boldsymbol{g}_N}{\partial \boldsymbol{r}_N} \right)^{\mathrm{T}} \boldsymbol{\lambda}_N \Bigg\} \mathrm{d}t = 0 \end{aligned} \tag{4.113}$$

where

$$\begin{aligned} \boldsymbol{M}_i^W &= \boldsymbol{M}_i^{123} + \boldsymbol{M}_i^{134} + \boldsymbol{M}_i^{145} + \boldsymbol{M}_i^{152} (i = 1, 2, 3) \\ \boldsymbol{F} &= \left[-F_T \boldsymbol{r}_1^{\mathrm{T}} / \|\boldsymbol{r}_1\|, \boldsymbol{F}_2^{\mathrm{T}}, \boldsymbol{F}_3^{\mathrm{T}}, \boldsymbol{F}_4^{\mathrm{T}}, \boldsymbol{F}_5^{\mathrm{T}} \right]^{\mathrm{T}} \end{aligned}$$

Considering the property of the variation operation, the dynamical equations of MSTN can be written as

$$\boldsymbol{M}\ddot{\boldsymbol{r}}_N + \boldsymbol{C}\dot{\boldsymbol{r}}_N + \boldsymbol{K}\boldsymbol{r}_N = \boldsymbol{F} + \left(\frac{\partial \boldsymbol{g}_N}{\partial \boldsymbol{r}_N}\right)^{\mathrm{T}} \boldsymbol{\lambda}_N \tag{4.114}$$

where

$$\begin{aligned}
\boldsymbol{M} &= \frac{m_W}{4}\boldsymbol{M}_1^W + m_M \mathrm{diag}(0,1,1,1,1) \otimes \mathrm{diag}(1,1,1) \\
\boldsymbol{C} &= \frac{m_W}{4}\boldsymbol{M}_2^W + m_M \mathrm{diag}(0,1,1,1,1) \otimes \boldsymbol{M}_{\dot{r}} \\
\boldsymbol{K} &= \frac{m_W}{4}\boldsymbol{M}_2^W + m_M \mathrm{diag}(0,1,1,1,1) \otimes \boldsymbol{M}_r
\end{aligned}$$

In order to solve the constraint forces, we take out the numbers of the constraints which are in the last two cases of Eq. (4.109), sort them from lowest to highest, and consequently get the number vector $\bar{\boldsymbol{I}}$. Sorting the corresponding interval functions and constraint forces in the same sequence yields the interval function vector $\bar{g}_N$ and the constraint force vector $\bar{\lambda}_N$. For the constraint whose number is not included in vector $\bar{\boldsymbol{I}}$, the corresponding constraint force can be directly decided as zero. For the remaining constraints, the second derivative of $\bar{g}_N$ can be calculated as

$$\ddot{\bar{\boldsymbol{g}}}_N = \frac{\partial \bar{\boldsymbol{g}}_N}{\partial \boldsymbol{r}_N}\ddot{\boldsymbol{r}}_N + \boldsymbol{h}(\boldsymbol{r}_N, \dot{\boldsymbol{r}}_N) \tag{4.115}$$

In addition, the relation between $\bar{\lambda}_N$ and λ_N can be written as

$$\boldsymbol{\lambda}_N = \boldsymbol{M}_R \bar{\boldsymbol{\lambda}}_N \tag{4.116}$$

where

$$\boldsymbol{M}_R(i,j) = \begin{cases} 1 & i = \bar{\boldsymbol{I}}(j) \\ 0 & \text{otherwise} \end{cases}$$

Inserting Eqs. (4.114) and (4.116) into Eq. (4.115) yields

$$\ddot{\bar{\boldsymbol{g}}}_N = \boldsymbol{A}\bar{\boldsymbol{\lambda}}_N + \boldsymbol{b} \tag{4.117}$$

where

$$\begin{aligned}
\boldsymbol{A} &= \frac{\partial \bar{\boldsymbol{g}}_N}{\partial \boldsymbol{r}_N}\boldsymbol{M}^{-1}\left(\frac{\partial \boldsymbol{g}_N}{\partial \boldsymbol{r}_N}\right)^{\mathrm{T}}\boldsymbol{M}_R \\
\boldsymbol{b} &= \frac{\partial \bar{\boldsymbol{g}}_N}{\partial \boldsymbol{r}_N}\boldsymbol{M}^{-1}(\boldsymbol{F} - \boldsymbol{C}\dot{\boldsymbol{r}}_N - \boldsymbol{K}\boldsymbol{r}_N) + \boldsymbol{h}(\boldsymbol{r}_N, \dot{\boldsymbol{r}}_N)
\end{aligned}$$

Considering the constraints conformed by $\bar{g}_N$, the problem to solve the constraint forces can

be written as

$$\begin{cases} \bar{\boldsymbol{g}}_N = \boldsymbol{0}, \dot{\bar{\boldsymbol{g}}}_N = \boldsymbol{0} \\ \ddot{\bar{\boldsymbol{g}}}_N = \boldsymbol{A}\bar{\boldsymbol{\lambda}}_N + \boldsymbol{b} \\ \ddot{\bar{\boldsymbol{g}}}_N^{\mathrm{T}} \bar{\boldsymbol{\lambda}}_N = 0,\ \ddot{\bar{\boldsymbol{g}}}_N \geq \boldsymbol{0},\ \bar{\boldsymbol{\lambda}}_N \geq \boldsymbol{0} \end{cases} \tag{4.118}$$

This is a typical linear complementarity problem. It can be solved by using the Lemke algorithm whose detailed procedures are investigated in [11].

According to the analysis above, when the position vector $\boldsymbol{r}_N$, the velocity vector $\dot{\mathbf{r}}_N$ and the control force vector $\boldsymbol{F}$ are known, we can obtain the constraint force vector λ_N through solving Eqs. (4.116) and (4.118). Then, the acceleration vector $\ddot{\boldsymbol{r}}_N$ can be obtained by solving the dynamical Eq. (4.114). Thus, the dynamics of MSTN is captured.

Considering the scenario shown in Fig. 4.34, at time t^-, the distance between point i and point j reaches the undeformed cord length L_{ij}. Meanwhile, the relative velocity along the cord between the two points is not zero, and they have a tendency to go away from each other. Thus, the cord becomes taut at time t and the tightening force is transferred through the cord, which is similar to the process of impact. Due to the high damping of the braided structure and the stiff fiber, at time t^+, the two points will get the same velocity along the cord.

In order to describe the instantaneous velocity jump, we integrate the overall dynamical Eq. (4.114) at the time interval $[t^-, t^+]$ and obtain

$$\int_{t^-}^{t^+} \boldsymbol{M}\ddot{\boldsymbol{r}}_N \mathrm{d}t + \int_{t^-}^{t^+} \boldsymbol{C}\dot{\boldsymbol{r}}_N \mathrm{d}t + \int_{t^-}^{t^+} \boldsymbol{K}\boldsymbol{r}_N \mathrm{d}t = \int_{t^-}^{t^+} \boldsymbol{F}\mathrm{d}t + \left(\frac{\partial \boldsymbol{g}_N}{\partial \boldsymbol{r}_N}\right)^{\mathrm{T}} \int_{t^-}^{t^+} \boldsymbol{\lambda}_N \mathrm{d}t \tag{4.119}$$

Since the position vector remains unchanged and the control forces are limited during the extremely transient period, the last two terms at the left side of Eq. (4.119) and the first term at the right side are equal to zero. For the constraint which satisfies the conditions of becoming taut, namely $g_i = 0$ and $\dot{g}_i < 0$, the constraint force goes to infinity over an infinitesimal period and its integral is not zero. However, for the constraint which does not satisfy the conditions, the constraint force is zero or limited, and therefore the integral remains zero. In order to facilitate the expression, we take out the numbers of those constraints satisfying the tightening condition, sort them from the lowest to the highest, and consequently get the column vector $\bar{\boldsymbol{I}}_{NT}$. Sorting the corresponding interval functions and constraint forces in the same sequence yields the interval function vector $\bar{\boldsymbol{g}}_{NT}$ and the constraint force $\bar{\boldsymbol{\lambda}}_{NT}$. Furthermore, we write the integrals of $\boldsymbol{\lambda}_{NT}$ and $\bar{\boldsymbol{\lambda}}_{NT}$ over $[t^-, t^+]$ as Λ_{NT} and $\bar{\Lambda}_{NT}$, respectively.

The relation between them can be written as

$$\boldsymbol{\Lambda}_{NT} = \boldsymbol{M}_{RT}\bar{\boldsymbol{\Lambda}}_{NT} \tag{4.120}$$

where

$$\boldsymbol{M}_{RT}(i,j) = \begin{cases} 1 & i = \bar{\boldsymbol{I}}_{NT}(j) \\ 0 & \text{otherwise} \end{cases}$$

Thus, Eq. (4.119) can be rewritten as

$$\boldsymbol{M}\left(\dot{\boldsymbol{r}}_N|_{t^+} - \dot{\boldsymbol{r}}_N|_{t^-}\right) = \left(\frac{\partial \boldsymbol{g}_N}{\partial \boldsymbol{r}_N}\right)^{\mathrm{T}} \boldsymbol{M}_{RT}\bar{\boldsymbol{\Lambda}}_{NT} \tag{4.121}$$

Besides, the interval function vector $\bar{\boldsymbol{g}}_{NT}$ must satisfy

$$\dot{\bar{\boldsymbol{g}}}_{NT}\big|_{t^+} = \boldsymbol{0} \tag{4.122}$$

Solving Eqs. (4.121) and (4.122) together yields the velocity vector at time t^+. Thus, the velocity jump is obtained.

4.3.2 Controller Design

When entering the capture phase, as long as the target is in the net mouth, it will be successfully captured by MSTN regardless of the net shape. Since the position and direction of the net mouth are completely decided by the position of the four maneuvering robots, we only need to require them to move along the expected trajectories when considering the control of MSTN. Thus, the position control vector $\boldsymbol{r}_c$ can be written as

$$\boldsymbol{r}_c = [\boldsymbol{r}_2^{\mathrm{T}}, \boldsymbol{r}_3^{\mathrm{T}}, \boldsymbol{r}_4^{\mathrm{T}}, \boldsymbol{r}_5^{\mathrm{T}}]^{\mathrm{T}} \tag{4.123}$$

Besides, we assume that the space platform employs the simplest passive mechanism to deploy the tether, which is similar to that used in the Small Expendable Deployer System mission [12]. Thus, the dragging force in the connecting force is uncontrollable in the approaching phase while it cannot be completely eliminated due to the existence of friction. Hence, the control force vector $\boldsymbol{F}_c$ of MSTN only includes the control forces acting on the maneuvering robots, that is,

$$\boldsymbol{F}_c = [\boldsymbol{F}_2^{\mathrm{T}}, \boldsymbol{F}_3^{\mathrm{T}}, \boldsymbol{F}_4^{\mathrm{T}}, \boldsymbol{F}_5^{\mathrm{T}}]^{\mathrm{T}} \tag{4.124}$$

The uncontrollable dragging force F_T will cause disturbances to the motion of MSTN and meanwhile can help the flexible net keep stable. Therefore, we split the dynamical Eq. (4.114) into two equations as

$$\begin{cases} \boldsymbol{M}_1\ddot{\boldsymbol{r}}_1 + \boldsymbol{M}_{1c}\ddot{\boldsymbol{r}}_c + \boldsymbol{C}_1\dot{\boldsymbol{r}}_N + \boldsymbol{K}_1\boldsymbol{r}_N = \frac{\boldsymbol{r}_1}{\|\boldsymbol{r}_1\|}F_T + \left(\frac{\partial \boldsymbol{g}_N}{\partial \boldsymbol{r}_1}\right)^{\mathrm{T}}\boldsymbol{\lambda}_N \\ \boldsymbol{M}_c\ddot{\boldsymbol{r}}_c + \boldsymbol{M}_{c1}\ddot{\boldsymbol{r}}_1 + \boldsymbol{C}_c\dot{\boldsymbol{r}}_N + \boldsymbol{K}_c\boldsymbol{r}_N = \boldsymbol{F}_c + \left(\frac{\partial \boldsymbol{g}_N}{\partial \boldsymbol{r}_c}\right)^{\mathrm{T}}\boldsymbol{\lambda}_N \end{cases} \tag{4.125}$$

where $\boldsymbol{M}_1$, $\boldsymbol{M}_{1c}$, $\boldsymbol{M}_{c1}$, and $\boldsymbol{M}_c$ are the corresponding partitions of $\boldsymbol{M}$; $\boldsymbol{C}_1$ and $\boldsymbol{C}_c$ are the corresponding partitions of $\boldsymbol{C}$; $\boldsymbol{K}_1$ and $\boldsymbol{K}_c$ are the corresponding partitions of $\boldsymbol{K}$. They satisfy

$$\boldsymbol{M} = \begin{bmatrix} \boldsymbol{M}_1 & \boldsymbol{M}_1 \\ \boldsymbol{M}_{c1} & \boldsymbol{M}_c \end{bmatrix}, C = \begin{bmatrix} \boldsymbol{C}_1 \\ \boldsymbol{C}_c \end{bmatrix}, \boldsymbol{K} = \begin{bmatrix} \boldsymbol{K}_1 \\ \boldsymbol{K}_c \end{bmatrix}.$$

According to Eq. (4.114), the four maneuvering robots will interact with each other. Besides, for constraints which are not in the first two cases of Eq. (4.109), whether they are in the third or fourth case is affected by the control forces required by the controller. Different control forces might lead to different structures of MSTN. In other words, the structure of MSTN is coupled with the control forces. This makes the control of MSTN remarkably different from the conventional space robot and the satellite formation. Referring to the contact control of rigid robots [9, 10], we present a coordinated controller with the structure as shown in Fig. 4.35. The PD adjuster is used to generate the acceleration correction vector according the following equation:

$$\Delta\ddot{\boldsymbol{r}}_c = \lambda_p\left(\boldsymbol{r}_c^* - \boldsymbol{r}_c\right) + \lambda_d\left(\dot{\boldsymbol{r}}_c^* - \dot{\boldsymbol{r}}_c\right) \tag{4.126}$$

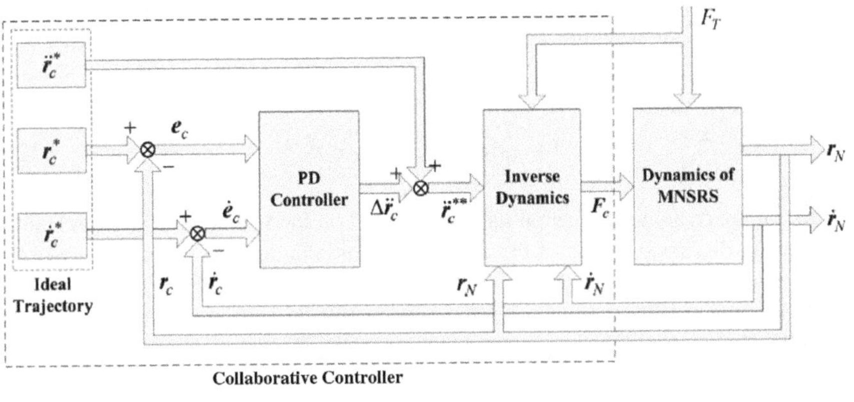

Fig. 4.35 Structure of the coordinated controller

where λ_p and λ_d are the proportional and differential coefficient, respectively. The expected acceleration $\ddot{\boldsymbol{r}}_c^{**}$ is obtained by adding the ideal acceleration $\ddot{\boldsymbol{r}}_c^{*}$ and the acceleration correction $\Delta\ddot{\boldsymbol{r}}_c$, that is,

$$\ddot{\boldsymbol{r}}_c^{**} = \ddot{\boldsymbol{r}}_c^{*} + \Delta\ddot{\boldsymbol{r}}_c \tag{4.127}$$

Then, the inverse dynamics is used to generate the required control force according to the expected acceleration. However, the problem faced by MSTN is a little different from the contact control of rigid robots, which makes the conventional Jacobian matrix and Moore–Penrose pseudoinverse not applicable for MSTN. Specifically, for the rigid robot, whether the arms will contact with objects can be inferred from the current state and the expected acceleration. Nevertheless, for MSTN, we do not give the expected acceleration of point 1. Even if the expected acceleration is given, requiring point 1 to gain the required acceleration is also very hard. This is because that there is no active control force acting on it. Its motion is affected by other points' motion, its own position and the uncontrollable dragging force. This situation is similar to when one robot arm gets out of control. Therefore, we need to get other methods to solve inverse dynamics.

As for the stability of the closed-loop controller, it is demonstrated in [9, 10] that the inverse dynamics module does not affect the stability of the whole system. The stability is mainly decided by the control law of $\Delta\ddot{\boldsymbol{r}}_c$, namely the PD adjuster. As long as the parameters of the PD adjuster are well selected, the closed-loop system will keep stable.

Considering that the original idea of the Moore–Penrose pseudo-inverse is to search the solution with minimal errors, we also transfer the problem of inverse dynamics into an optimization problem. The goal of optimization is to solve for the control force and the acceleration which conform to the dynamical constraints and meanwhile can make the error between the actual and expected accelerations minimal. However, since the system structure is affected by the control force, we have to traverse all cases and solve problems with all the possible structures one by one. The final solution is the best one among all solutions. Thus, the inverse dynamics is finally transferred to a double-level optimization problem.

No matter what structure MSTN is, the performance function J can always be written as

$$J = \frac{1}{2}\left(\boldsymbol{r}_c - \ddot{\boldsymbol{r}}_c^{**}\right)^{\mathrm{T}}\left(\boldsymbol{r}_c - \ddot{\boldsymbol{r}}_c^{**}\right) \tag{4.128}$$

As for the dynamical constraints, they depend on the structure of MSTN. Hence, we need to discuss the state of the connecting cords in detail. Corresponding to the different cases of Eq. (4.109), the number vector $\bar{\boldsymbol{I}}$ will be different, and there will be different constraints. Since we assume the controller can get the position vector $\boldsymbol{r}_N$ and the velocity vector $\dot{\boldsymbol{r}}_N$ from the sensors, the first two cases can be directly distinguished and the vector $\bar{\boldsymbol{I}}$ can also be decided. However, to which case the remaining constraints belong to is decided by the control force which is waiting to

Table 4.3 Optimization problems corresponding to $length(\bar{\boldsymbol{I}}) = 1$

Number		1	2
Problem	Objective	$\min J$	
	Constraint 1	$\boldsymbol{M}\ddot{\boldsymbol{r}}_N + \boldsymbol{C}\dot{\boldsymbol{r}}_N + \boldsymbol{K}\boldsymbol{r}_N = \boldsymbol{F} + \left(\frac{\partial \boldsymbol{g}_N}{\partial \boldsymbol{r}_1}\right)^{\mathrm{T}} \boldsymbol{M}_R \bar{\boldsymbol{\lambda}}_N$	
	Constraint 2	$\lambda_{i1} = 0,\ \ddot{g}_{i1} = 0$	$\lambda_{i1} \geq 0,\ \ddot{g}_{i1} = 0$

be solved for. Hence, when designing the method to solve the inverse dynamics, we must discuss all situations corresponding to the different length of vector $\bar{\boldsymbol{I}}$:

(i) If $length(\bar{\boldsymbol{I}}) = 0$, then $\boldsymbol{\lambda}_N = \boldsymbol{0}$ and $\ddot{\boldsymbol{r}}_c = \ddot{\boldsymbol{r}}_c^{**}$. Thus, Eq. (4.125) can be rewritten as

$$\begin{cases} \boldsymbol{M}_1\ddot{\boldsymbol{r}}_1 = \frac{\boldsymbol{r}_1}{\|\boldsymbol{r}_1\|}F_T - \left(\boldsymbol{M}_{1c}\ddot{\boldsymbol{r}}_c^{**} + \boldsymbol{C}_1\dot{\boldsymbol{r}}_N + \boldsymbol{K}_1\boldsymbol{r}_N\right) \\ -\boldsymbol{F}_c + \boldsymbol{M}_{c1}\ddot{\boldsymbol{r}}_1 = -\left(\boldsymbol{M}_c\ddot{\boldsymbol{r}}_c^{**} + \boldsymbol{C}_c\dot{\boldsymbol{r}}_N + \boldsymbol{K}_c\boldsymbol{r}_N\right) \end{cases} \quad (4.129)$$

The control force $\boldsymbol{F}_c$ can be obtained by solving the two equations together.

(ii) If $length(\bar{\boldsymbol{I}}) = 1$ and $\bar{\boldsymbol{I}} = [i_1]$, then there will be 2 linear programming problems in the first level as shown in Table 4.3. The final optimal solution is the best one of these 2 problems.

(iii) If $length(\bar{\boldsymbol{I}}) = 2$ and $\bar{\boldsymbol{I}} = [i_1, i_2]$, then there will be 4 linear programming problems in the first level as shown in Table 4.4. The final optimal solution is the best one of these 4 problems.

…

(n + 1) If $length(\bar{\boldsymbol{I}}) = n(1 \leq n \leq 8)$ and $\bar{\boldsymbol{I}} = [i_1, i_2, \ldots, i_n]$, then there will be 2^n linear programming problems in the first level as shown in Table 4.5. The final optimal solution is the best one of these 2^n problems.

Table 4.4 Optimization problems corresponding to $length(\bar{\boldsymbol{I}}) = 2$

Number		1	2	3	4
Problem	Objective	$\min J$			
	Constraint 1	$\boldsymbol{M}\ddot{\boldsymbol{r}}_N + \boldsymbol{C}\dot{\boldsymbol{r}}_N + \boldsymbol{K}\boldsymbol{r}_N = \boldsymbol{F} + \left(\frac{\partial \boldsymbol{g}_N}{\partial \boldsymbol{r}_1}\right)^{\mathrm{T}} \boldsymbol{M}_R \bar{\boldsymbol{\lambda}}_N$			
	Constraint 2	$\lambda_{i1} = 0,$ $\ddot{g}_{i1} > 0$	$\lambda_{i1} \geq 0,$ $\ddot{g}_{i1} = 0$	$\lambda_{i1} = 0,$ $\ddot{g}_{i1} > 0$	$\lambda_{i1} \geq 0,$ $\ddot{g}_{i1} = 0$
	Constraint 3	$\lambda_{i2} = 0,$ $\ddot{g}_{i2} > 0$	$\lambda_{i2} = 0,$ $\ddot{g}_{i2} > 0$	$\lambda_{i2} \geq 0,$ $\ddot{g}_{i2} = 0$	$\lambda_{i2} \geq 0,$ $\ddot{g}_{i2} = 0$

Table 4.5 Optimization problems corresponding to $length(\bar{\boldsymbol{I}}) = n$

Number		1	2	…	$2^n - 1$	2^n
Problem	Objective		$\min J$			
	Constraint 1		$\boldsymbol{M}\ddot{\boldsymbol{r}}_N + \boldsymbol{C}\dot{\boldsymbol{r}}_N + \boldsymbol{K}\boldsymbol{r}_N = \boldsymbol{F} + \left(\frac{\partial g_N}{\partial \boldsymbol{r}_1}\right)^{\mathrm{T}} \boldsymbol{M}_R \bar{\boldsymbol{\lambda}}_N$			
	Constraint 2	$\lambda_{i1} = 0,$ $\ddot{g}_{i1} > 0$	$\lambda_{i1} \geq 0,$ $\ddot{g}_{i1} = 0$	…	$\lambda_{i1} = 0,$ $\ddot{g}_{i1} > 0$	$\lambda_{i1} \geq 0,$ $\ddot{g}_{i1} = 0$
	Constraint 3	$\lambda_{i2} = 0,$ $\ddot{g}_{i2} > 0$	$\lambda_{i2} = 0,$ $\ddot{g}_{i2} > 0$	…	$\lambda_{i2} \geq 0,$ $\ddot{g}_{i2} = 0$	$\lambda_{i2} \geq 0,$ $\ddot{g}_{i2} = 0$
	…	…	…	…	…	…
	Constraint n + 1	$\lambda_{in} = 0,$ $\ddot{g}_{in} > 0$	$\lambda_{in} = 0,$ $\ddot{g}_{in} > 0$	…	$\lambda_{in} \geq 0,$ $\ddot{g}_{in} = 0$	$\lambda_{in} \geq 0,$ $\ddot{g}_{in} = 0$

4.3.3 Numerical Simulation

A representative system has been selected to analyze the dynamical characteristics of MSTN and demonstrate the performance of the proposed coordinated controller. The basic parameters of the selected system are shown in Table 4.6. The detailed dimension of the flexible net is shown in Fig. 4.36. The target orbital debris is assumed to be in the +V-bar direction, namely the +x-axis of the orbital coordinate system. Hence, similar to the rendezvous and docking of conventional spacecraft, the approaching process is mainly conducted in the orbital plane, that is the xoz plane of the orbital frame. In order to facilitate the analysis, we assume the approaching process begins when point 1 is 1 m away from the platform, as shown in Fig. 4.37. If the ejecting process is ideal, the net is fully unfolded and all points have the same velocity when the approaching process begins. Therefore, the ideal initial states of MSTN can be written as

$$r_1 = \boldsymbol{M}_{IS} \begin{bmatrix} 1 \\ 0 \\ 0 \end{bmatrix}, r_2 = \boldsymbol{M}_{IS} \begin{bmatrix} 11 \\ 0 \\ 10 \end{bmatrix}, r_3 = \boldsymbol{M}_{IS} \begin{bmatrix} 11 \\ 10 \\ 0 \end{bmatrix}, r_4 = \boldsymbol{M}_{IS} \begin{bmatrix} 11 \\ 0 \\ -10 \end{bmatrix}, r_5 = \boldsymbol{M}_{IS} \begin{bmatrix} 11 \\ -10 \\ 0 \end{bmatrix}$$

$$\boldsymbol{v}_1 = \boldsymbol{v}_2 = \boldsymbol{v}_3 = \boldsymbol{v}_4 = \boldsymbol{v}_5 = \boldsymbol{M}_{IS} \begin{bmatrix} 1 \\ 0 \\ 0 \end{bmatrix}$$

Table 4.6 Parameters of MSTN

Parameter	Value
Orbital angle velocity ω	0.0011 rad/s
Mass of the maneuvering robot m_M	10 kg
Mass of the flexible net m_W	8 kg
Undeformed cord length between any connecting points L_{ij}	$10\sqrt{2}$ m

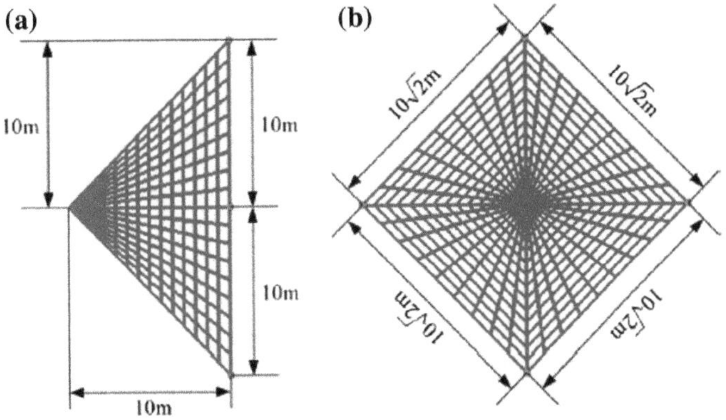

Fig. 4.36 Dimension of the undeformed net: **a** front view; **b** side view

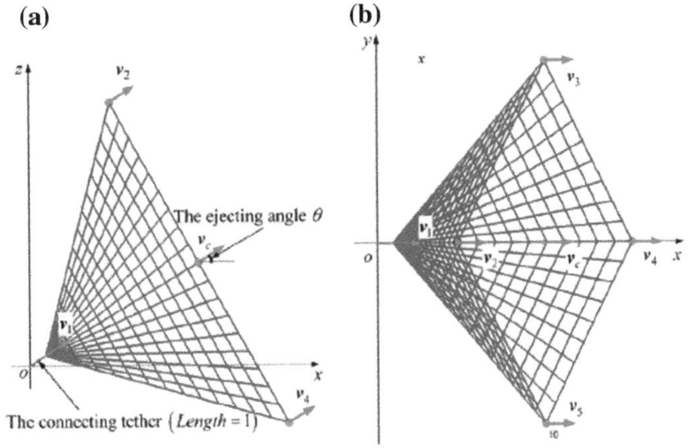

Fig. 4.37 Initial state of MSTN: **a** front view; **b** top view

where $\boldsymbol{M}_{IS}$ is the rotating matrix. It can be calculated as

$$\boldsymbol{M}_{IS} = \begin{bmatrix} \cos\theta & 0 & -\sin\theta \\ 0 & 1 & 0 \\ \sin\theta & 0 & \cos\theta \end{bmatrix}$$

Here, θ is the ejecting angle. Besides, in Fig. 4.37, $\boldsymbol{v}_c$ stands for the velocity of the center of the net mouth. It can be calculated as

$$v_c = \frac{1}{4}(v_2 + v_3 + v_4 + v_5)$$

Non-control capture

If MSTN is without active control, it is the same as the conventional Tethered net system, and its ideal motion should be like the fishing net thrown by the fisherman. However, due to the low damping property of the space environment, the dragging force in the connecting tether can cause great disturbances to the motion of MSTN. Therefore, we discuss the ideal case without the dragging force at first. Then, we will investigate the effects of the dragging force.

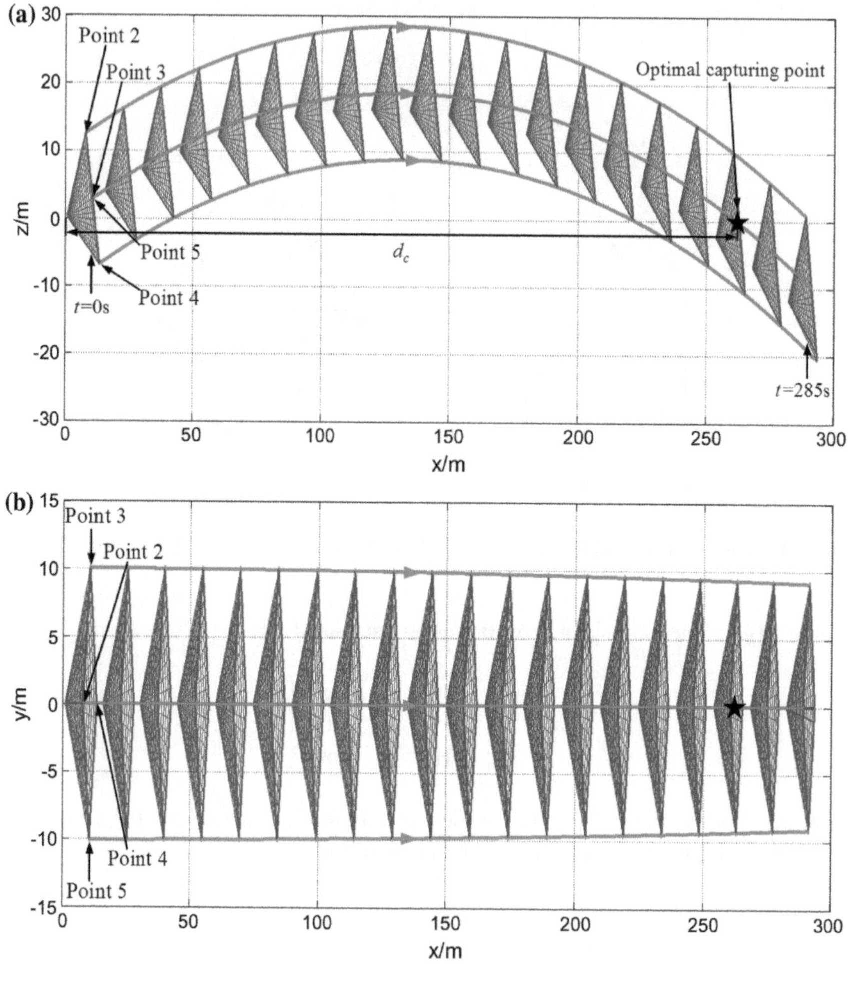

Fig. 4.38 Ideal uncontrolled motion of MSTN: **a** front view; **b** top view

Case 1: ideal uncontrolled motion

Different ejecting angles will lead to different trajectories of MSTN. In Fig. 4.38, we show the motion of MSTN corresponding to the case $\theta = 15°$. From the figure we can conclude that if there is no external nonconservative force, the flexible net will not experience significant deformation. On the other hand, Fig. 4.38b reveals that the net mouth has the tendency to shrink along the y-axis. This is because of the ω^2 term of M_r in Eq. (4.104) which causes negative force when the point is in the +y-axis, positive force in the −y-axis.

When capturing the orbital debris, we hope the target can be at the center of the net mouth and the velocity of the net is perpendicular to the net mouth. Hence, as shown in Fig. 4.38, we define the intersection of the x-axis and the trajectory of the mouth center as the optimal capturing point. The distance between the optimal capturing point and the origin is defined as the optimal capturing distance d_c. Besides, we introduce the effective net mouth area A_c to evaluate the ability of MSTN to capture the target. It stands for the projected area of the net mouth on the plane which is perpendicular to the velocity vector $\boldsymbol{v}_c$. When the net is fully unfolded, A_c reaches the maximum value, namely $200\,\text{m}^2$. When A_c is lower than a certain threshold, it shows the net has deformed significantly or the direction of the net mouth has greatly deviated from the direction of A_c. In both cases, MSTN might be unable to capture the target. Therefore, such cases should be avoided. In this paper, we set the threshold as the half of the maximum value, namely $100\,\text{m}^2$.

Figure 4.39 shows the link between the ejecting angle θ and the optimal capturing distance d_c. The linear fitting demonstrates that dc will increase about 17 m for every 1° increase in θ. This reveals that in the case of uncontrolled capture, the optimal capturing distance is very sensitive to the ejecting angle, which requires the

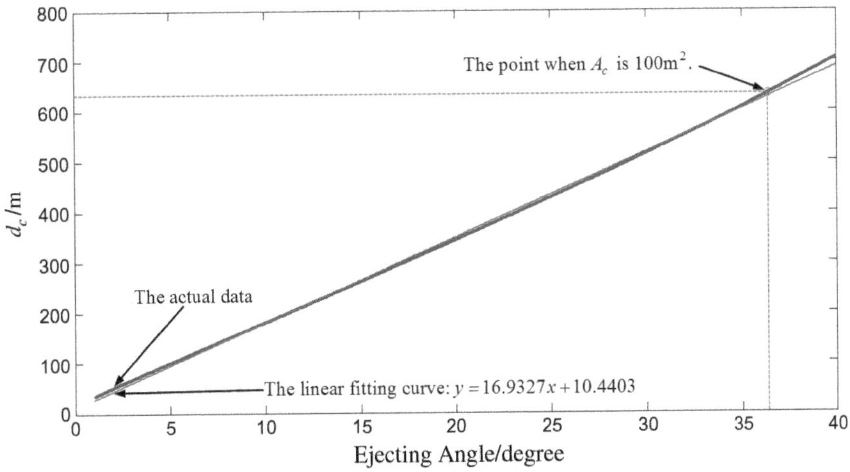

Fig. 4.39 The optimal capturing distance versus ejecting angle

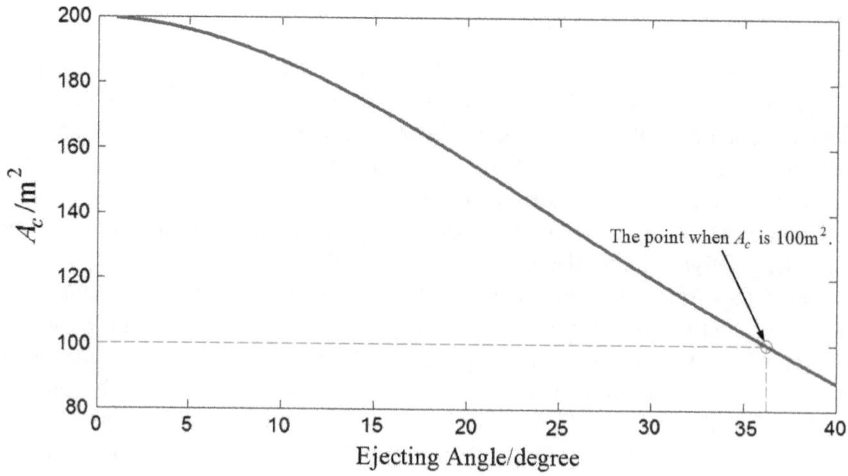

Fig. 4.40 The effective net mouth area at the optimal capturing point versus ejecting angle

ejecting mechanism to have very high precision. Figure 4.40 shows that as the ejecting angle increases, the effective area A_c at the optimal capturing point will decrease remarkably. When the ejecting angle reaches about 36.4° and the optimal capturing distance reaches 633 m, the effective capturing area at the optimal capturing point will decrease to 100 m^2. Therefore, even if all external disturbances are neglected, when the ejecting velocity is 1 m/s, the maximum operating distance of the uncontrolled MSTN is only 633 m.

Case 2: uncontrolled motion with dragging force

As shown in Fig. 4.41, when the dragging force is 200 mN and the ejecting angle is 30°, the net mouth will close rapidly after the ejection, and the uncontrolled MSTN cannot fulfill the objective of capturing the orbital debris in the +V-bar direction. This is because the dragging force in the connecting tether will be transferred to the maneuvering robots through the cord between point 1 and these robots. The component of the tensile force in the cord will drive the maneuvering robots to approach each other. Since the motion damping of the space environment is extremely small, the accumulative effect of the dragging force will be very significant. In order to analyze the effect of the dragging force more clearly, we demonstrate the variation of the effective net mouth area Ac with different dragging forces in Fig. 4.42. From the figure, we know that even though the dragging force is very small, the effect will also be very obvious. Since the ejecting velocity is 1 m/s, the target cannot be more than 60 m away from the space platform when the

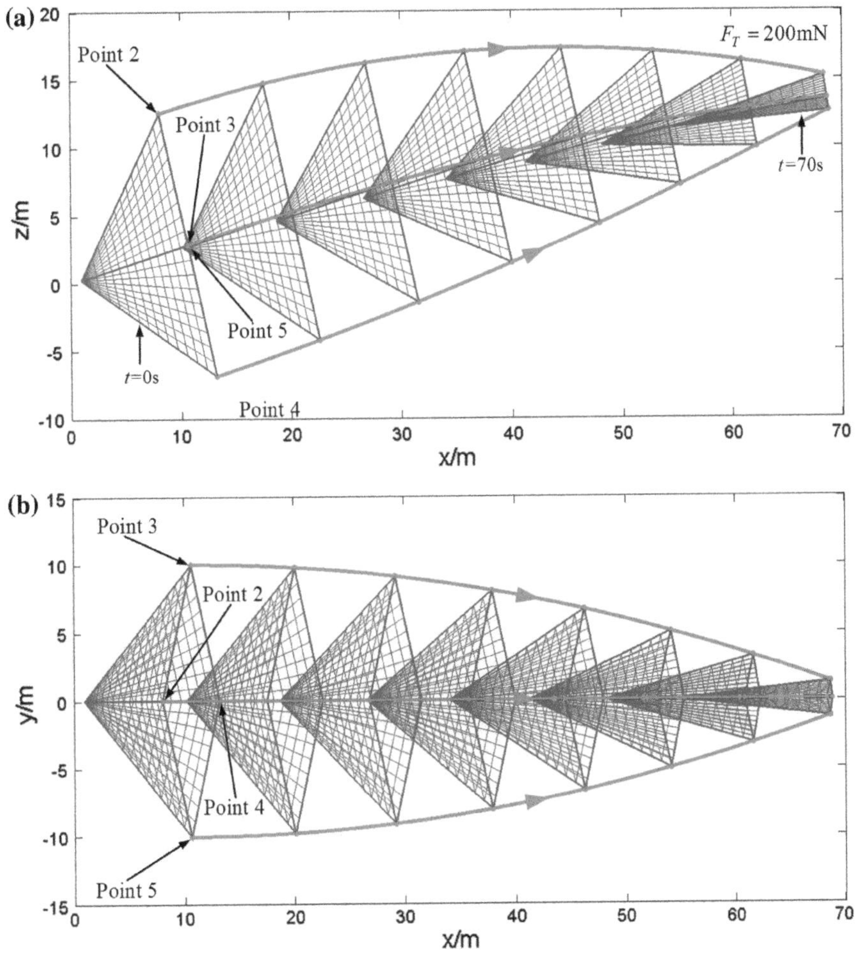

Fig. 4.41 Uncontrolled motion of MSTN with dragging force: **a** front view; **b** top view

dragging force is more than 100 mN. If the dragging force is more than 500 mN, the operating range of uncontrolled MSTN is further narrowed to be less than 30 m. Thus, we can conclude that the conventional uncontrolled MSTN can only be used to capture targets which are extremely close to the space platform. This will bring great limitations for the application of MSTN.

Controlled capture

Since the uncontrolled motion of MSTN is very sensitive to the ejecting angle and the dragging force, it is of great necessity to introduce the closed-loop controller. In

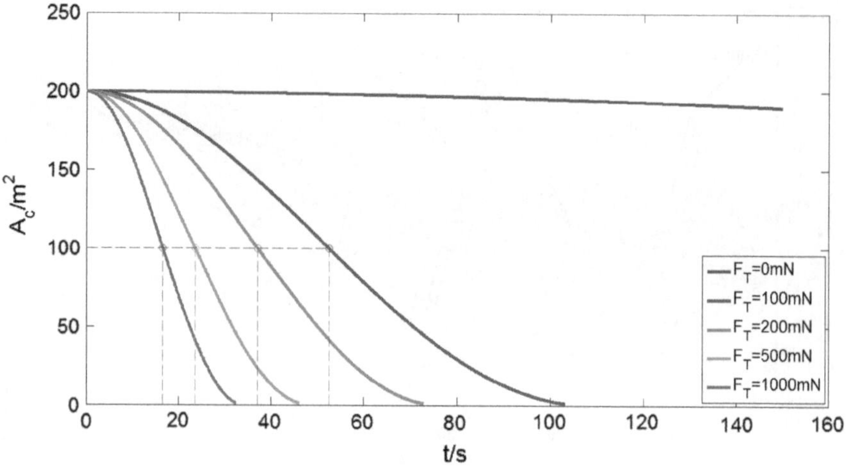

Fig. 4.42 The effective net mouth area versus time

this section, we will demonstrate the effectiveness of the proposed coordinated controller and its abilities to resist the effects of various disturbances and errors. In order to simplify the analysis, we assume the target debris is at $[250, 0, 0]^T$ and the dragging force in the connecting tether is 200 mN. Besides, in order to get the optimal capture performance, we expect the center of the net mouth to move along the x-axis, and the net mouth can be perpendicular to the x-axis. Thus, we require the four maneuvering robots to move along straight lines which are parallel to the x-axis, and the ideal ejecting angle is 0°. Furthermore, we set the expected velocity and acceleration of the four robots to be $[1, 0, 0]^T$ and $[0, 0, 0]^T$, respectively. The parameters of PD adjuster, λ_p and λ_d, are both set to be 1.

Case 1: ideal controlled motion

When all initial states are precisely equal to the expected values and all states can be accurately measured, the proposed coordinated controller should be able to ensure that MSTN moves along the expected trajectory. In Figs. 4.43 and 4.44, we show the controlled motion of MSTN and the control commands corresponding to this case, respectively. According to Fig. 4.43, we can conclude that the proposed coordinated controller has the ability to help MSTN fulfil the goal of capturing the target when there are no external disturbances and parameter errors. Besides, from Fig. 4.44, we find that the control commands constantly fluctuate at two levels. One is at about 50 mN and the other is at about 200 mN. This is mainly caused by the structure change of MSTN. When the connecting cord with point 1 is slack, the maneuvering robot will only need to overcome the effect of the Coriolis force and

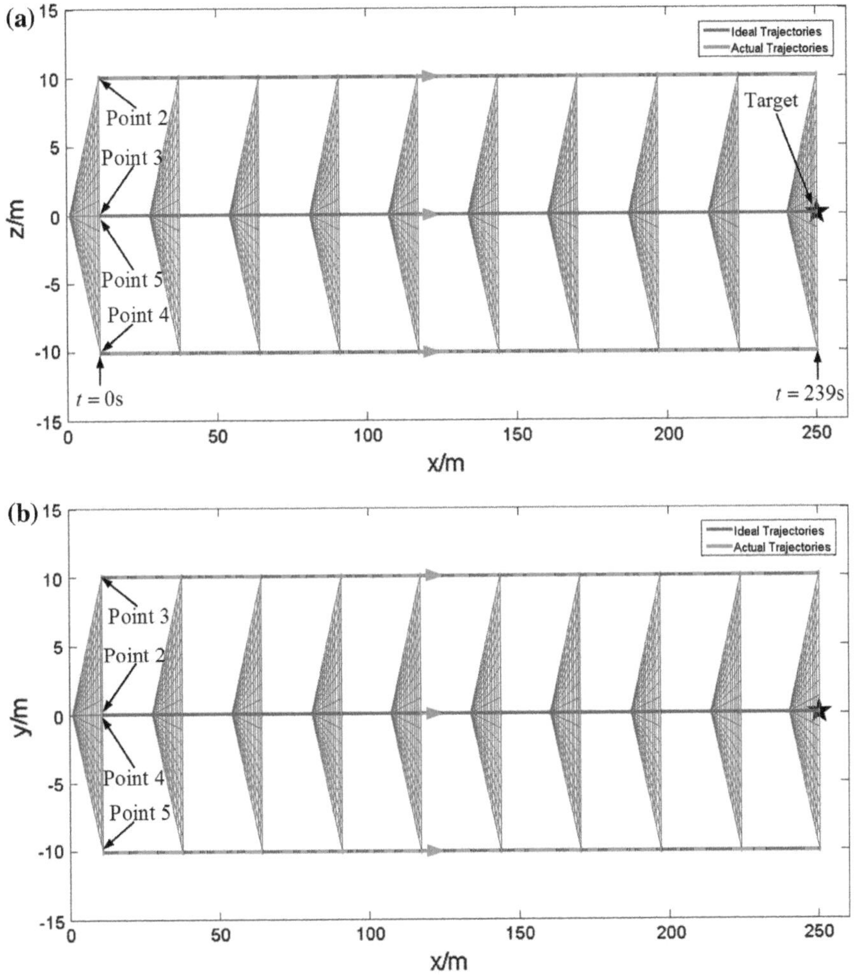

Fig. 4.43 Ideal controlled motion of MSTN: **a** front view; **b** top view

the gravitational force. Nevertheless, when the connecting cord with point 1 is tent, thrusters will also need to counterbalance the effect of the dragging force.

Case 2: controlled motion with initial state errors

In this case, we want to discuss the performance of the coordinated controller when the actual ejecting angle significantly deviates from the expected value. As mentioned before, the expected ejecting angle is 0°. However, we set the actual ejecting

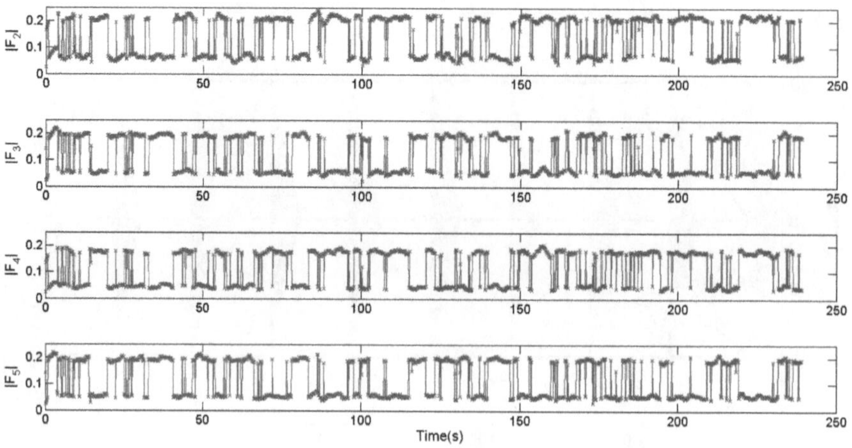

Fig. 4.44 Control commands of the coordinated controller

angle as 30°, and the motion of MSTN is shown in Fig. 4.45. From the figure, we find that the flexible net will experience remarkable deformation: at first, the center of the net is behind the net mouth; then, since the mouth net needs to adjust its direction, the four maneuvering robots obtain different velocities, and the net center gradually goes ahead of the net mouth; finally, the robots follow the ideal trajectories steadily, and the net gradually returns to the ideal shape due to the dragging force in the connecting tether. Through the analysis of the whole process, we can conclude that the proposed coordinated controller has the ability to eliminate a large ejecting angle error and the dragging force is helpful for the net to keep the ideal net shape.

Case 3: controlled motion with measurement noise

In order to demonstrate the coordinated controller's ability to resist the effect of measurement noise, we add Gaussian white noise with zero mean to all measurements. The standard deviation of the added term to the position vector $\boldsymbol{r}_N$, the velocity vector $\dot{\boldsymbol{r}}_N$ and the dragging force F_T is set as 1 m, 0.2 m/s and 50 mN, respectively. Figures 4.46 and 4.47 show the position deviation of point 2 and point 3, respectively. (Due to the kinetic symmetry, the deviation of point 4 and point 5 is similar to that of point 2 and point 3, respectively.) From these two figures, we can conclude that the proposed controller is able to effectively weaken the effect of the measurement noise.

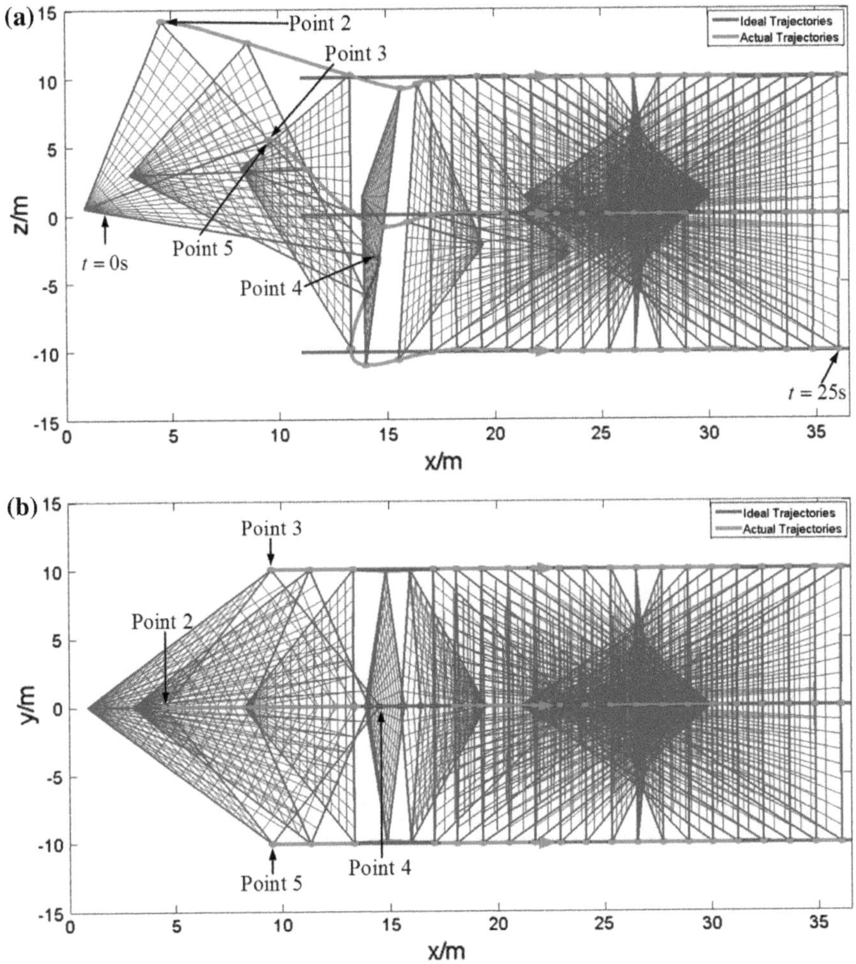

Fig. 4.45 Controlled motion of MSTN with initial state errors: **a** front view; **b** top view (the figure only shows the motion from 0 to 25 s since errors are basically eliminated during this period and the motion after 25 s is similar to that in Fig. 4.43)

Case 4: controlled motion with kinetic parameter errors

In the ideal case, the parameters in the inverse dynamics module are equal to those in the dynamics module. However, in order to test the sensitivity of the coordinated controller to the error of kinetic parameters, we introduce a parameter error of 30%

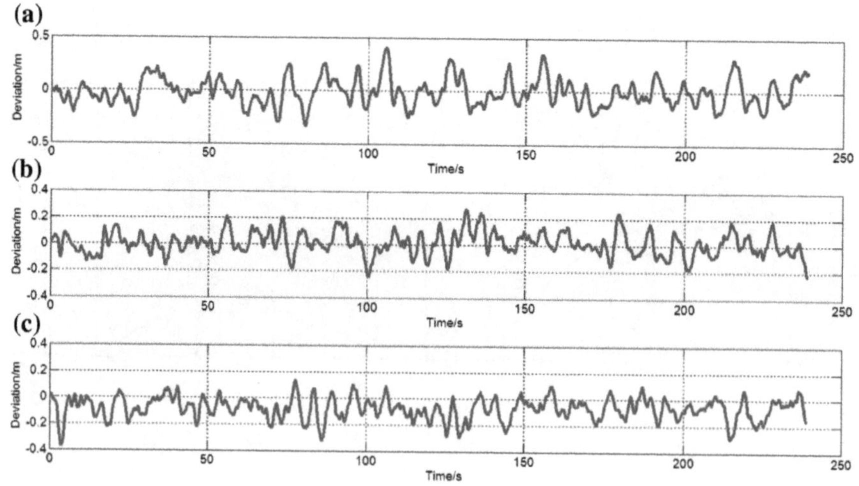

Fig. 4.46 Position deviation of point 2: **a** the x component; **b** the y component; **c** the z component

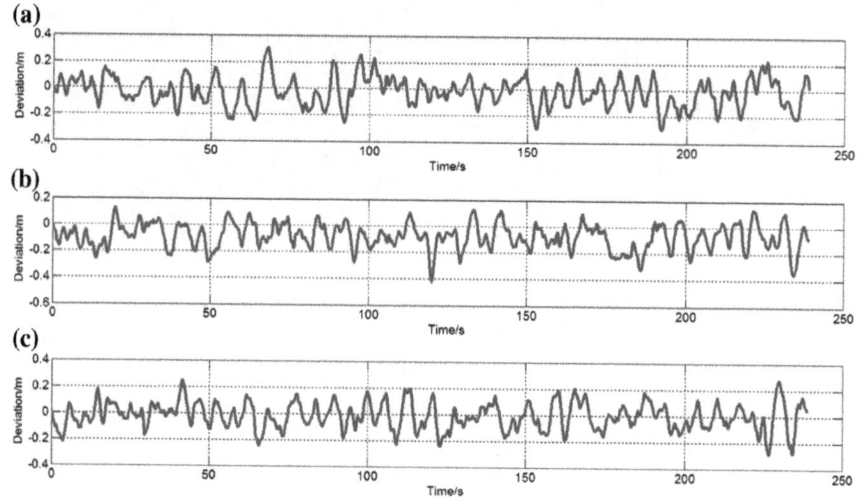

Fig. 4.47 Position deviation of point 3: **a** the x component; **b** the y component; **c** the z component

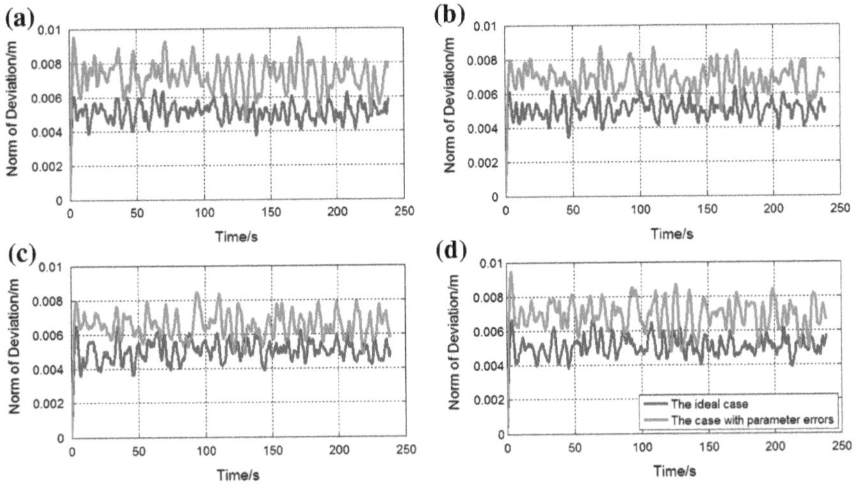

Fig. 4.48 Position deviation comparisons between the ideal case and the case with structure parameter errors: **a** point 1; **b** point 2; **c** point 3; **d** point 4

and set those used in the former to be different from those used in the latter. Specifically, the mass of the flexible net and the maneuvering robot when calculating the dynamics is 8 and 10 kg, respectively. However, when calculating the inverse dynamics, they are set as 10.4 and 7 kg. Comparisons between the ideal case and the case with parameter errors are shown in Fig. 4.48. From the figure, we can conclude that the parameter error of 30% causes an increase of about 30% in the position deviation of all four robots. However, the deviations of the four points are still very small and the stability of the whole system can be ensured.

References

1. Khalil H (1996) Nonlinear control. Prentice-Hall, New Jersey
2. Yan XG, Spurgeon SK, Edwards C (2003) Decentralized sliding mode control for multimachine power systems using only output information. In: Conference of the IEEE industrial electronics society. IECON '03, vol 2. IEEE, pp 1944–1949
3. Moreno J, Osorio M (2012) Strict Lyapunov functions for the super-twisting algorithm. IEEE Trans Autom Control 57(4):1035–1040
4. Nagesh I, Edwards C (2014) Technical communique: a multivariable super-twisting sliding mode approach. Automatica 50(3):984–988
5. Wang Z (2015) Adaptive smooth second-order sliding mode control method with application to missile guidance. Trans Inst Meas Control 0142331215621616
6. Bachtler J, Begg I (2006) Fuzzy controller design: theory and applications. Macromol Chem Phys 205(17):23–29
7. Forg M, Pfeiffer F, Ulbrich H (2005) Simulation of unilateral constrained systems with many bodies. Multibody Syst Dyn 14(2):137–154

8. Mankala KK, Agrawal SK (2004) Dynamic modeling and simulation of impact in tether net/gripper systems. Multibody Syst Dyn 11(3):235–250
9. Mansard N, Khatib O, Kheddar A (2009) A unified approach to integrate unilateral constraints in the stack of tasks. IEEE Trans Robot 25(3):670–685
10. Kanoun O, Lamiraux F, Wieber PB, Kanehiro F, Yoshida E, Laumond JP (2009) Prioritizing linear equality and inequality systems: application to local motion planning for redundant robots. In: Proceedings of IEEE international conference on robotics and automation
11. Klarbring A (1994) Mathematical programming in contact problems. In: Aliabadi MH (ed) Computational methods for contact problems. Elsevier, Amsterdam
12. Leamy MJ, Noor AK, Wasfy TM (2001) Dynamic simulation of a tethered satellite system using finite elements and fuzzy sets. Comput Methods Appl Mech Eng 190(37–38):4847–4870

Chapter 5
Distributed Deployment Control

The TSNR is a typical distributed system with the requirement of synchronization of multi-agent (the Maneuverable Units). Meanwhile, the configurations of the communication subsystem and the sensor subsystem of four MUs are different in practical applications. Therefore, these requirements and features necessitate distributed control method for the TSNR. During the whole approaching phase, there exists the relative distance constraint between arbitrary adjacent MUs connected with the net. Like the assembly of a team of spacecraft, the distance between arbitrary MUs is very small. Therefore, the minimal physical separation among the MUs for collision avoidance must be involved to ensure the approaching phase. Equally, the aggressive elasticity of the braid tether induces the periodical transition from slack to taut and slack again, namely bounce effect [1], which will yield the collision of two end masses. Therefore, to avoid the explosive elasticity of the connecting net, the maximal separation between arbitrary adjacent MUs should be less than a certain value. Thus, the approaching phase is not trivial. The controllers should be suitable for compensating the disturbances due to the net tensile force while achieving other control objectives like the relative distance constraint assurance. Many strategies have been proposed so far for formation control problems with collision avoidance and maintaining network connectivity. Two major approaches are rule-based methods and optimization-based methods, respectively [2]. Artificial potential field method is a kind of rule-based methods which has been used extensively for its analysis and continuity [3]. The way this method is applied is to assume that the agents are immersed in a virtual potential field. This field generates a repulsive force that prevents the agents from colliding with each other when they are too close; the field may also generate an attractive force between a pair of agents when they are too distant from each other.

P. Huang and F. Zhang, *Theory and Applications of Multi-Tethers in Space*,
Springer Tracts in Mechanical Engineering,
https://doi.org/10.1007/978-981-15-0387-0_5

5.1 Distributed Configuration-Maneuvering Control After Releasing

5.1.1 Orbital Dynamics

As shown in Fig. 5.1, the Earth-Centered Inertial (ECI) frame, *EXYZ*, is used as inertial frame to define the positions of platform satellite and TSNR in space. The Local-Vertical Local-Horizontal (LVLH) frame, *Oxyz*, originated in the centroid of the platform satellite (marked by *O*) is used to describe the relative motion between the platform satellite and TSNR.

The mass–spring model is utilized to establish the dynamics model of the connecting net in this research. As shown in Fig. 5.2, the side of a mesh is called knit cable with the natural length l_0. L is the side length of the connecting net. The mass–spring modeling method for connecting net is a kind of simplified modeling method by lumping mass of a knit cable between two knots at the ends of knit cable, and the knit cables are abstracted as massless spring–damper elements. The position of knot ij is expressed as

$$\boldsymbol{\rho}_{ij} = \mathbf{r}_{ij} - \mathbf{r}_o \tag{5.1}$$

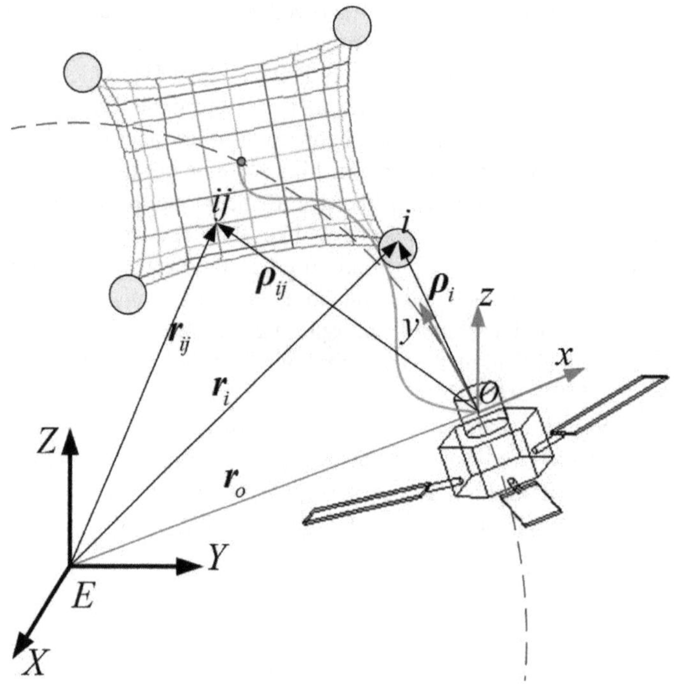

Fig. 5.1 Illustration of TSNR model and frames

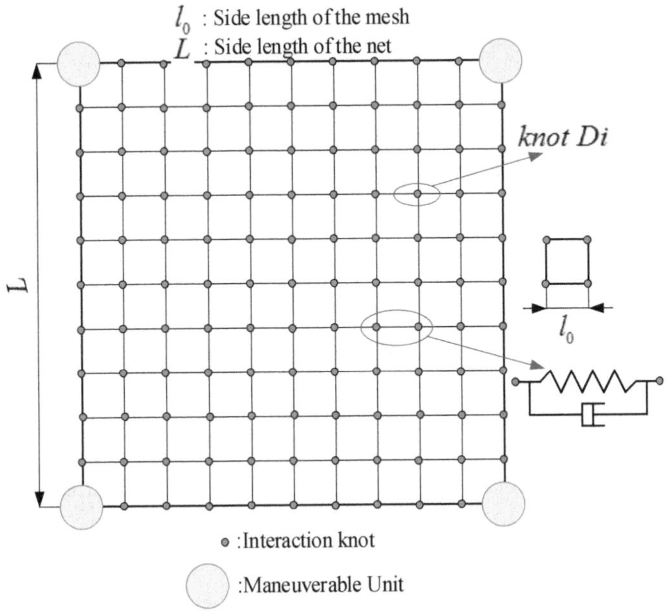

Fig. 5.2 Schematic diagram of TSNR

where $\mathbf{r}_{ij}$ is the position vector of knot ij in ECI frame, $\mathbf{r}_o|_{\mathrm{LVLH}} = [r_o, 0, 0]^{\mathrm{T}}$ represents the positive position vector of centroid O expressed in LVLH frame, and $\boldsymbol{\rho}_{ij}$ is the vector from O to knot ij in ECI frame, where $i = A, B, \ldots, K$ is the symbol of the rows of connecting net's knot, and $j = a, b, \ldots, k$ represents the columns (as shown in Fig. 5.2).

The position vector $\boldsymbol{\rho}_{ij}$ can be written as $\boldsymbol{\rho}_{ij}|_{\mathrm{LVLH}} = [x_{ij}, y_{ij}, z_{ij}]^{\mathrm{T}}$ in LVLH frame. The second derivative with respect to Eq. (5.1) without considering orbital perturbations is

$$\ddot{\boldsymbol{\rho}}_{ij} = \frac{\mu \mathbf{r}_o}{r_o^3} - \frac{\mu(\boldsymbol{\rho}_{ij} + \mathbf{r}_o)}{\left((r_o + x_{ij})^2 + y_{ij}^2 + z_{ij}^2\right)^{\frac{3}{2}}} + \frac{\sum_{mn\in\Omega_{ij}} \mathbf{F}_{ij-mn} + \mathbf{u}_{ij}}{m_{ij}} - \mathbf{d} \tag{5.2}$$

where μ is the earth gravitational constant, $\mathbf{u}_{ij}|_{\mathrm{LVLH}} = [u_{x,ij}, u_{y,ij}, u_{z,ij}]^{\mathrm{T}}$ is the control input of knot ij, $\mathbf{d}|_{\mathrm{LVLH}} = [d_x, d_y, d_z]^{\mathrm{T}}$ is the external acceleration of the platform caused by the platform thrust and the tension acting on platform, m_{ij} represents the mass of knot ij, Ω_{ij} is the set of the adjacent knots which connect to knot ij by an adjacent knit cable, and $\mathbf{F}_{ij-mn}$ is the tension acting on knot ij exerted by the neighbor knot mn.

$$\mathbf{F}_{ij-mn} = \begin{cases} 0 & d_{ij-mn} \le l_0 \\ \left(\frac{EA}{l_0}\left(d_{ij-mn} - l_0\right) + \zeta \dot{d}_{ij-mn}\right)\hat{\mathbf{d}}_{ij-mn} & d_{ij-mn} > l_0 \end{cases} \tag{5.3}$$

where $\mathbf{F}_{ij-mn}$ can be formulated as $\mathbf{F}_{ij-mn}\big|_{\mathrm{LVLH}} = \left[f_{x,ij-mn}, f_{y,ij-mn}, f_{z,ij-mn}\right]^{\mathrm{T}}$, E represents Young's modulus of the flexible knit cable material, A is the cross section of the knit cable, d_{ij-mn} is actual length between the adjacent knots ij and mn. The damping coefficient of the knit cable is defined as $\zeta = 2\xi\sqrt{mEA/l_0}$, in which ξ represents its damping ratio, m is the nominal mass of a knit cable. $\hat{\mathbf{d}}_{ij-mn}$ is the unit vector from knot ij to knot mn.

The relationship between $\boldsymbol{\rho}_{ij}$ and $\boldsymbol{\rho}_{ij}\big|_{\mathrm{LVLH}} = \left[x_{ij}, y_{ij}, z_{ij}\right]^{\mathrm{T}}$ is held as

$$\boldsymbol{\rho}_{ij} = {}^{E}\mathbf{R}_{L}\boldsymbol{\rho}_{ij}\big|_{\mathrm{LVLH}} \tag{5.4}$$

where ${}^{E}\mathbf{R}_{L}$ is the transformation matrix from LVLH to ECI frame. The second derivative with respect to Eq. (5.4) is

$$\begin{aligned} \ddot{\boldsymbol{\rho}}_{ij}\big|_{\mathrm{LVLH}} = {} & {}^{E}\mathbf{R}_{L}^{-1}\ddot{\boldsymbol{\rho}}_{ij} - \dot{\boldsymbol{\omega}}_{p}\big|_{\mathrm{LVLH}} \times \boldsymbol{\rho}_{ij}\big|_{\mathrm{LVLH}} \\ & - \boldsymbol{\omega}_{p}\big|_{\mathrm{LVLH}} \times \left(\boldsymbol{\omega}_{p}\big|_{\mathrm{LVLH}} \times \boldsymbol{\rho}_{ij}\big|_{\mathrm{LVLH}}\right) - 2\boldsymbol{\omega}_{p}\big|_{\mathrm{LVLH}} \times \dot{\boldsymbol{\rho}}_{ij}\big|_{\mathrm{LVLH}} \end{aligned} \tag{5.5}$$

where $\boldsymbol{\omega}_{p}\big|_{\mathrm{LVLH}} = [0, 0, \omega]^{\mathrm{T}}$ is the angular velocity of the circular reference orbit. Considering Eqs. (5.2) and (5.5), the relative dynamics model of knot ij can be derived as Eq. (5.6).

$$\begin{cases} \ddot{x}_{ij} - 2\omega\dot{y}_{ij} - \omega^2 x_{ij} - \dot{\omega} y_{ij} + \dfrac{\mu\left(r_o + x_{ij}\right)}{\left(\left(r_o + x_{ij}\right)^2 + y_{ij}^2 + z_{ij}^2\right)^{\frac{3}{2}}} - \dfrac{\mu}{r_o^2} = \dfrac{\sum\limits_{mn\in\Omega_{ij}} f_{x,ij-mn} + u_{x,ij}}{m_{ij}} - d_x \\ \ddot{y}_{ij} + 2\omega\dot{x}_{ij} + \dot{\omega} x_{ij} - \omega^2 y_{ij} + \dfrac{\mu y_{ij}}{\left(\left(r_o + x_{ij}\right)^2 + y_{ij}^2 + z_{ij}^2\right)^{\frac{3}{2}}} = \dfrac{\sum\limits_{mn\in\Omega_{ij}} f_{y,ij-mn} + u_{y,ij}}{m_{ij}} - d_y \\ \ddot{z}_{ij} + \dfrac{\mu z_{ij}}{\left(\left(r_o + x_{ij}\right)^2 + y_{ij}^2 + z_{ij}^2\right)^{\frac{3}{2}}} = \dfrac{\sum\limits_{mn\in\Omega_{ij}} f_{z,ij-mn} + u_{z,ij}}{m_{ij}} - d_z \end{cases} \tag{5.6}$$

When $ij = Aa, Ak, Ka, Kk$ which represent four-corner MUs of TSNR, Eq. (5.6) is the relative dynamics model of MUs expressed in LVLH frame. For convenience, rewrite Eq. (5.6) as the following form:

$$\begin{cases} \dot{\boldsymbol{\rho}}_i = \mathbf{v}_i \\ \mathbf{M}_i\dot{\mathbf{v}}_i + \mathbf{C}_i\mathbf{v}_i + \mathbf{g}_i = \mathbf{u}_i + \mathbf{d}_i \end{cases} \tag{5.7}$$

where the relative position vector of MU i relative to the platform satellite is denoted as $\boldsymbol{\rho}_i|_{\mathrm{LVLH}} = [x_i, y_i, z_i]^{\mathrm{T}}$ in LVLH frame, and $\mathbf{v}_i$ is the relative velocity vector. $\mathbf{M}_i$ $(\mathrm{i} = 1, 2, \ldots, \mathrm{n})$ denotes the positive-definite diagonal inertia matrix. $\mathbf{C}_i$ is a skew matrix. $\mathbf{g}_i$ is a nonlinear term. $\mathbf{u}_i$ is the control force, and $\mathbf{d}_i$ is the disturbance on MU i.

Assumption 1 The disturbance vector $\mathbf{d}_i$ $(\forall i = 1, 2, \ldots, n)$ is bounded which appends the additional constraint $\|\mathbf{d}_i\| \le \bar{d}_i$.

Property 1 The matrix $\mathbf{C}_i$ is a skew-symmetric matrix which satisfies $\mathbf{x}^{\mathrm{T}}\mathbf{C}_i\mathbf{x} = 0$ for any vector $\mathbf{x}$ with corresponding dimensions.

5.1.2 Problem Statement

The configuration maintenance of the TSNR during approaching phase requires that the MUs should hold accurate geometric construction during maneuvers. This subsection is aimed at studying the cooperative relative tracking control of four MUs to realize the shape maintenance of the net during deployment.

5.1.3 Controller Design

The configuration coordination of the flexible net requires that the MUs should maintain accurate geometric configuration during the TSNR maneuvers. In this subsection, a distributed coordination control scheme is designed for the MUs by employing the behavior-based control strategy and consensus algorithms.

Assumption 2 The relative position and velocity information of all the MUs with respect to the platform satellite are available.

Before controller design, to simplify the presentation, the error variables are defined as follows:

$$\tilde{\boldsymbol{\rho}}_i = \boldsymbol{\rho}_i - \boldsymbol{\rho}_i^d, \quad \tilde{\mathbf{v}}_i = \mathbf{v}_i - \mathbf{v}_i^d \tag{5.8}$$

where the notations $\boldsymbol{\rho}_i^d$ and $\mathbf{v}_i^d$ are the desired position and velocity to be tracked by MU i.

Inspired by [1] and [4], we define the neighbor errors for MU i as

$$\mathbf{e}_i = \sum_{j \in N_i} a_{ij}\left(\tilde{\boldsymbol{\rho}}_i - \tilde{\boldsymbol{\rho}}_j\right) + b_i\tilde{\boldsymbol{\rho}}_i \tag{5.9}$$

$$\dot{\mathbf{e}}_i = \sum_{j \in N_i} a_{ij}\left(\tilde{\mathbf{v}}_i - \tilde{\mathbf{v}}_j\right) + b_i\, \tilde{\mathbf{v}}_i \tag{5.10}$$

where N_i is the set of indices corresponding to the information neighbors of the MU i, a_{ij} is the ith row and jth column entry of the weighted adjacency matrix $\mathbf{A}$, and b_i represents the ith principal diagonal element of matrix $\mathbf{B}$.

Further, Eqs. (5.9) and (5.10) can be rewritten in a vector form as

$$\mathbf{e} = ((\mathbf{L} + \mathbf{B}) \otimes \mathbf{I})\tilde{\boldsymbol{\rho}} = (\mathbf{H} \otimes \mathbf{I})\tilde{\boldsymbol{\rho}} \tag{5.11}$$

$$\dot{\mathbf{e}} = ((\mathbf{L} + \mathbf{B}) \otimes \mathbf{I})\tilde{\mathbf{v}} = (\mathbf{H} \otimes \mathbf{I})\tilde{\mathbf{v}} \tag{5.12}$$

where $\mathbf{e} = \left[\mathbf{e}_1^{\mathrm{T}} \ \ldots \ \mathbf{e}_n^{\mathrm{T}}\right]^{\mathrm{T}}$, $\tilde{\boldsymbol{\rho}} = \left[\tilde{\boldsymbol{\rho}}_1^{\mathrm{T}} \ \ldots \ \tilde{\boldsymbol{\rho}}_n^{\mathrm{T}}\right]^{\mathrm{T}}$ and $\tilde{\mathbf{v}} = \left[\tilde{\mathbf{v}}_1^{\mathrm{T}} \ \ldots \ \tilde{\mathbf{v}}_n^{\mathrm{T}}\right]^{\mathrm{T}}$ are stack vectors, the symbol $\otimes$ represents the Kronecker product, and $\mathbf{I}$ denotes an identity matrix with corresponding dimensions.

Remark 1 The configuration maintenance of the TSNR during approaching phase requires that the MUs should hold accurate geometric construction during maneuvers. It means when the station keeping is achieved, the formation keeping is also realized, namely $\boldsymbol{\rho}_i \to \boldsymbol{\rho}_i^d$, $\mathbf{v}_i \to \mathbf{v}_i^d$, $\tilde{\boldsymbol{\rho}}_i \to \tilde{\boldsymbol{\rho}}_j$ and $\tilde{\mathbf{v}}_i \to \tilde{\mathbf{v}}_j$, $\forall i, j \in N_i$.

Motivated by [5] and [6], we introduce the following auxiliary variable:

$$\mathbf{s}_i = k_1\tilde{\boldsymbol{\rho}}_i + \tilde{\mathbf{v}}_i + k_2\mathbf{e}_i \tag{5.13}$$

where k_1 and k_2 are positive constant. Then, from Eqs. (5.5), (5.8), (5.9), and (5.13), we have

$$\begin{aligned} &\mathbf{M}_i\dot{\mathbf{s}}_i + \mathbf{M}_i\dot{\mathbf{v}}_i^d - k_1\mathbf{M}_i\tilde{\mathbf{v}}_i - k_2\mathbf{M}_i\left[\sum_{j \in N_i} a_{ij}\left(\tilde{\mathbf{v}}_i - \tilde{\mathbf{v}}_j\right) + b_i\tilde{\mathbf{v}}_i\right] + \mathbf{C}_i\mathbf{s}_i \\ &\quad + \mathbf{C}_i\mathbf{v}_i^d - k_1\mathbf{C}_i\tilde{\boldsymbol{\rho}}_i - k_2\mathbf{C}_\mathbf{i}\left[\sum_{j \in N_i} a_{ij}\left(\tilde{\boldsymbol{\rho}}_i - \tilde{\boldsymbol{\rho}}_j\right) + b_i\tilde{\boldsymbol{\rho}}_i\right] + \mathbf{g}_i \\ &\quad = \mathbf{u}_i + \mathbf{d}_i \end{aligned} \tag{5.14}$$

It is obvious that MUs are always subject to various disturbances during maneuvers such as tension strain, gravitational perturbations. In our research, a hyperbolic tangent functions is introduced to render the controller robust to disturbance $\mathbf{d}_i$.

$$\boldsymbol{\psi}_i(\mathbf{s}_i) = \tanh(\mathbf{s}_i/\varepsilon) \tag{5.15}$$

where constant $\varepsilon > 0$. Then, inspired by [7], the robust coordination controller is presented as

$$\begin{aligned}\mathbf{u}_i = {} & \mathbf{M}_i\dot{\mathbf{v}}_i^d + \mathbf{C}_i\mathbf{v}_i^d - k_1\mathbf{C}_i\tilde{\boldsymbol{\rho}}_i - k_1\mathbf{M}_i\tilde{\mathbf{v}}_i - k_2\mathbf{C}_i\left[\sum_{j\in N_i} a_{ij}\left(\tilde{\boldsymbol{\rho}}_i - \tilde{\boldsymbol{\rho}}_j\right) + b_i\tilde{\boldsymbol{\rho}}_i\right] \\ & - k_2\mathbf{M}_i\left[\sum_{j\in N_i} a_{ij}\left(\tilde{\mathbf{v}}_i - \tilde{\mathbf{v}}_j\right) + b_i\tilde{\mathbf{v}}_i\right] + \mathbf{g}_i - k_3\mathbf{s}_i - \hat{\beta}_i\boldsymbol{\psi}_i(\mathbf{s}_i)\end{aligned} \tag{5.16}$$

$$\dot{\hat{\beta}}_i = \mathbf{s}_i^{\mathrm{T}}\tanh(\mathbf{s}_i/\varepsilon) - \gamma_i\left(\hat{\beta}_i - \beta_{i0}\right) \tag{5.17}$$

where k_3, γ_i and β_{i0} are positive constants.

Remark 2 The first to seventh items in Eq. (5.16) are used to eliminate the same terms on the left-hand side of Eq. (5.14). The item $-k_3\mathbf{s}_i$ is the proportional feedback term, which is used to regulate the MUs to the desired state. The robustness signal $-\hat{\beta}_i\boldsymbol{\psi}_i(\mathbf{s}_i)$ with the adaptive updating law of the varying gain $\hat{\beta}_i$ in Eq. (5.17) is used to eliminate the effect of external bounded disturbance $\mathbf{d}_i$. The ideal robustness gain $\beta_i = \hat{\beta}_i + \tilde{\beta}_i$ is chosen to satisfy $\beta_i \geq \bar{d}_i$, in which $\hat{\beta}_i$ is the estimate of the ideal constant β_i, and $\tilde{\beta}_i$ is the estimation error. In Eq. (5.16), obviously, the control input of each MU is determined by the information from itself and the neighbors. Thus, the control algorithm is distributed.

5.1.4 Stability Analysis of the Closed-Loop System

Lemma 1 When the weighted undirected simple graph G is connected and the platform satellite has a directed path to all of the MUs, each diagonal entry of matrix $\mathbf{B}$ satisfies $b_i \neq 0$ and matrix $\mathbf{H}$ is a strictly diagonally dominant matrix with the positive principal diagonal elements. By the Gershgorin law, it can be obtained that all of the eigenvalues of matrix $\mathbf{H}$ have positive real parts.

Lemma 2 When the graph G contains an undirected spanning tree and the diagonal matrix $\mathbf{B}$ is a positive-definite matrix, the matrix $\mathbf{Q} = \mathbf{PH} + \mathbf{H}^{\mathrm{T}}\mathbf{P}$ is a symmetric positive matrix, where $\mathbf{P}$ is defined as $\mathbf{P} = diag(p_1, p_2, \ldots, p_n)$ with $p_i > 0$, $i = 1, 2, \ldots, n$ [8].

Lemma 3 It is obtained that the following inequality holds for any $\varepsilon > 0$ and $x \in \mathbb{R}$:

$$0 \leq |x| - x \cdot \tanh(x/\varepsilon) \leq \kappa\varepsilon \tag{5.18}$$

where κ is a positive constant which satisfies $\kappa = e^{-(\kappa+1)}$; i.e., $\kappa = 0.2785$ [9].

Remark 3 The multi-agent system in this paper is composed of n MUs and one platform satellite, which is considered as a time-varying leader. Suppose that the undirected graph G contains an undirected spanning tree and all of the MUs can get information from the platform satellite.

Theorem 1 By employing the control scheme in Eqs. (5.16) and (5.17) for systems (5.7), when $\boldsymbol{\rho}_i^d$, $\mathbf{v}_i^d$, and $\dot{\mathbf{v}}_i^d$ are all bounded, we can derive that $\tilde{\boldsymbol{\rho}}_i$, $\tilde{\mathbf{v}}_i$, and $\hat{\beta}_i$ are also all bounded. Furthermore, the position tracking errors are uniformly ultimate boundedness.

Proof By substituting Eq. (5.16) into (5.14), the closed-loop system is derived as follows:

$$\mathbf{M}_i\dot{\mathbf{s}}_i + (\mathbf{C}_i + k_3\mathbf{I})\mathbf{s}_i = -\hat{\beta}_i\tanh(\mathbf{s}_i/\varepsilon) + \mathbf{d}_i \tag{5.19}$$

where $\mathbf{I}$ denotes an identity matrix with corresponding dimensions. We choose the Lyapunov function candidate as

$$V(t) = \sum_{i=1}^{n}\left[\frac{1}{2}\mathbf{s}_i^{\mathrm{T}}\mathbf{M}_i\mathbf{s}_i + \frac{1}{2}\tilde{\beta}_i^2\right] + \frac{1}{2}\mathbf{e}^{\mathrm{T}}(\mathbf{P}\otimes\mathbf{I})\mathbf{e} \tag{5.20}$$

Note that $V(t)$ is positive definite and radially unbounded with respect to $\mathbf{s}_i$ and $\tilde{\beta}_i$. The time derivative of V along Eqs. (5.7) and (5.16) is given by

$$\begin{aligned}
\dot{V}(t) &= \sum_{i=1}^{n}\left[\mathbf{s}_i^{\mathrm{T}}\mathbf{M}_i\dot{\mathbf{s}}_i + \tilde{\beta}_i\dot{\tilde{\beta}}_i\right] + \mathbf{e}^{\mathrm{T}}(\mathbf{P}\otimes\mathbf{I})\dot{\mathbf{e}} \\
&= \sum_{i=1}^{n}\left\{\mathbf{s}_i^{\mathrm{T}}\left[-(\mathbf{C}_i + k_3\mathbf{I})\mathbf{s}_i - \hat{\beta}_i\tanh(\mathbf{s}_i/\varepsilon) + \mathbf{d}_i\right] - \tilde{\beta}_i\left[\mathbf{s}_i^{\mathrm{T}}\tanh(\mathbf{s}_i/\varepsilon) - \gamma_i\left(\hat{\beta}_i - \beta_{i0}\right)\right]\right\} \\
&\quad + \mathbf{e}^{\mathrm{T}}(\mathbf{P}\otimes\mathbf{I})\dot{\mathbf{e}} \\
&= \sum_{i=1}^{n}\left\{-k_3\mathbf{s}_i^{\mathrm{T}}\mathbf{s}_i - \left(\beta_i - \tilde{\beta}_i\right)\mathbf{s}_i^{\mathrm{T}}\tanh(\mathbf{s}_i/\varepsilon) + \mathbf{s}_i^{\mathrm{T}}\mathbf{d}_i - \tilde{\beta}_i\mathbf{s}_i^{\mathrm{T}}\tanh(\mathbf{s}_i/\varepsilon) + \gamma_i\tilde{\beta}_i\left(\hat{\beta}_i - \beta_{i0}\right)\right\} \\
&\quad + \mathbf{e}^{\mathrm{T}}(\mathbf{P}\otimes\mathbf{I})\dot{\mathbf{e}} \\
&\leq \sum_{i=1}^{n}\left\{-k_3\mathbf{s}_i^{\mathrm{T}}\mathbf{s}_i - \beta_i\mathbf{s}_i^{\mathrm{T}}\tanh(\mathbf{s}_i/\varepsilon) + \beta_i\|\mathbf{s}_i\| + \gamma_i\tilde{\beta}_i\left(\hat{\beta}_i - \beta_{i0}\right)\right\} + \mathbf{e}^{\mathrm{T}}(\mathbf{P}\otimes\mathbf{I})\dot{\mathbf{e}}
\end{aligned} \tag{5.21}$$

Furthermore, the following inequality holds:

$$\begin{aligned}\tilde{\beta}_i\left(\hat{\beta}_i-\beta_{i0}\right)&=-\frac{1}{2}\tilde{\beta}_i^2-\frac{1}{2}\left(\hat{\beta}_i-\beta_{i0}\right)^2+\frac{1}{2}\left(\beta_i-\beta_{i0}\right)^2\\&\le-\frac{1}{2}\tilde{\beta}_i^2+\frac{1}{2}\left(\beta_i-\beta_{i0}\right)^2\end{aligned}\tag{5.22}$$

By substituting Eq. (5.22) into (5.21), it can be deduced that

$$\begin{aligned}\dot{V}(t)\le&\sum_{i=1}^{n}\left\{-k_3\mathbf{s}_i^{\mathrm{T}}\mathbf{s}_i-\beta_i\mathbf{s}_i^{\mathrm{T}}\tanh(\mathbf{s}_i/\varepsilon)+\beta_i\|\mathbf{s}_i\|-\frac{1}{2}\gamma_i\tilde{\beta}_i^2+\frac{1}{2}\gamma_i\left(\beta_i-\beta_{i0}\right)^2\right\}\\&+\mathbf{e}^{\mathrm{T}}(\mathbf{P}\otimes\mathbf{I})\dot{\mathbf{e}}\end{aligned}\tag{5.23}$$

With the item $\beta_i\|\mathbf{s}_i\|$ caused by the disturbance $\mathbf{d}_i$ in Eq. (5.23), it can't be concluded that $\dot{V}(t)\le-\lambda V(t)+c$ or $\dot{V}(t)<0$ without the robustness signal, where constant $\lambda>0$. However, the item $\beta_i\|\mathbf{s}_i\|$ will be eliminated by the robustness signal introduced in Eq. (5.15). By Lemma 3, it can be derived that if $\beta_i\ge 0$, then

$$\beta_i\mathbf{s}_i^{\mathrm{T}}\tanh(\mathbf{s}_i/\varepsilon)\ge\beta_i\|\mathbf{s}_i\|-\kappa\varepsilon\beta_i\tag{5.24}$$

The auxiliary variable in Eq. (5.13) can be rewritten in a stack vector form $\mathbf{s}=k_1\tilde{\boldsymbol{\rho}}+\tilde{\mathbf{v}}+k_2\mathbf{e}$ with $\mathbf{s}=\left(\mathbf{s}_1^{\mathrm{T}},\mathbf{s}_2^{\mathrm{T}},\ldots,\mathbf{s}_n^{\mathrm{T}}\right)^{\mathrm{T}}$. Considering Lemma 2 and Eq. (5.12), we carry out the following equation:

$$\begin{aligned}\mathbf{e}^{\mathrm{T}}(\mathbf{P}\otimes\mathbf{I})\dot{\mathbf{e}}&=\mathbf{e}^{\mathrm{T}}(\mathbf{PH}\otimes\mathbf{I})(\mathbf{s}-k_1\tilde{\boldsymbol{\rho}}-k_2\mathbf{e})\\&\le\|\mathbf{PH}\|_{\mathrm{F}}\|\mathbf{e}\|\|\mathbf{s}\|-\frac{k_1}{2}\underline{\rho}(\mathbf{Q})\|\mathbf{e}\|\|\tilde{\boldsymbol{\rho}}\|-\frac{k_2}{2}\underline{\rho}(\mathbf{Q})\|\mathbf{e}\|^2\\&\le\|\mathbf{PH}\|_{\mathrm{F}}\|\mathbf{e}\|\|\mathbf{s}\|-\frac{k_2}{2}\underline{\rho}(\mathbf{Q})\|\mathbf{e}\|^2\end{aligned}\tag{5.25}$$

where notation $\underline{\rho}(\mathbf{Q})$ represents the minimum eigenvalue of matrix $\mathbf{Q}$ being $\underline{\rho}(\mathbf{Q})>0$ by Lemma 2. The products $\|\cdot\|_{\mathrm{F}}$ and $\|\cdot\|$ refer to the Frobenius norm of a matrix and the Euclidean norm of a vector, respectively.

The following inequality is established according to Young's inequality:

$$\|\mathbf{PH}\|_{\mathrm{F}}\|\mathbf{e}\|\|\mathbf{s}\|\le\frac{\|\mathbf{PH}\|_{\mathrm{F}}^2\|\mathbf{e}\|^2}{2\delta_1}+\frac{\delta_1\|\mathbf{s}\|^2}{2}\tag{5.26}$$

where $\delta_1>0$ is a constant. Upon substituting Eqs. (5.24)–(5.26) into (5.23), we then arrive at

$$
\begin{aligned}
\dot{V}(t) \leq & \sum_{i=1}^{n} \left\{ -k_3 \mathbf{s}_i^{\mathrm{T}} \mathbf{s}_i + \kappa \varepsilon \beta_i - \frac{1}{2} \gamma_i \tilde{\beta}_i^2 + \frac{1}{2} \gamma_i (\beta_i - \beta_{i0})^2 \right\} \\
& + \left(\frac{\|\mathbf{PH}\|_{\mathrm{F}}^2 \|\mathbf{e}\|^2}{2\delta_1} + \frac{\delta_1 \|\mathbf{s}\|^2}{2} \right) - \frac{k_2}{2} \underline{\rho}(\mathbf{Q}) \|\mathbf{e}\|^2 \\
= & -\frac{1}{2} \left(k_2 \underline{\rho}(\mathbf{Q}) - \frac{\|\mathbf{PH}\|_{\mathrm{F}}^2}{\delta_1} \right) \|\mathbf{e}\|^2 - \left(k_3 - \frac{\delta_1}{2} \right) \|\mathbf{s}\|^2 \\
& - \sum_{i=1}^{n} \frac{1}{2} \gamma_i \tilde{\beta}_i^2 + \sum_{i=1}^{n} \left[\kappa \varepsilon \beta_i + \frac{1}{2} \gamma_i (\beta_i - \beta_{i0})^2 \right]
\end{aligned}
\tag{5.27}
$$

It should be mentioned that k_2 and k_3 can be chosen as $k_2 > \|\mathbf{PH}\|_{\mathrm{F}}^2 / \underline{\rho}(\mathbf{Q}) \delta_1$ and $k_3 > \delta_1 / 2$ such that $\left(k_2 \underline{\rho}(\mathbf{Q}) - \|\mathbf{PH}\|_{\mathrm{F}}^2 / \delta_1 \right) > 0$ and $(k_3 - \delta_1/2) > 0$. Furthermore, Eq. (5.27) can be written as

$$
\begin{aligned}
\dot{V}(t) \leq & -\frac{1}{2} \left(k_2 \underline{\rho}(\mathbf{Q}) - \frac{\|\mathbf{PH}\|_{\mathrm{F}}^2}{\delta_1} \right) \|\mathbf{e}\|^2 - \left(k_3 - \frac{\delta_1}{2} \right) \|\mathbf{s}\|^2 \\
& - \sum_{i=1}^{n} \frac{1}{2} \gamma_i \tilde{\beta}_i^2 + \sum_{i=1}^{n} \left[\kappa \varepsilon \beta_i + \frac{1}{2} \gamma_i (\beta_i - \beta_{i0})^2 \right] \\
= & -\frac{1}{2\bar{\rho}(\mathbf{P})} \left(k_2 \underline{\rho}(\mathbf{Q}) - \frac{\|\mathbf{PH}\|_{\mathrm{F}}^2}{\delta_1} \right) \bar{\rho}(\mathbf{P}) \|\mathbf{e}\|^2 \\
& - \frac{1}{2\bar{\rho}(\mathbf{M}_i)} (2k_3 - \delta_1) \bar{\rho}(\mathbf{M}_i) \|\mathbf{s}\|^2 - \sum_{i=1}^{n} \frac{1}{2} \gamma_i \tilde{\beta}_i^2 + c
\end{aligned}
\tag{5.28}
$$

where $c = \sum_{i=1}^{n} \left[\kappa \varepsilon \beta_i + \gamma_i (\beta_i - \beta_{i0})^2 / 2 \right]$, $\bar{\rho}(\mathbf{P})$ and $\bar{\rho}(\mathbf{M}_i)$ represent the maximum eigenvalues of matrix $\mathbf{P}$ and $\mathbf{M}_i$, respectively.

Let $\lambda = \min\left\{ \left(k_2 \underline{\rho}(\mathbf{Q}) - \|\mathbf{PH}\|_{\mathrm{F}}^2 / \delta_1 \right) / \bar{\rho}(\mathbf{P}), (2k_3 - \delta_1)/\bar{\rho}(\mathbf{M}_i), \gamma_i \right\}$. It thus follows from Eq. (5.28) that

$$
\dot{V}(t) \leq -\lambda V(t) + c \tag{5.29}
$$

where $\lambda > 0$, c are constants, from [10], we can conclude that $V(t)$ is bounded. Then, with any initial compact set, we can conclude that

$$
0 \leq V(t) \leq \left[V(0) - \frac{c}{\lambda} \right] e^{-\lambda t} + \frac{c}{\lambda} \leq V(0) + \frac{c}{\lambda} \tag{5.30}
$$

where $V(0)$ is the bounded initial value of Lyapunov function candidate $V(t)$. From Eq. (5.20), we have

$$\frac{1}{2}\underline{\rho}(\mathbf{M}_i)\|\mathbf{s}_i\|^2 \leq \frac{1}{2}\mathbf{s}_i^{\mathrm{T}}\mathbf{M}_i\mathbf{s}_i \leq V(t) \tag{5.31}$$

$$\frac{1}{2}\underline{\rho}(\mathbf{P})\|\mathbf{e}\|^2 \leq \frac{1}{2}\mathbf{e}^{\mathrm{T}}(\mathbf{P}\otimes\mathbf{I})\mathbf{e} \leq V(t) \tag{5.32}$$

where $\underline{\rho}(\mathbf{M}_i)$ and $\underline{\rho}(\mathbf{P})$ denote the minimum eigenvalues of matrices $\mathbf{M}_i$ and $\mathbf{P}$, respectively. From Eqs. (5.30) to (5.32), we carry out the following equations:

$$\|\mathbf{s}_i\| \leq \sqrt{\frac{2\left[V(0)-\frac{c}{\lambda}\right]e^{-\lambda t}+\frac{2c}{\lambda}}{\underline{\rho}(\mathbf{M}_i)}} \tag{5.33}$$

$$\|\mathbf{e}(t)\| \leq \sqrt{\frac{2\left[V(0)-\frac{c}{\lambda}\right]e^{-\lambda t}+\frac{2c}{\lambda}}{\underline{\rho}(\mathbf{P})}} \tag{5.34}$$

Further, from Eqs. (5.33) and (5.34), we have

$$\lim_{t\to\infty}\|\mathbf{s}_i\| = \sqrt{\frac{2c}{\lambda\underline{\rho}(\mathbf{M}_i)}} \tag{5.35}$$

$$\lim_{t\to\infty}\|\mathbf{e}(t)\| = \sqrt{\frac{2c}{\lambda\underline{\rho}(\mathbf{P})}} \tag{5.36}$$

Remark 4 According to Eqs. (5.3)–(5.34), the auxiliary variables $\mathbf{s}_i(t)$ and the neighbor errors $\mathbf{e}_i(t)(i \in \{1,2,\ldots,n\})$ in the closed-loop system will remain in a bounded compact set Ω. From Eqs. (5.35) to (5.36), $\mathbf{s}_i(t)$ and $\mathbf{e}_i(t)$ will eventually converge to another steady-state compact set Ω_s, where $\Omega_s \subseteq \Omega$, and Ω_s can be made smaller by changing the appropriate parameters. From Eq. (5.13), it can be derived that $\tilde{\boldsymbol{\rho}}_i$ and $\tilde{\mathbf{v}}_i$ are also bounded. Thus, it follows that the station keeping of MUs is achieved, meanwhile, the formation keeping is also realized, meaning that the configuration of the TSNR is maintained during approaching phase.

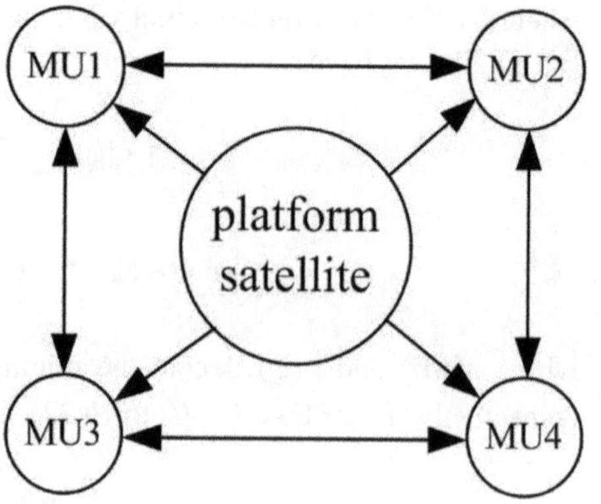

Fig. 5.3 Communication topology of TSNR

5.1.5 Numerical Simulation

5.1.5.1 Simulation Environment

TSNR is stored in a platform satellite before shooting, and the net is folded in a square pattern with sideline 0.5 m [11]. To validate the effectiveness of the distributed coordination controllers proposed in this paper, two different cases are simulated in this section. Case 1 is a symmetric configuration case without an initial velocity error, whereas case 2 is an asymmetric configuration case. In case 1, the initial velocities of four MUs are $(0.1 \quad 0.1 \quad 0.1)$ m/s, $(-0.1 \quad 0.1 \quad 0.1)$ m/s, $(0.1 \quad 0.1 \quad -0.1)$ m/s, and $(-0.1 \quad 0.1 \quad -0.1)$ m/s, respectively. In case 2, a velocity error is imposed on MU1, where the initial velocity of MU1 is $(0.2 \quad 0.1 \quad 0.1)$ m/s. The orbital angular velocity is $\omega = 9.2430 \times 10^{-5}$ rad/s. When any component of the position tracking error of any MU is less than 0.01 m, the controllers are switched on. The controller is opened at $t = 20.221$ s in this section, while the instantaneous positions of four MUs are $(2.49 \quad 2.0119 \quad 2.4878)$, $(-2.4824 \quad 2.0189 \quad 2.4856)$, $(2.49 \quad 2.0119 \quad -2.4878)$, and $(-2.4824 \quad 2.0189 \quad -2.4856)$, respectively. The desired trajectories of four MUs in the orbital coordinate frame are $(2.5 \quad 2.0189 + 0.1(t - 20.221) \quad 2.5)$, $(-2.5 \quad 2.0189 + 0.1(t - 20.221) 2.5)$, $(2.5 \quad 2.0189 + 0.1(\mathrm{t} - 20.221) \quad -2.5)$, $(-2.5 \quad 2.0189 + 0.1(t - 20.221) \quad -2.5)$, respectively, and the desired velocities are $(0 \quad 0.1 \quad 0)$ m/s for all MUs. Parameters of the controllers are $k_1 = k_2 = k_3 = 0.1$, $\beta_{i0} = 0.1$, $\varepsilon = 0.01$, $\gamma_i = 10$ $(i = 1, 2, 3, 4)$. The adjacency matrix $\mathbf{A}$ is $\mathbf{A} = [0 \quad 1 \quad 1.5 \quad 0;\ 1 \quad 0 \quad 0 \quad 1;\ 1.5 \quad 0 \quad 0 \quad 2;\ 0 \quad 1 \quad 2 \quad 0]$, $\mathbf{B} = \mathbf{I}_4$, in which $\mathbf{I}_4$ is an four-dimensional identity matrix. The relationship of communication between the platform satellite and the Maneuverable Units is shown in Fig. 5.3. Other parameters of the system are shown in Table 5.1.

Table 5.1 Parameters of the system

Parameters	Value
Mass of each MU (kg)	10
Mass of the net (kg)	1
Side length of the net (m)	5
Side length of the mesh (m)	0.5
Material Young's modulus (MPa)	445.6
Diameter of the net's thread (mm)	1
Damping ratio of the net's thread	0.106
Orbit altitude (km)	36000
Orbit inclination (rad)	0

5.1.5.2 Simulation Results and Discussion

The position and velocity states of four MUs in both cases are plotted in Figs. 5.4 and 5.5, respectively. The controller inputs, disturbances on four MUs in symmetric configuration and asymmetric configuration are shown in Figs. 5.6, 5.7, and 5.8, respectively. Figure 5.9 shows the motions of sidelines of the net in case 2. Finally, a shape of the TSNR is presented in Fig. 5.10.

From Figs. 5.4 and 5.5, the simulation results show that the MUs can simultaneously track the desired trajectories with the wanted velocities in both case 1 and case 2. Position and velocity errors of MUs have been attenuated under the designed controllers. It is obvious that the control strategy can suppress the errors more rapidly in case 1. After 30 s, the position and velocity states of four MUs vibrate slightly and periodically near the desired values in case 1, but the same phenomenon occurs after 80 s in case 2. The results indicate the designed controllers work well in both symmetric and asymmetric cases, but the convergence property of case 1 is superior to case 2.

In Fig. 5.5, the amplitude and frequency of vibration in the y-axis are slighter than that of the x-axis and the z-axis, which shows that the oscillation of the MUs along the flight direction is weaker than the oscillation along the other two directions. The same conclusion can be obtained from Figs. 5.7 and 5.8, in which the disturbance in y-axis is smaller than that of the x-axis and the z-axis. Compared with case 1, the overshoots of the positions and velocities of four MUs in case 2 are larger due to the initial velocity error. But the vibration amplitudes are feasible. In this respect, the proposed strategy is effective for relative position control of four MUs.

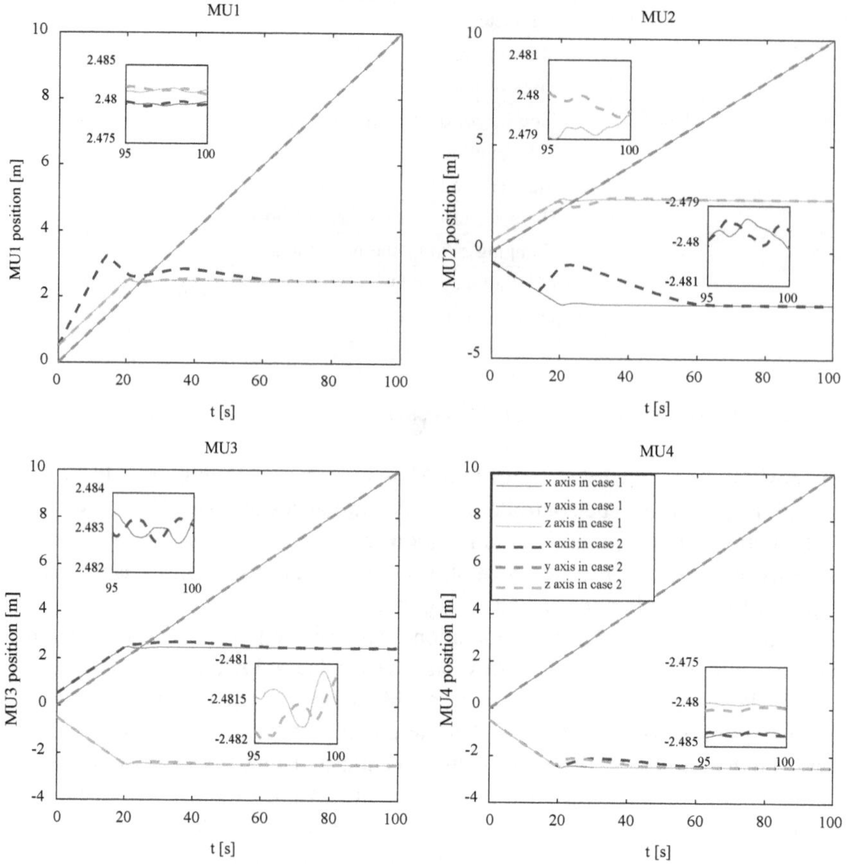

Fig. 5.4 Position states of four MUs

When the controllers are activated, the controller inputs (Fig. 5.6) and disturbances (Figs. 5.7 and 5.8) on the MUs instantly increase to a peak, which is caused by the elastic force of the net. As shown in Figs. 5.6, 5.7, and 5.8, the instantaneous peaks of the controller inputs and disturbances on the MUs in case 2 are larger than that in case 1. We can draw the conclusion that the asymmetric configuration leads to the greater disturbances on the MUs, which adds challenges to the controller design.

Figure 5.9 shows that when the MUs are ejected with the initial velocities from the platform satellite, the net begins to unfold. The motions of sidelines of the net are disorderly and irregular before the controllers are activated, while the motions of sidelines of the net become more orderly after the controllers are activated. It implies that the controllers are effective in the maintenance control for the shape of the net.

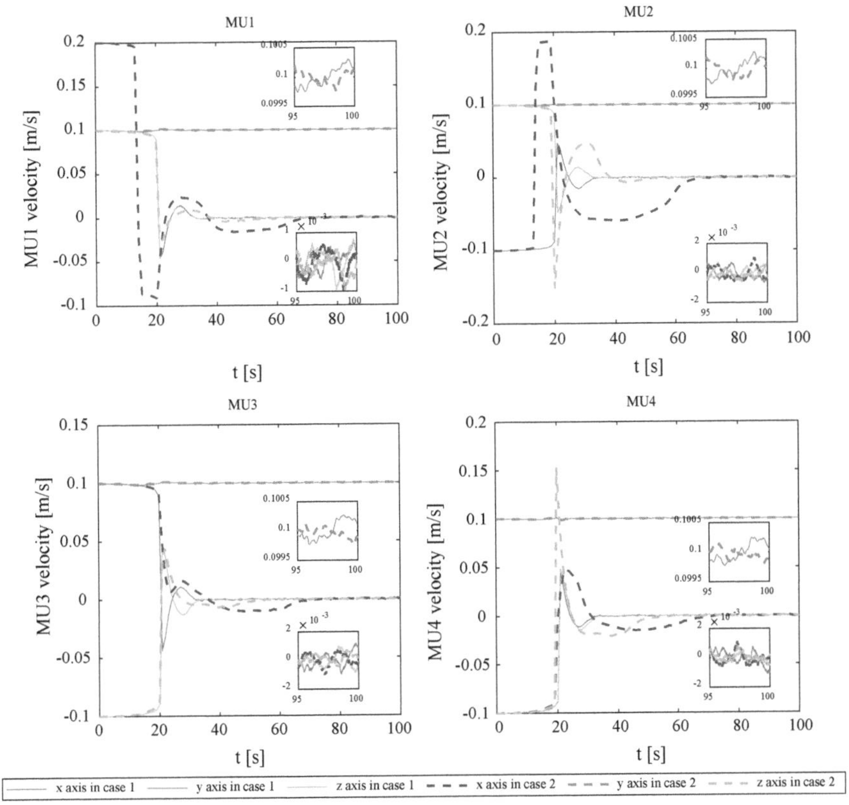

Fig. 5.5 Velocities of four MUs

5.2 Distributed Configuration-Maneuvering Control with Constraints

5.2.1 Problem Statement

To avoid the snapping elasticity of the connecting net, the maximal separation between arbitrary adjacent MUs should be less than a certain value. Thus, the constraints of the MUs are obtained as

$$\Theta = \left\{ \begin{array}{l} \boldsymbol{\rho}_{II} | \boldsymbol{\rho}_{II} = \left[\boldsymbol{\rho}_1^{\mathrm{T}}, \boldsymbol{\rho}_2^{\mathrm{T}}, \ldots \boldsymbol{\rho}_n^{\mathrm{T}}\right]^{\mathrm{T}}, \left\|\boldsymbol{\rho}_i - \boldsymbol{\rho}_j\right\| < L + \delta \\ , \text{for } i = 1, 2, \ldots, n, \text{ and } j \in \Omega_i \end{array} \right. \tag{5.37}$$

where δ is the maximum elastic elongation of the net side, n is the number of MUs. Ω_i is the set of indices corresponding to the information neighbors of MU i.

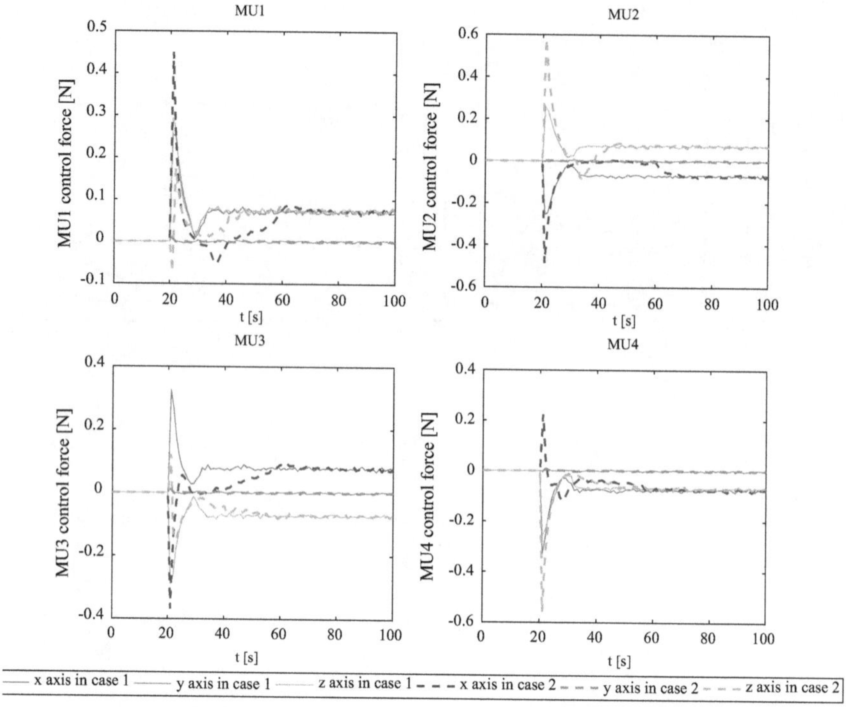

Fig. 5.6 Controller inputs of four MUs

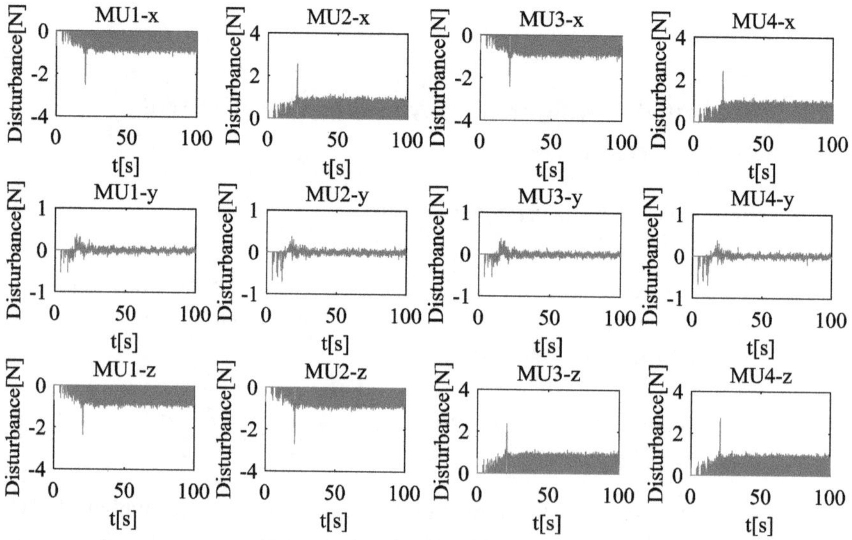

Fig. 5.7 Disturbances on four MUs in three directions in case 1

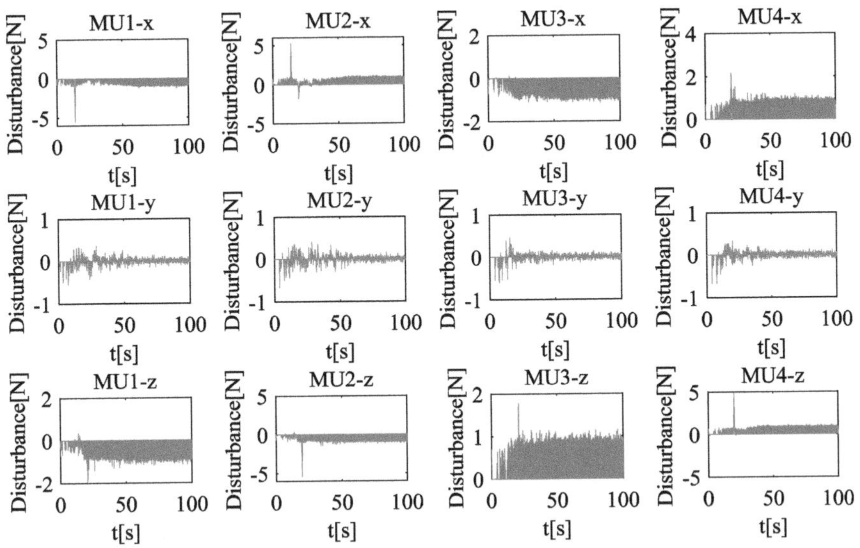

Fig. 5.8 Disturbances on four MUs in three directions in case 2

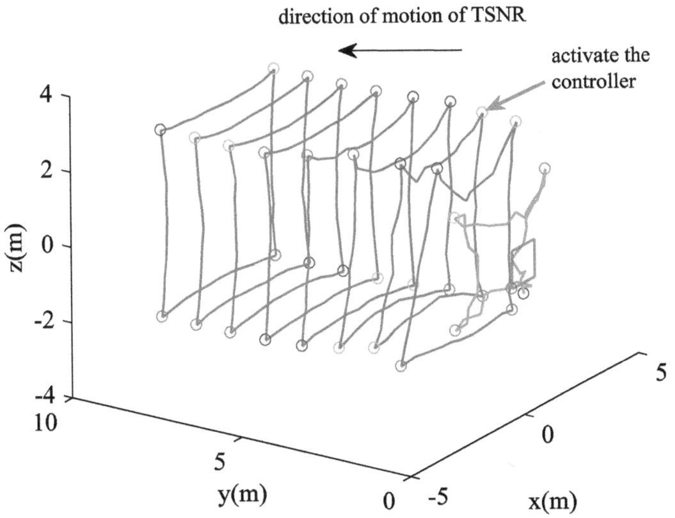

Fig. 5.9 The motions of sidelines of the net in case 2

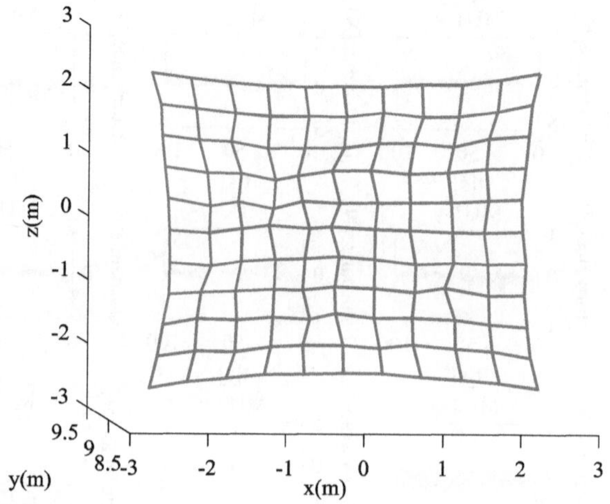

Fig. 5.10 Configuration of TSNR

To avoid collision between adjacent MUs, define the following avoidance sets for each pair of MUs as

$$\Pi = \begin{cases} \boldsymbol{\rho}_{II} | \boldsymbol{\rho}_{II} = \left[\boldsymbol{\rho}_1^{\mathrm{T}}, \boldsymbol{\rho}_2^{\mathrm{T}}, \ldots \boldsymbol{\rho}_n^{\mathrm{T}}\right]^{\mathrm{T}}, \left\|\boldsymbol{\rho}_i - \boldsymbol{\rho}_j\right\| > \ell_0 \\ \text{, for } i = 1, 2, \ldots, n, \text{ and } j \in \Omega_i \end{cases} \tag{5.38}$$

where ℓ_0 is the minimum safety distance that guarantees no collisions between MUs.

Remark 5 Based on the model dynamics (5.7) and constraints (5.37) and (5.38), the configuration maintenance of TSNR is stated formally as follows: Design robust adaptive distributed formation control for the system (5.7) based on artificial potential function and neural network such that the controlled MUs achieve that $\boldsymbol{\rho}_i(t)$ and $\mathbf{v}_i(t)$ $(i = 1, 2, \ldots, n)$ asymptotically converge to $\boldsymbol{\rho}_i^d(t)$ and $\mathbf{v}_i^d(t)$, respectively, as $t \to \infty$ in the presence of relative distance constraints and external disturbances.

Remark 6 In this paper, the research is aimed at achieving the configuration maintenance of multi-agent systems with the relative distance constraints between agents, with application to the TSNR system during approaching phase. Thus, a distributed control strategy for MUs based on artificial potential functions is proposed. As we can know, most of the finished work about TSNR does not consider the relative distance constraints between the MUs. Consequently, this research is a spread of current achievement. For simplicity, we only consider the fixed topology systems. At the same time, it is assumed that the communication topology of the system is connected. It is a rational assumption since the fixed connected communication graph is the base of the research on switching communication topologies.

5.2.2 Controller Design

5.2.2.1 Artificial Potential Function

To satisfy the requirements in Eqs. (5.37) and (5.38), artificial potential functions are introduced in order to construct an attractive or repulsive force between adjacent MUs. The potential function that will be used for the controller design in this work must satisfy both tracking and formation specification. In other words, the potential function is composed of a tracking part and a formation control part [12]. With the designed potential function, each MU is forced to track the desired trajectory while undergoing the relative distance constraints. Consequently, inspired by [13], the artificial potential function can be represented by

$$V\left(\boldsymbol{\rho},\boldsymbol{\rho}^d\right)=\sum_{i=1}^{n}V_i^d\left(\left\|\boldsymbol{\rho}_i-\boldsymbol{\rho}_i^d\right\|\right)+\sum_{i=1}^{n-1}\sum_{j>i\text{ and }j\in\Omega_i}V_{ij}\left(\left\|\boldsymbol{\rho}_i-\boldsymbol{\rho}_j\right\|\right) \tag{5.39}$$

where

$$V_{ij}\left(\left\|\boldsymbol{\rho}_i-\boldsymbol{\rho}_j\right\|\right)=\begin{cases}\dfrac{1}{\sin\left(\dfrac{\pi\left\|\boldsymbol{\rho}_i-\boldsymbol{\rho}_j\right\|}{2(L-\ell_0)}-\dfrac{\pi\ell_0}{2(L-\ell_0)}\right)} & \ell_0<\left\|\boldsymbol{\rho}_i-\boldsymbol{\rho}_j\right\|\le L\\[2ex] \dfrac{1}{\sin\left(\dfrac{\pi\left\|\boldsymbol{\rho}_i-\boldsymbol{\rho}_j\right\|}{2\delta}-\dfrac{\pi(L-\delta)}{2\delta}\right)} & L<\left\|\boldsymbol{\rho}_i-\boldsymbol{\rho}_j\right\|<L+\delta\end{cases} \tag{5.40}$$

$$V_i^d\left(\left\|\boldsymbol{\rho}_i-\boldsymbol{\rho}_i^d\right\|\right)=\left\|\boldsymbol{\rho}_i-\boldsymbol{\rho}_i^d\right\|^2 \tag{5.41}$$

where $\boldsymbol{\rho}=\left[\boldsymbol{\rho}_1^{\mathrm{T}},\boldsymbol{\rho}_2^{\mathrm{T}},\ldots,\boldsymbol{\rho}_n^{\mathrm{T}}\right]^{\mathrm{T}}$ and $\boldsymbol{\rho}^d=\left[\boldsymbol{\rho}_1^{d\mathrm{T}},\boldsymbol{\rho}_2^{d\mathrm{T}},\ldots,\boldsymbol{\rho}_n^{d\mathrm{T}}\right]^{\mathrm{T}}$ are the stack vectors. $\boldsymbol{\rho}_i^d$ with $i=1,2,\ldots,n$ is the desired trajectory of MU i.

Here, $V_i^d\left(\left\|\boldsymbol{\rho}_i-\boldsymbol{\rho}_i^d\right\|\right)$ is the potential function between MU i and virtual target i which is assumed to be compatible with the expected trajectory $\boldsymbol{\rho}_i^d$ of MU i. $V_i^d\left(\left\|\boldsymbol{\rho}_i-\boldsymbol{\rho}_i^d\right\|\right)$ which has a unique minimum equal to zero at $\boldsymbol{\rho}_i=\boldsymbol{\rho}_i^d$ is used for tracking mission. It always satisfies $V_i^d\left(\left\|\boldsymbol{\rho}_i-\boldsymbol{\rho}_i^d\right\|\right)\ge 0$. Therefore, this function can be viewed as a Lyapunov function. $V_{ij}\left(\left\|\boldsymbol{\rho}_i-\boldsymbol{\rho}_j\right\|\right)$ is the potential function between MU i and MU j. It is to make $V_{ij}\left(\left\|\boldsymbol{\rho}_i-\boldsymbol{\rho}_j\right\|\right)\to\infty$ as $\left\|\boldsymbol{\rho}_i-\boldsymbol{\rho}_j\right\|\to\ell_0$ or $\left\|\boldsymbol{\rho}_i-\boldsymbol{\rho}_j\right\|\to L+\delta$, so that each MU can achieve relative distance constraints in the whole time. The function $V_{ij}\left(\left\|\boldsymbol{\rho}_i-\boldsymbol{\rho}_j\right\|\right)$ has a unique minimum at $\left\|\boldsymbol{\rho}_i-\boldsymbol{\rho}_j\right\|=L$.

Assumption 3 The reference trajectory $\boldsymbol{\rho}_i^d$ and the desired relative velocity $\mathbf{v}_i^d$ $(i = 1, 2, \ldots, n)$ of MUs are assumed to be bounded for all time.

The gradients of the potential function at $\boldsymbol{\rho}_i$ and $\boldsymbol{\rho}_i^d$ are given by

$$\nabla_{\boldsymbol{\rho}_i} V(\boldsymbol{\rho}, \boldsymbol{\rho}^d) = \left\{ \begin{array}{l} \nabla_{\boldsymbol{\rho}_i} V_i^d\left(\left\|\boldsymbol{\rho}_i - \boldsymbol{\rho}_i^d\right\|\right) \\ + \sum\limits_{\substack{j \in \Omega_i \\ j > i}} \nabla_{\boldsymbol{\rho}_i} V_{ij}\left(\left\|\boldsymbol{\rho}_i - \boldsymbol{\rho}_j\right\|\right) \\ + \sum\limits_{\substack{j \in \Omega_i \\ j < i}} \nabla_{\boldsymbol{\rho}_i} V_{ji}\left(\left\|\boldsymbol{\rho}_j - \boldsymbol{\rho}_i\right\|\right) \end{array} \right\} \tag{5.42}$$

and

$$\nabla_{\boldsymbol{\rho}_i^d} V(\boldsymbol{\rho}, \boldsymbol{\rho}^d) = -\nabla_{\boldsymbol{\rho}_i} V_i^d\left(\left\|\boldsymbol{\rho}_i - \boldsymbol{\rho}_i^d\right\|\right) \tag{5.43}$$

Combining Eqs. (5.42) and (5.43), we obtain

$$\begin{aligned} &\sum_{i=1}^{n} \nabla_{\boldsymbol{\rho}_i^d} V(\boldsymbol{\rho}, \boldsymbol{\rho}^d) \\ &= -\sum_{i=1}^{n} \nabla_{\boldsymbol{\rho}_i} V(\boldsymbol{\rho}, \boldsymbol{\rho}^d) + \sum_{i=1}^{n} \left(\begin{array}{l} \sum\limits_{\substack{j \in \Omega_i \\ j > i}} \nabla_{\boldsymbol{\rho}_i} V_{ij}\left(\left\|\boldsymbol{\rho}_i - \boldsymbol{\rho}_j\right\|\right) \\ + \sum\limits_{\substack{j \in \Omega_i \\ j < i}} \nabla_{\boldsymbol{\rho}_i} V_{ji}\left(\left\|\boldsymbol{\rho}_j - \boldsymbol{\rho}_i\right\|\right) \end{array} \right) \end{aligned} \tag{5.44}$$

From the defining of $V_{ij}\left(\left\|\boldsymbol{\rho}_i - \boldsymbol{\rho}_j\right\|\right)$ in Eq. (5.40), the following reciprocity property is held:

$$\nabla_{\boldsymbol{\rho}_i} V_{ij}\left(\left\|\boldsymbol{\rho}_i - \boldsymbol{\rho}_j\right\|\right) = -\nabla_{\boldsymbol{\rho}_j} V_{ij}\left(\left\|\boldsymbol{\rho}_i - \boldsymbol{\rho}_j\right\|\right) \tag{5.45}$$

According to Eq. (5.45), the formation part of the potential function satisfies

$$\sum_{i=1}^{n} \left(\sum_{\substack{j \in \Omega_i \\ j > i}} \nabla_{\boldsymbol{\rho}_i} V_{ij}\left(\left\|\boldsymbol{\rho}_i - \boldsymbol{\rho}_j\right\|\right) + \sum_{\substack{j \in \Omega_i \\ j < i}} \nabla_{\boldsymbol{\rho}_i} V_{ji}\left(\left\|\boldsymbol{\rho}_j - \boldsymbol{\rho}_i\right\|\right) \right) = 0 \tag{5.46}$$

Rearranging Eq. (5.44) by using this result, we have

$$\sum_{i=1}^{n} \nabla_{\boldsymbol{\rho}_i^d} V(\boldsymbol{\rho}, \boldsymbol{\rho}^d) = -\sum_{i=1}^{n} \nabla_{\boldsymbol{\rho}_i} V(\boldsymbol{\rho}, \boldsymbol{\rho}^d) \tag{5.47}$$

5.2.2.2 Robust Adaptive Distributed Controller

Due to the hard nonlinear dynamics of the system, we introduce the Radial Basis Function Neural Network (RBFNN) to approximate the complicated nonlinear item. Dragged simultaneously by the net and the disturbances in space environment, the MUs struggle to the desired trajectories. Accordingly, the controller should have strong robustness. The configuration coordination of the connecting net requires that the MUs should maintain accurate geometric configuration during the TSNR maneuvers. Therefore, we will design robust adaptive distributed control scheme based on the RBFNN to overcome the external disturbances' influence and maintain the shape of the TSNR in this paper.

Assumption 4 The measurement of the relative position and velocity of all the MUs with respect to the LVLH frame are accurately accessible.

Assumption 5 The undirected graph G is connected and there exists a directed path from the platform satellite to each MU.

We introduce the following auxiliary variable:

$$\mathbf{s}_i = \tilde{\mathbf{v}}_i + k_{1i}\mathbf{e}_i + \alpha\, \nabla_{\boldsymbol{\rho}_i} V\left(\boldsymbol{\rho}, \boldsymbol{\rho}^d\right) \tag{5.48}$$

where k_{1i} and α are positive constants. Then, from Eqs. (5.7), (5.8), and (5.48), we have

$$\begin{aligned}
&\mathbf{M}_i(\boldsymbol{\rho}_i)\dot{\mathbf{s}}_i + \mathbf{C}_i(\boldsymbol{\rho}_i, \dot{\boldsymbol{\rho}}_i)\mathbf{s}_i \\
&= \left\{\begin{array}{l}
\mathbf{u}_i + \mathbf{d}_i - \mathbf{M}_i(\boldsymbol{\rho}_i)\ddot{\boldsymbol{\rho}}_i^d + \mathbf{M}_i(\boldsymbol{\rho}_i)k_{1i}\dot{\mathbf{e}}_i \\
+\alpha\mathbf{M}_i(\boldsymbol{\rho}_i)\frac{d}{dt}\left(\nabla_{\boldsymbol{\rho}_i} V\left(\boldsymbol{\rho}, \boldsymbol{\rho}^d\right)\right) - \mathbf{C}_i(\boldsymbol{\rho}_i, \dot{\boldsymbol{\rho}}_i)\dot{\boldsymbol{\rho}}_i^d \\
+\mathbf{C}_i(\boldsymbol{\rho}_i, \dot{\boldsymbol{\rho}}_i)k_{1i}\mathbf{e}_i + \alpha\mathbf{C}_i(\boldsymbol{\rho}_i, \dot{\boldsymbol{\rho}}_i)\nabla_{\boldsymbol{\rho}_i} V\left(\boldsymbol{\rho}, \boldsymbol{\rho}^d\right) - \mathbf{g}_i(\boldsymbol{\rho}_i)
\end{array}\right\}
\end{aligned} \tag{5.49}$$

Here, we define the nonlinear item $\mathbf{f}_i(\mathbf{x}_i)$

$$\begin{aligned}
\mathbf{f}_i(\mathbf{x}_i) = &-\mathbf{M}_i(\boldsymbol{\rho}_i)\ddot{\boldsymbol{\rho}}_i^d + \mathbf{M}_i(\boldsymbol{\rho}_i)k_{1i}\dot{\mathbf{e}}_i \\
&- \mathbf{C}_i(\boldsymbol{\rho}_i, \dot{\boldsymbol{\rho}}_i)\dot{\boldsymbol{\rho}}_i^d + \mathbf{C}_i(\boldsymbol{\rho}_i, \dot{\boldsymbol{\rho}}_i)k_{1i}\mathbf{e}_i - \mathbf{g}_i(\boldsymbol{\rho}_i)
\end{aligned} \tag{5.50}$$

where $\mathbf{x}_i = \left[\boldsymbol{\rho}_i^{\mathrm{T}}, \dot{\boldsymbol{\rho}}_i^{\mathrm{T}}, \mathbf{e}_i^{\mathrm{T}}, \dot{\mathbf{e}}_i^{\mathrm{T}}\right]^{\mathrm{T}} \in \mathsf{R}^{12}$. In this paper, the continuous function $\mathbf{f}_i(\mathbf{x}_i)$ can be represented as follows:

$$\mathbf{f}_i(\mathbf{x}_i) = \mathbf{W}_i^{*\mathrm{T}}\mathbf{h}_i(\mathbf{x}_i) + \boldsymbol{\varepsilon}_i \tag{5.51}$$

where $\mathbf{W}_i^* \in \mathsf{R}^{p_i \times 3}$ represents the ideal constant weight matrix, p_i denotes the number of neurons, $\boldsymbol{\varepsilon}_i \in \mathsf{R}^3$ is the approximation error which satisfying $\|\boldsymbol{\varepsilon}_i\| \leq \bar{\varepsilon}_i$. W_i is a positive constant which is picked out to suffice $\mathrm{tr}\left(\mathbf{W}_i^{*\mathrm{T}}\mathbf{W}_i^*\right) \leq W_i$, and $\mathbf{h}_i(\mathbf{x}_i) = \left[h_{i1}(\mathbf{x}_i), h_{i2}(\mathbf{x}_i), \ldots, h_{ip_i}(\mathbf{x}_i)\right]^{\mathrm{T}}$ is the activation function

$$h_{ij}(\mathbf{x}_i) = \exp\left(\frac{-\left\|\mathbf{x}_i - \mathbf{c}_{ij}\right\|^2}{2\theta_{ij}^2}\right) \tag{5.52}$$

where $\mathbf{c}_{ij} \in \mathsf{R}^{12}$ is the basis function centers, $\theta_{ij} > 0$ is the width of the Gaussian function. However, the ideal weight matrix $\mathbf{W}_i^*$ is only for theoretical analysis. Its estimation $\hat{\mathbf{W}}_i$ is used for the controller design. The estimation of $\mathbf{f}_i(\mathbf{x}_i)$ is given as

$$\hat{\mathbf{f}}_i(\mathbf{x}_i) = \hat{\mathbf{W}}_i^{\mathrm{T}} \mathbf{h}_i(\mathbf{x}_i) \tag{5.53}$$

For the purpose of guaranteeing that the weight matrices retain bounded, the projection algorithm [14] is adopted to design the parameters' updating tune.

$$\dot{\hat{\mathbf{W}}}_i = \begin{cases} \boldsymbol{\Gamma}_i \mathbf{h}_i(\mathbf{x}_i)\mathbf{s}_i^{\mathrm{T}} & \text{if } \operatorname{tr}\left(\hat{\mathbf{W}}_i^{\mathrm{T}}\hat{\mathbf{W}}_i\right) < W_i \\ \boldsymbol{\Gamma}_i \mathbf{h}_i(\mathbf{x}_i)\mathbf{s}_i^{\mathrm{T}} - \boldsymbol{\Gamma}_i \hat{\mathbf{W}}_i & \text{if } \operatorname{tr}\left(\hat{\mathbf{W}}_i^{\mathrm{T}}\hat{\mathbf{W}}_i\right) = W_i \text{ and } \mathbf{s}_i^{\mathrm{T}}\hat{\mathbf{W}}_i^{\mathrm{T}}\mathbf{h}_i(\mathbf{x}_i) < 0 \\ \boldsymbol{\Gamma}_i \mathbf{h}_i(\mathbf{x}_i)\mathbf{s}_i^{\mathrm{T}} - \boldsymbol{\Gamma}_i \frac{\mathbf{s}_i^{\mathrm{T}}\hat{\mathbf{W}}_i^{\mathrm{T}}\mathbf{h}_i(\mathbf{x}_i)}{\operatorname{tr}\left(\hat{\mathbf{W}}_i^{\mathrm{T}}\hat{\mathbf{W}}_i\right)}\hat{\mathbf{W}}_i & \text{if } \operatorname{tr}\left(\hat{\mathbf{W}}_i^{\mathrm{T}}\hat{\mathbf{W}}_i\right) = W_i \text{ and } \mathbf{s}_i^{\mathrm{T}}\hat{\mathbf{W}}_i^{\mathrm{T}}\mathbf{h}_i(\mathbf{x}_i) \geq 0 \end{cases} \tag{5.54}$$

where $\boldsymbol{\Gamma}_i$ is a positive-definite matrix, which regulates the updating tune of $\hat{\mathbf{W}}_i$.

Lemma 4 If the updating tune for NN weight matrices $\hat{\mathbf{W}}_i$ where $i = 1, 2, \ldots, n$ are chosen as Eq. (5.54) and the initial value of $\hat{\mathbf{W}}_i$ satisfies $\operatorname{tr}\left(\hat{\mathbf{W}}_i^{\mathrm{T}}(0)\hat{\mathbf{W}}_i(0)\right) \leq W_i$, where $\boldsymbol{\Gamma}_i$ is a diagonal positive matrix, then the boundedness $\forall t \geq 0$, $\operatorname{tr}\left(\hat{\mathbf{W}}_i^{\mathrm{T}}(t)\hat{\mathbf{W}}_i(t)\right) \leq W_i$ can be guaranteed.

Proof Consider the Lyapunov function candidate

$$V_{\hat{\mathbf{W}}_i} = \operatorname{tr}\left(\hat{\mathbf{W}}_i^{\mathrm{T}}\hat{\mathbf{W}}_i\right)$$

Its time derivative is

$$\dot{V}_{\hat{\mathbf{W}}_i} = 2\operatorname{tr}\left(\hat{\mathbf{W}}_i^{\mathrm{T}}\dot{\hat{\mathbf{W}}}_i\right)$$

By Eq. (5.54), the following cases are considered:

(1) If $V_{\hat{\mathbf{W}}_i} < W_i$, the conclusion has already held.

(2) If $V_{\hat{\mathbf{W}}_i} = W_i$ and $\mathbf{s}_i^{\mathrm{T}}\hat{\mathbf{W}}_i^{\mathrm{T}}\mathbf{h}_i(\mathbf{x}_i) < 0$, we have

$$\dot{V}_{\hat{\mathbf{W}}_i} = 2\operatorname{tr}\left(\hat{\mathbf{W}}_i^{\mathrm{T}}\boldsymbol{\Gamma}_i\mathbf{h}_i(\mathbf{x}_i)\mathbf{s}_i^{\mathrm{T}} - \hat{\mathbf{W}}_i^{\mathrm{T}}\boldsymbol{\Gamma}_i\hat{\mathbf{W}}_i\right) < 0$$

(3) If $V_{\hat{\mathbf{W}}_i} = W_i$ and $\mathbf{s}_i^{\mathrm{T}}\hat{\mathbf{W}}_i^{\mathrm{T}}\mathbf{h}_i(\mathbf{x}_i) \geq 0$, we obtain

$$\dot{V}_{\hat{\mathbf{W}}_i} = 2\mathrm{tr}\left(\hat{\mathbf{W}}_i^{\mathrm{T}}\boldsymbol{\Gamma}_i\mathbf{h}_i(\mathbf{x}_i)\mathbf{s}_i^{\mathrm{T}} - \hat{\mathbf{W}}_i^{\mathrm{T}}\boldsymbol{\Gamma}_i\frac{\mathbf{s}_i^{\mathrm{T}}\hat{\mathbf{W}}_i^{\mathrm{T}}\mathbf{h}_i(\mathbf{x}_i)}{\mathrm{tr}\left(\hat{\mathbf{W}}_i^{\mathrm{T}}\hat{\mathbf{W}}_i\right)}\hat{\mathbf{W}}_i\right) = 0$$

Hence, $\forall t \geq 0$, $\mathrm{tr}\left(\hat{\mathbf{W}}_i^{\mathrm{T}}(t)\hat{\mathbf{W}}_i(t)\right) \leq W_i$, where $i = 1, 2, \ldots, n$, always holds. Thus, this concludes the proof.

Consider the following robust distributed control algorithm:

$$\mathbf{u}_i = \left\{\begin{array}{l} -k_{2i}\mathbf{s}_i - \hat{\mathbf{W}}_i^{\mathrm{T}}\mathbf{h}_i(\mathbf{x}_i) - \hat{\phi}_i\mathbf{sgn}(\mathbf{s}_i) \\ -\alpha\mathbf{M}_i(\boldsymbol{\rho}_i)\frac{d}{dt}\left(\nabla_{\boldsymbol{\rho}_i}V(\boldsymbol{\rho}, \boldsymbol{\rho}^d)\right) \\ -\alpha\mathbf{C}_i(\boldsymbol{\rho}_i, \dot{\boldsymbol{\rho}}_i)\nabla_{\boldsymbol{\rho}_i}V(\boldsymbol{\rho}, \boldsymbol{\rho}^d) \end{array}\right\} \tag{5.55}$$

where k_{2i} is a positive constant, $\hat{\phi}_i$ is the estimate of the optimal robustness gain ϕ_i which is chosen to satisfy $\phi_i \geq \bar{\varepsilon}_i + \bar{d}_i$. The adaptive updating law for the robustness gain $\hat{\phi}_i$ is given as

$$\dot{\hat{\phi}}_i = \hbar_i\|\mathbf{s}_i\|_1 \tag{5.56}$$

where $\hbar_i$ is positive constant, and $\|\cdot\|_1$ denotes the 1-norm of vector.

Remark 7 The robust adaptive distributed control scheme for MU i comprises consensus algorithm, artificial potential auxiliary system, RBFNN, and robust control term, as shown in Fig. 5.11. The consensus algorithm is designed for

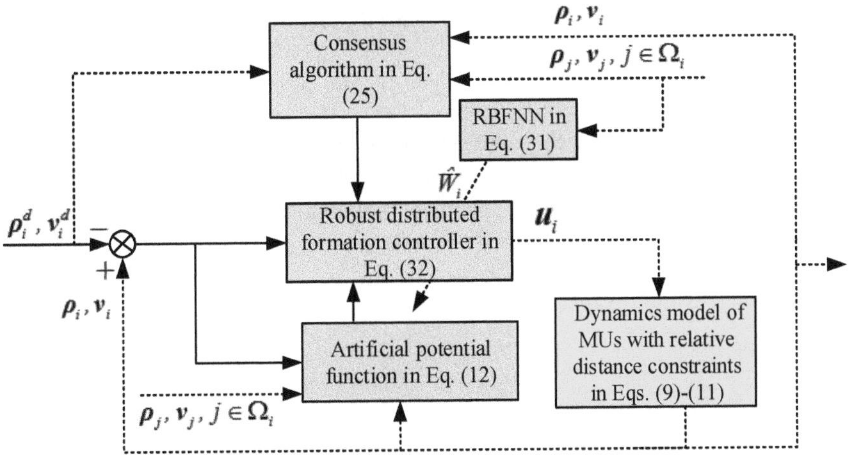

Fig. 5.11 Schematic diagram of the proposed control system structure

guaranteeing the MUs to track the desired trajectory simultaneously maintaining the net's configuration. The artificial potential auxiliary system is utilized to limit the amplitude of the relative distance between adjacent MUs. The RBFNN is designed to approximate the nonlinear term, where the robust term is utilized to offset the negative influence of the estimation error and disturbances.

5.2.3 Stability Analysis of the Closed-Loop System

In this subsection, some lemmas will be employed in the asymptotical stability proof. For convenience, they are repeated here.

Lemma 5 Assume that $f(t)$ is an uniformly continuous function and the result of the mathematical operation $\lim_{t\to\infty}\int_0^t f(\tau)d\tau$ is a finite value, then $f(t)\to 0$ [15].

Corollary 1 Consider the function $f(t)$, if $f(t)$, $\dot{f}(t)\in L_\infty$ and $f(t)\in L_p$, $p\in[1,\infty)$, then $\lim_{t\to\infty}f(t)=0$ [16].

Theorem 2 By employing the control scheme in Eq. (5.55) for systems (5.7) with constraints (5.37) and (5.38), when $\boldsymbol{\rho}_i^d$ and $\mathbf{v}_i^d$ are all bounded, we can derive that the states of MUs are asymptotically reaching to the desired value, which guarantees that a formation is reached asymptotically.

Proof Choose the Lyapunov function as

$$\begin{aligned} V(t) =& \frac{1}{2}\sum_{i=1}^{n}\mathbf{s}_i^{\mathrm{T}}\mathbf{M}_i\mathbf{s}_i + \frac{1}{2}\sum_{i=1}^{n}\mathrm{tr}\left(\tilde{\mathbf{W}}_i^{\mathrm{T}}\boldsymbol{\Gamma}_i^{-1}\tilde{\mathbf{W}}_i\right) \\ &+ \frac{1}{2}\sum_{i=1}^{n}\frac{1}{\hbar_i}\tilde{\phi}_i^2 + \frac{1}{2}\mathbf{e}^{\mathrm{T}}(\mathbf{P}\otimes\mathbf{I})\mathbf{e} + \alpha V\left(\boldsymbol{\rho},\boldsymbol{\rho}^d\right) \end{aligned} \tag{5.57}$$

where $\tilde{\mathbf{W}}_i=\mathbf{W}_i^*-\hat{\mathbf{W}}_i$ is the approximation error of $\mathbf{W}_i^*$, $\tilde{\phi}_i=\phi_i-\hat{\phi}_i$ is the approximation error of ϕ_i. $\mathbf{M}_i$ is a positive define inertia matrix, $\boldsymbol{\Gamma}_i$ and $\mathbf{P}$ are positive define design matrices, $V\left(\boldsymbol{\rho},\boldsymbol{\rho}^d\right)>0$, therefore, $V(t)$ is a positive function. The time derivative of $V(t)$ along Eqs. (5.7) and (5.55) is given by

$$\begin{aligned}
&\dot{V}(t)\\
&=\sum_{i=1}^{n}\mathbf{s}_i^{\mathrm{T}}\mathbf{M}_i\dot{\mathbf{s}}_i+\frac{1}{2}\sum_{i=1}^{n}\mathbf{s}_i^{\mathrm{T}}\dot{\mathbf{M}}_i\mathbf{s}_i-\sum_{i=1}^{n}\mathrm{tr}\left(\tilde{\mathbf{W}}_i^{\mathrm{T}}\boldsymbol{\Gamma}_i^{-1}\dot{\hat{\mathbf{W}}}_i\right)\\
&\quad-\sum_{i=1}^{n}\frac{1}{\hbar_i}\tilde{\phi}_i\dot{\hat{\phi}}_i+\mathbf{e}^{\mathrm{T}}(\mathbf{P}\otimes\mathbf{I})\dot{\mathbf{e}}+\alpha\dot{V}\left(\boldsymbol{\rho},\boldsymbol{\rho}^d\right)\\
&=\sum_{i=1}^{n}\mathbf{s}_i^{\mathrm{T}}\left(-k_{2i}\mathbf{s}_i-\hat{\mathbf{W}}_i^{\mathrm{T}}\mathbf{h}_i(\mathbf{x}_i)-\hat{\phi}_i\mathbf{sgn}(\mathbf{s}_i)+\mathbf{d}_i+\mathbf{f}_i(\mathbf{x}_i)\right)\\
&\quad-\sum_{i=1}^{n}\mathrm{tr}\left(\tilde{\mathbf{W}}_i^{\mathrm{T}}\boldsymbol{\Gamma}_i^{-1}\dot{\hat{\mathbf{W}}}_i\right)-\sum_{i=1}^{n}\frac{1}{\hbar_i}\tilde{\phi}_i\dot{\hat{\phi}}_i+\mathbf{e}^{\mathrm{T}}(\mathbf{P}\otimes\mathbf{I})\dot{\mathbf{e}}+\alpha\dot{V}\left(\boldsymbol{\rho},\boldsymbol{\rho}^d\right)
\end{aligned}\tag{5.58}$$

Substitute Eq. (5.51) into $\dot{V}(t)$, it can be derived as

$$\begin{aligned}
\dot{V}(t)&=\sum_{i=1}^{n}\mathbf{s}_i^{\mathrm{T}}\left(-k_{2i}\mathbf{s}_i-\hat{\phi}_i\mathbf{sgn}(\mathbf{s}_i)+\mathbf{d}_i+\tilde{\mathbf{W}}_i^{\mathrm{T}}\mathbf{h}_i(\mathbf{x}_i)+\boldsymbol{\varepsilon}_i\right)\\
&\quad-\sum_{i=1}^{n}\mathrm{tr}\left(\tilde{\mathbf{W}}_i^{\mathrm{T}}\boldsymbol{\Gamma}_i^{-1}\dot{\hat{\mathbf{W}}}_i\right)-\sum_{i=1}^{n}\frac{1}{\hbar_i}\tilde{\phi}_i\dot{\hat{\phi}}_i+\mathbf{e}^{\mathrm{T}}(\mathbf{P}\otimes\mathbf{I})\dot{\mathbf{e}}+\alpha\dot{V}\left(\boldsymbol{\rho},\boldsymbol{\rho}^d\right)
\end{aligned}\tag{5.59}$$

Considering the adaptive law in Eq. (5.54), it can be obtained

(1) If $\dot{\hat{\mathbf{W}}}_i=\boldsymbol{\Gamma}_i\mathbf{h}_i(\mathbf{x}_i)\mathbf{s}_i^{\mathrm{T}}$, then

$$\begin{aligned}
&\mathbf{s}_i^{\mathrm{T}}\tilde{\mathbf{W}}_i^{\mathrm{T}}\mathbf{h}_i(\mathbf{x}_i)-\mathrm{tr}\left(\tilde{\mathbf{W}}_i^{\mathrm{T}}\boldsymbol{\Gamma}_i^{-1}\dot{\hat{\mathbf{W}}}_i\right)\\
&=\mathrm{tr}\left(\tilde{\mathbf{W}}_i^{\mathrm{T}}\mathbf{h}_i(\mathbf{x}_i)\mathbf{s}_i^{\mathrm{T}}-\tilde{\mathbf{W}}_i^{\mathrm{T}}\boldsymbol{\Gamma}_i^{-1}\dot{\hat{\mathbf{W}}}_i\right)=0
\end{aligned}$$

(2) If $\mathrm{tr}\left(\hat{\mathbf{W}}_i^{\mathrm{T}}\hat{\mathbf{W}}_i\right)=W_i$ and $\mathbf{s}_i^{\mathrm{T}}\hat{\mathbf{W}}_i^{\mathrm{T}}\mathbf{h}_i(\mathbf{x}_i)<0$, $\dot{\hat{\mathbf{W}}}_i=\boldsymbol{\Gamma}_i\mathbf{h}_i(\mathbf{x}_i)\mathbf{s}_i^{\mathrm{T}}-\boldsymbol{\Gamma}_i\hat{\mathbf{W}}_i$, then

$$\begin{aligned}
&\mathbf{s}_i^{\mathrm{T}}\tilde{\mathbf{W}}_i^{\mathrm{T}}\mathbf{h}_i(\mathbf{x}_i)-\mathrm{tr}\left(\tilde{\mathbf{W}}_i^{\mathrm{T}}\boldsymbol{\Gamma}_i^{-1}\dot{\hat{\mathbf{W}}}_i\right)\\
&=\mathrm{tr}\left(\tilde{\mathbf{W}}_i^{\mathrm{T}}\mathbf{h}_i(\mathbf{x}_i)\mathbf{s}_i^{\mathrm{T}}-\tilde{\mathbf{W}}_i^{\mathrm{T}}\boldsymbol{\Gamma}_i^{-1}\dot{\hat{\mathbf{W}}}_i\right)=\mathrm{tr}\left(\tilde{\mathbf{W}}_i^{\mathrm{T}}\hat{\mathbf{W}}_i\right)
\end{aligned}$$

$$\begin{aligned}
&\mathrm{tr}\left(\tilde{\mathbf{W}}_i^{\mathrm{T}}\hat{\mathbf{W}}_i\right)\\
&=\frac{1}{2}\mathrm{tr}\left(\tilde{\mathbf{W}}_i^{\mathrm{T}}\tilde{\mathbf{W}}_i+\tilde{\mathbf{W}}_i^{\mathrm{T}}\hat{\mathbf{W}}_i+\hat{\mathbf{W}}_i^{\mathrm{T}}\tilde{\mathbf{W}}_i+\hat{\mathbf{W}}_i^{\mathrm{T}}\hat{\mathbf{W}}_i-\tilde{\mathbf{W}}_i^{\mathrm{T}}\tilde{\mathbf{W}}_i-\hat{\mathbf{W}}_i^{\mathrm{T}}\hat{\mathbf{W}}_i\right)\\
&=\frac{1}{2}\mathrm{tr}\left(\mathbf{W}_i^{*\mathrm{T}}\mathbf{W}_i^{*}\right)-\frac{1}{2}\mathrm{tr}\left(\tilde{\mathbf{W}}_i^{\mathrm{T}}\tilde{\mathbf{W}}_i\right)-\frac{1}{2}\mathrm{tr}\left(\hat{\mathbf{W}}_i^{\mathrm{T}}\hat{\mathbf{W}}_i\right)\\
&\le\frac{1}{2}W_i-\frac{1}{2}W_i-\frac{1}{2}\mathrm{tr}\left(\tilde{\mathbf{W}}_i^{\mathrm{T}}\tilde{\mathbf{W}}_i\right)\\
&\le 0
\end{aligned}\tag{5.60}$$

Therefore, $\mathbf{s}_i^{\mathrm{T}}\tilde{\mathbf{W}}_i^{\mathrm{T}}\mathbf{h}_i(\mathbf{x}_i)-\mathrm{tr}\left(\tilde{\mathbf{W}}_i^{\mathrm{T}}\boldsymbol{\Gamma}_i^{-1}\dot{\hat{\mathbf{W}}}_i\right)\le 0$.

(3) If $\mathrm{tr}\left(\hat{\mathbf{W}}_i^{\mathrm{T}}\hat{\mathbf{W}}_i\right)=W_i$ and $\mathbf{s}_i^{\mathrm{T}}\hat{\mathbf{W}}_i^{\mathrm{T}}\mathbf{h}_i(\mathbf{x}_i)\ge 0$, $\dot{\hat{\mathbf{W}}}_i=\boldsymbol{\Gamma}_i\mathbf{h}_i(\mathbf{x}_i)\mathbf{s}_i^{\mathrm{T}}-\boldsymbol{\Gamma}_i\mathbf{s}_i^{\mathrm{T}}\hat{\mathbf{W}}_i^{\mathrm{T}}\mathbf{h}_i(\mathbf{x}_i)\hat{\mathbf{W}}_i/\mathrm{tr}\left(\hat{\mathbf{W}}_i^{\mathrm{T}}\hat{\mathbf{W}}_i\right)$, then

$$\begin{aligned}
&\mathbf{s}_i^{\mathrm{T}}\tilde{\mathbf{W}}_i^{\mathrm{T}}\mathbf{h}_i(\mathbf{x}_i)-\mathrm{tr}\left(\tilde{\mathbf{W}}_i^{\mathrm{T}}\boldsymbol{\Gamma}_i^{-1}\dot{\hat{\mathbf{W}}}_i\right)\\
&=\mathbf{s}_i^{\mathrm{T}}\hat{\mathbf{W}}_i^{\mathrm{T}}\mathbf{h}_i(\mathbf{x}_i)\mathrm{tr}\left(\tilde{\mathbf{W}}_i^{\mathrm{T}}\hat{\mathbf{W}}_i\right)/\mathrm{tr}\left(\hat{\mathbf{W}}_i^{\mathrm{T}}\hat{\mathbf{W}}_i\right)\\
&\le 0
\end{aligned}$$

Hence, it is obvious that $\mathbf{s}_i^{\mathrm{T}}\tilde{\mathbf{W}}_i^{\mathrm{T}}\mathbf{h}_i(\mathbf{x}_i)\le\mathrm{tr}\left(\tilde{\mathbf{W}}_i^{\mathrm{T}}\boldsymbol{\Gamma}_i^{-1}\dot{\hat{\mathbf{W}}}_i\right)$ is satisfied in all the cases.

Thus, Eq. (5.59) can be rewritten as

$$\begin{aligned}
\dot{V}(t)=&\sum_{i=1}^{n}\mathbf{s}_i^{\mathrm{T}}\left(-k_{2i}\mathbf{s}_i-\hat{\phi}_i\mathbf{sgn}(\mathbf{s}_i)+\mathbf{d}_i+\boldsymbol{\varepsilon}_i\right)\\
&-\sum_{i=1}^{n}\frac{1}{\hbar_i}\tilde{\phi}_i\dot{\hat{\phi}}_i+\mathbf{e}^{\mathrm{T}}(\mathbf{P}\otimes\mathbf{I})\dot{\mathbf{e}}+\alpha\dot{V}\left(\boldsymbol{\rho},\boldsymbol{\rho}^d\right)
\end{aligned}\tag{5.61}$$

Furthermore, the following inequality holds:

$$\begin{aligned}
&\mathbf{s}_i^{\mathrm{T}}(\mathbf{d}_i+\boldsymbol{\varepsilon}_i)\\
&\le\left\|\mathbf{s}_i^{\mathrm{T}}\mathbf{d}_i\right\|_1+\left\|\mathbf{s}_i^{\mathrm{T}}\boldsymbol{\varepsilon}_i\right\|_1\\
&\le\|\mathbf{s}_i\|_1\|\mathbf{d}_i\|_1+\|\mathbf{s}_i\|_1\|\boldsymbol{\varepsilon}_i\|_1\\
&\le\phi_i\|\mathbf{s}_i\|_1
\end{aligned}\tag{5.62}$$

Substitute Eqs. (5.56) and (5.62) into (5.61), it can be obtained as

$$\dot{V}(t) \leq -\sum_{i=1}^{n} \mathbf{s}_i^{\mathrm{T}} k_{2i} \mathbf{s}_i + \mathbf{e}^{\mathrm{T}}(\mathbf{P} \otimes \mathbf{I})\dot{\mathbf{e}} + \alpha \dot{V}(\boldsymbol{\rho}, \boldsymbol{\rho}^d) \tag{5.63}$$

By Eqs. (5.47) and (5.48), we can derive that

$$\begin{aligned} \dot{V}(\boldsymbol{\rho}, \boldsymbol{\rho}^d) &= \sum_{i=1}^{n} \nabla_{\boldsymbol{\rho}_i} V(\boldsymbol{\rho}, \boldsymbol{\rho}^d)\dot{\boldsymbol{\rho}}_i + \sum_{i=1}^{n} \nabla_{\boldsymbol{\rho}_i^d} V(\boldsymbol{\rho}, \boldsymbol{\rho}^d)\dot{\boldsymbol{\rho}}_i^d \\ &= \sum_{i=1}^{n} \nabla_{\boldsymbol{\rho}_i} V(\boldsymbol{\rho}, \boldsymbol{\rho}^d)\left(\mathbf{s}_i - k_{1i}\mathbf{e}_i - \alpha \nabla_{\boldsymbol{\rho}_i} V(\boldsymbol{\rho}, \boldsymbol{\rho}^d)\right) \end{aligned} \tag{5.64}$$

According to Eqs. (5.11), (5.12), and (5.48), we have

$$\mathbf{e}^{\mathrm{T}}(\mathbf{P} \otimes \mathbf{I})\dot{\mathbf{e}} = \mathbf{e}^{\mathrm{T}}(\mathbf{PH} \otimes \mathbf{I})(\mathbf{s} - \mathbf{k}_1\mathbf{e} - \alpha\boldsymbol{\chi}) \tag{5.65}$$

where $\mathbf{s} = \left[\mathbf{s}_1^{\mathrm{T}}, \mathbf{s}_2^{\mathrm{T}}, \ldots, \mathbf{s}_n^{\mathrm{T}}\right]^{\mathrm{T}}$, $\mathbf{k}_1 = \mathrm{diag}(k_{11}\mathbf{I}, k_{12}\mathbf{I}, \ldots, k_{1n}\mathbf{I})$, $\boldsymbol{\chi} = \left[\nabla_{\boldsymbol{\rho}_1} V(\boldsymbol{\rho}, \boldsymbol{\rho}^d), \nabla_{\boldsymbol{\rho}_2} V(\boldsymbol{\rho}, \boldsymbol{\rho}^d), \ldots, \nabla_{\boldsymbol{\rho}_n} V(\boldsymbol{\rho}, \boldsymbol{\rho}^d)\right]^{\mathrm{T}}$.

Substitute Eqs. (5.64) and (5.65), it can be obtained as

$$\begin{aligned} &\dot{V}(t) \\ &\leq -\sum_{i=1}^{n} \mathbf{s}_i^{\mathrm{T}} k_{2i}\mathbf{s}_i + \mathbf{e}^{\mathrm{T}}(\mathbf{PH} \otimes \mathbf{I})(\mathbf{s} - \mathbf{k}_1\mathbf{e} - \alpha\boldsymbol{\chi}) \\ &\quad + \alpha \sum_{i=1}^{n} \nabla_{\boldsymbol{\rho}_i} V(\boldsymbol{\rho}, \boldsymbol{\rho}^d)\left(\mathbf{s}_i - k_{1i}\mathbf{e}_i - \alpha \nabla_{\boldsymbol{\rho}_i} V(\boldsymbol{\rho}, \boldsymbol{\rho}^d)\right) \\ &\leq -\mathbf{s}^{\mathrm{T}}\mathbf{k}_2\mathbf{s} - \alpha^2\boldsymbol{\chi}^{\mathrm{T}}\boldsymbol{\chi} - \mathbf{e}^{\mathrm{T}}(\mathbf{PH} \otimes \mathbf{I})\mathbf{k}_1\mathbf{e} + \alpha\mathbf{s}^{\mathrm{T}}\boldsymbol{\chi} \\ &\quad + \mathbf{e}^{\mathrm{T}}(\mathbf{PH} \otimes \mathbf{I})\mathbf{s} - \alpha\mathbf{e}^{\mathrm{T}}((\mathbf{PH} \otimes \mathbf{I}) + \mathbf{k}_1)\boldsymbol{\chi} \\ &\leq -\begin{bmatrix} \mathbf{s}^{\mathrm{T}} & \boldsymbol{\chi}^{\mathrm{T}} \end{bmatrix} \begin{bmatrix} \mathbf{A} & -\frac{\alpha}{2}\mathbf{I} \\ -\frac{\alpha}{2}\mathbf{I} & \mathbf{B} \end{bmatrix} \begin{bmatrix} \mathbf{s} \\ \boldsymbol{\chi} \end{bmatrix} \\ &\quad - \begin{bmatrix} \mathbf{s}^{\mathrm{T}} & \mathbf{e}^{\mathrm{T}} \end{bmatrix} \begin{bmatrix} \mathbf{C} & -\frac{(\mathbf{PH}\otimes\mathbf{I})}{2} \\ -\frac{(\mathbf{PH}\otimes\mathbf{I})}{2} & \mathbf{D} \end{bmatrix} \begin{bmatrix} \mathbf{s} \\ \mathbf{e} \end{bmatrix} \\ &\quad - \begin{bmatrix} \mathbf{e}^{\mathrm{T}} & \boldsymbol{\chi}^{\mathrm{T}} \end{bmatrix} \begin{bmatrix} \mathbf{E} & \mathbf{I} \\ \alpha((\mathbf{PH} \otimes \mathbf{I}) + \mathbf{k}_1) - \mathbf{I} & \mathbf{F} \end{bmatrix} \begin{bmatrix} \mathbf{e} \\ \boldsymbol{\chi} \end{bmatrix} \end{aligned} \tag{5.66}$$

The matrices **A**, **B**, **C**, **D**, **E**, and **F** satisfy the following relationships:

$$\begin{cases} \mathbf{A} + \mathbf{C} = \mathbf{k}_2 \\ \mathbf{B} + \mathbf{F} = \alpha^2\mathbf{I} \\ \mathbf{D} + \mathbf{E} = (\mathbf{PH} \otimes \mathbf{I})\mathbf{k}_1 \end{cases} \tag{5.67}$$

Now, for simplicity, we choose the matrices as

$$\begin{cases} \mathbf{A} = \mathbf{C} = \frac{\mathbf{k}_2}{2} \\ \mathbf{B} = \mathbf{F} = \frac{\alpha^2}{2}\mathbf{I} \\ \mathbf{D} = \mathbf{E} = \frac{(\mathbf{PH} \otimes \mathbf{I})\mathbf{k}_1}{2} \end{cases} \tag{5.68}$$

To guarantee the stability of the closed-loop system, that is $\dot{V}(t) \leq 0$, the following relationships should be also held:

$$\begin{cases} \mathbf{AB} - \frac{\alpha^2}{4}\mathbf{I} > \mathbf{0} \\ \mathbf{CD} - \frac{(\mathbf{PH} \otimes \mathbf{I})(\mathbf{PH} \otimes \mathbf{I})}{4} > \mathbf{0} \\ \mathbf{EF} - (\alpha((\mathbf{PH} \otimes \mathbf{I}) + \mathbf{k}_1) - \mathbf{I}) > \mathbf{0} \end{cases} \tag{5.69}$$

The actual values of designed parameters α, $\mathbf{k}_1$, $\mathbf{k}_2$, and $\mathbf{P}$ will be displayed in the simulation section. With the apposite values, the following inequalities hold:

$$\begin{cases} \begin{bmatrix} \mathbf{A} & -\frac{\alpha}{2}\mathbf{I} \\ -\frac{\alpha}{2}\mathbf{I} & \mathbf{B} \end{bmatrix} > 0 \\ \begin{bmatrix} \mathbf{C} & -\frac{(\mathbf{PH} \otimes \mathbf{I})}{2} \\ -\frac{(\mathbf{PH} \otimes \mathbf{I})}{2} & \mathbf{D} \end{bmatrix} > 0 \\ \begin{bmatrix} \mathbf{E} & \mathbf{I} \\ \alpha((\mathbf{PH} \otimes \mathbf{I}) + \mathbf{k}_1) - \mathbf{I} & \mathbf{F} \end{bmatrix} > 0 \end{cases} \tag{5.70}$$

It can be derived that $\dot{V}(t) \leq 0$ holds.

Remark 8 Hence, it is obvious that $\dot{V}(t)$ is semi-negative. We can deduce that $\lim_{t \to \infty} V(t)$ exists and is finite, and $\left(\mathbf{s}_i, \tilde{\mathbf{W}}_i, \tilde{\phi}_i, \mathbf{e}, V\left(\boldsymbol{\rho}, \boldsymbol{\rho}^d\right)\right)$ are globally uniformly bounded. By the auxiliary variable $\mathbf{s}_i = \tilde{\mathbf{v}}_i + k_{1i}\mathbf{e}_i + \alpha \nabla_{\boldsymbol{\rho}_i} V\left(\boldsymbol{\rho}, \boldsymbol{\rho}^d\right)$, the boundedness of $\mathbf{s}_i$ implies that both $\tilde{\mathbf{v}}_i$ and $\nabla_{\boldsymbol{\rho}_i} V\left(\boldsymbol{\rho}, \boldsymbol{\rho}^d\right)$ are bounded. Furthermore, $V\left(\boldsymbol{\rho}, \boldsymbol{\rho}^d\right)$ is bounded all the time, and $d\left(\nabla_{\boldsymbol{\rho}_i} V\left(\boldsymbol{\rho}, \boldsymbol{\rho}^d\right)\right)/dt \in L_\infty$ are satisfied. $\nabla_{\boldsymbol{\rho}_i} V\left(\boldsymbol{\rho}, \boldsymbol{\rho}^d\right)$ is uniformly continuous in time. It can be deduced that $\forall\, t \geq 0,\ l_0 < \left\|\boldsymbol{\rho}_i - \boldsymbol{\rho}_j\right\| < l + \delta$ holds, which implies that both the connecting constraint and the collision avoidance constraint are not violated in the whole time. According to Eq. (5.58), it is obvious that $\dot{V}(t) \in L_\infty$ and $\dot{\mathbf{s}} \in L_\infty$. Integrating both sides of Eq. (5.64) on time interval $[0, \infty)$, we have

$$\begin{aligned} &\int_0^\infty \dot{V}\left(\boldsymbol{\rho}, \boldsymbol{\rho}^d\right) dt \\ &= \sum_{i=1}^{n} \int_0^\infty \left(\nabla_{\boldsymbol{\rho}_i} V\left(\boldsymbol{\rho}, \boldsymbol{\rho}^d\right)\dot{\boldsymbol{\rho}}_i + \nabla_{\boldsymbol{\rho}_i^d} V\left(\boldsymbol{\rho}, \boldsymbol{\rho}^d\right)\dot{\boldsymbol{\rho}}_i^d\right) dt \\ &= V\left(\boldsymbol{\rho}, \boldsymbol{\rho}^d\right)\Big|_{t=\infty} - V\left(\boldsymbol{\rho}, \boldsymbol{\rho}^d\right)\Big|_{t=0} < \infty \end{aligned} \tag{5.71}$$

Thus, $\int_0^\infty \left(\nabla_{\boldsymbol{\rho}_i} V\left(\boldsymbol{\rho}, \boldsymbol{\rho}^d\right)\right) dt$ is bounded. According to Lemma 5, we can get the conclusion that $\lim_{t\to\infty} \nabla_{\boldsymbol{\rho}_i} V\left(\boldsymbol{\rho}, \boldsymbol{\rho}^d\right) = 0$, which means that the MUs will maintain accurate geometric configuration. Integrating both sides of Eq. (5.63) on time interval $[0, \infty)$, we have

$$\int_0^\infty \left(\sum_{i=1}^n \mathbf{s}_i^{\mathrm{T}} k_{2i} \mathbf{s}_i - \mathbf{e}^{\mathrm{T}} (\mathbf{P} \otimes \mathbf{I}) \dot{\mathbf{e}} - \alpha \dot{V}\left(\boldsymbol{\rho}, \boldsymbol{\rho}^d\right) \right) dt \leq V(0) - V(\infty) \tag{5.72}$$

Then, we conclude $\mathbf{s}_i \in L_2$. Following Corollary 1, we can have $\lim_{t\to\infty} \mathbf{s}_i(t) = 0$. According to Eq. (5.48), $\lim_{t\to\infty}\left(\dot{\boldsymbol{\rho}}_i - \dot{\boldsymbol{\rho}}_i^d + \mathbf{k}_{1i}\mathbf{e}_i\right) = \lim_{t\to\infty}\left(\mathbf{s}_i - \alpha \nabla_{\boldsymbol{\rho}_i} V\left(\boldsymbol{\rho}, \boldsymbol{\rho}^d\right)\right) = 0$, so $\dot{\boldsymbol{\rho}}_i \to \dot{\boldsymbol{\rho}}_i^d$, $\boldsymbol{\rho}_i \to \boldsymbol{\rho}_i^d$.

Therefore, Theorem 2 is proved.

5.2.4 *Numerical Simulation*

5.2.4.1 Simulation Environment

TSNR is stored in a platform satellite before shooting, and the net is folded in a square pattern with sideline 0.5 m. The connecting net is passively deployed after the catapulting of four-corner MUs. The initial velocities of four MUs are $(0.1; 0.1; 0.1)$ m/s, $(-0.1; 0.1; 0.1)$ m/s; $(0.1; 0.1; -0.1)$ m/s, and $(-0.1; 0.1; -0.1)$ m/s, respectively. The free-flying phase is a transient period from the moment the MUs are catapulting and the control strategy is involved. In this paper, the controller is involved at $t = 20$ s, while the relative positions between adjacent MUs meet the constraints in Eqs. (5.37) and (5.38). Other parameters of the system are shown in Table 5.2. The relationship of communication between the platform satellite and the MUs is shown in Fig. 5.3.

5.2.4.2 Simulation Results

The parameters in the stabilization controller are selected as shown in Table 5.3.

When the controlled flying phase is involved, the initial positions of MUs are $\boldsymbol{\rho}_1 = (2.4697, 1.9901, 2.4690)^{\mathrm{T}}$, $\boldsymbol{\rho}_2 = (-2.4625, 1.9966, 2.4665)^{\mathrm{T}}$, $\boldsymbol{\rho}_3 = (2.4697, 1.9901, -2.4690)^{\mathrm{T}}$, and $\boldsymbol{\rho}_4 = (-2.4625, 1.9966, -2.4665)^{\mathrm{T}}$, respectively. The desired positions of MUs are $\boldsymbol{\rho}_1^d = (2.5, 1.9966 + 0.1(t-20), 2.5)^{\mathrm{T}}$, $\boldsymbol{\rho}_2^d = (-2.5, 1.9966 + 0.1(t-20), 2.5)^{\mathrm{T}}$, $\boldsymbol{\rho}_3^d = (2.5, 1.9966 + 0.1(t-20), -2.5)^{\mathrm{T}}$,

Table 5.2 Parameters of the TSNR system

Parameters	Value
Mass of each MU (kg)	10
Mass of connecting net (kg)	1
Side length of the net (m)	5
Side length of the mesh (m)	0.5
Maximum elastic elongation (m)	0.01
Minimum safety distance (m)	0.9
Material Young's modulus (MPa)	445.6
Diameter of the net's thread (mm)	1
Damping ratio of the net's thread	0.106
Orbit altitude (km)	36000
Orbit inclination (rad)	0

Table 5.3 Stabilization controller parameters

Parameters	Values
$\mathbf{k}_1$	$0.01 \times \mathrm{diag}(1, 1, \ldots, 1)_{12\times12}$
α	0.000001
$\mathbf{k}_2$	$\mathrm{diag}(2, 2, \ldots, 2)_{12\times12}$
$\hbar_i$, for $i = 1, 2, 3, 4$	0.05

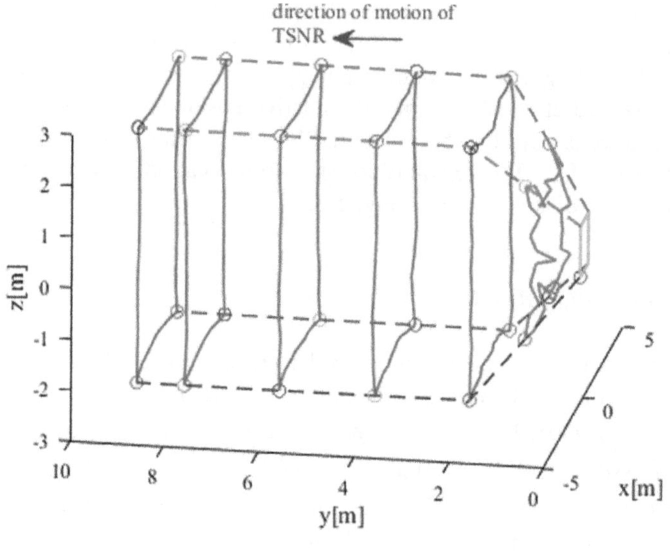

Fig. 5.12 The motions of sidelines of the net

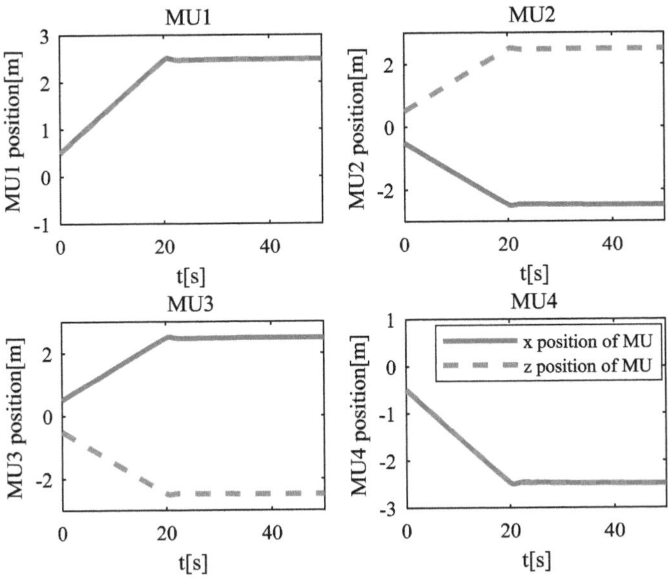

Fig. 5.13 Positions of MUs on x- and z-directions

$\boldsymbol{\rho}_4^d = (-2.5, 1.9966 + 0.1(t-20), -2.5)^{\mathrm{T}}$, respectively. The information exchange among MUs and platform satellite is represented by a graph shown in Fig. 5.3 with the following adjacency matrix:

$$\mathbf{A} = \begin{bmatrix} 0 & 1 & 1 & 0 \\ 1 & 0 & 0 & 1 \\ 1 & 0 & 0 & 1 \\ 0 & 1 & 1 & 0 \end{bmatrix}$$

and $\mathbf{B} = \mathbf{I}_4$. The four MUs have the identical configurations about RBFNN. Each of the RBFNN estimator has nine neurons in this paper. Centers of the activation function $h_{ij}(\mathbf{x}_i)$ are dispersed uniformly in the scope $[-3, 3] \times [-3, 3]$. The other parameters are chosen as $W_i = 10$, $\boldsymbol{\Gamma}_i = \mathbf{I}_3$, and $\hat{W}_i(0) = \mathbf{0}$.

The detailed simulation results are provided as follows. A three-dimensional view of the TSNR maneuver process in the LVLH frame is shown in Fig. 5.12, where the right solid square shape denotes the initial position of the folded net; 'o' represents the MUs at the corners of a square connecting net; the dashed lines show the trajectories of four MUs. The figure shows that when the MUs are ejected with the initial velocities from the platform satellite, the net begins to unfold. The motions of sidelines of the net become more orderly after the controllers are activated. It clearly shows that the designed control scheme can obtain good configuration keeping performance in the presence of relative distance constraints.

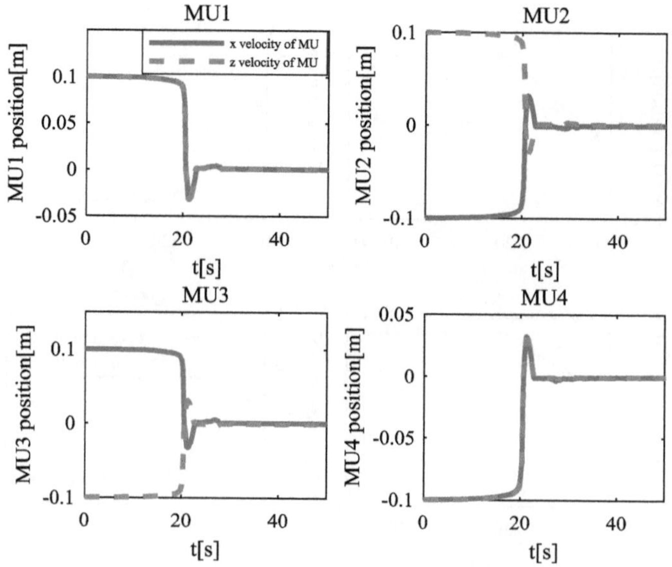

Fig. 5.14 Velocities of MUs on x- and z-directions

In addition, the positions and velocities of MUs are shown in Figs. 5.13 and 5.14, respectively. As we can see, the positions and velocities of MUs jumping at the moment the controller's opening are controlled well and rapidly, that is the MUs can track the desired trajectories with high accuracy and small overshoot. The slightly chattering of the positions and velocities of four MUs, which is still acceptable is induced by the bounce effect of the connecting net. It clearly shows that the proposed coordinated formation control scheme can obtain high accuracy of position tracking.

(1) *Case A*: Using the control scheme designed as Eq. (5.55) and the chosen parameters, Fig. 5.15 shows the relative distance between arbitrary adjacent MUs, which are immersed in a virtual potential field. The maximum and minimum relative distance are 5.0097 m and 1 m, respectively. The figure illustrates that the relative distances between arbitrary adjacent MUs always stay in the constraint of 0.9 $\sim$ 5.01 m, which implies that the artificial potential function works well on relative distance constraints between MUs. When the controllers are switching on, the MUs move further away between each other with the velocities on x- and z-directions. The connecting net is quickly taut. The virtual potential field generates a repulsive force that urges the MUs to reach the desired trajectories. However, while the distance between MUs reaches to the designed value, the attractive force generated by the potential field is imposed on the MUs. The distance between arbitrary adjacent MUs is quickly getting smaller and smaller owing to the elastic force of net and attraction force caused by the artificial potential function. The connecting net

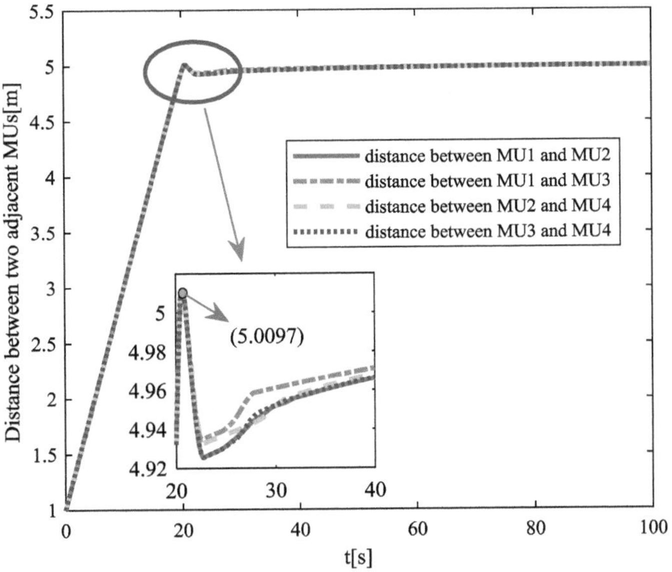

Fig. 5.15 Distance between arbitrary adjacent MUs with virtual potential field

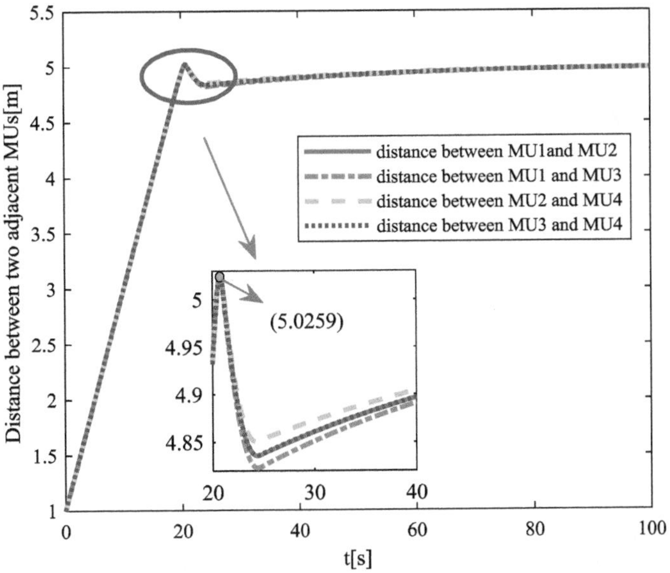

Fig. 5.16 Distance between arbitrary adjacent MUs without virtual potential field

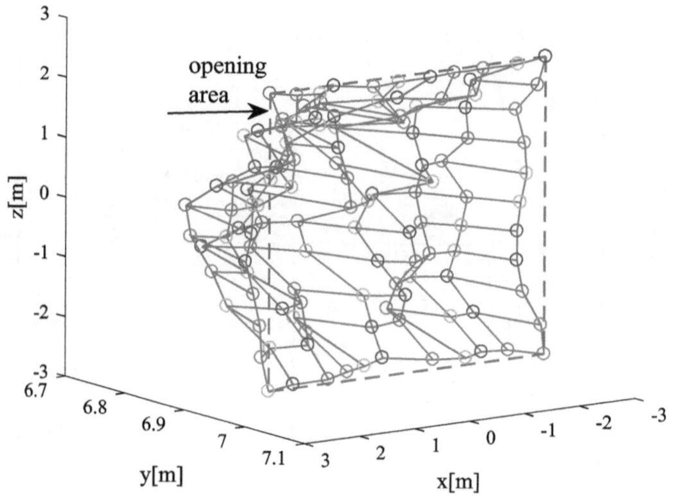

Fig. 5.17 The final shape of the TSNR

transforms from taut to slack. Finally, within the limitation, the MUs move further away again to track the desired trajectories.

(2) *Case B*: This case is utilized to demonstrate the relative distance constraint capability of artificial potential function. By the control scheme defined by Eq. (5.55) with $\alpha = 0$, Fig. 5.16 shows the relative distance between arbitrary adjacent MUs without virtual potential field. The aggressive elasticity of the connecting net induces the transition from slack to taut and slack again. Compared with case A, the chattering is stronger. The maximum relative distance is 5.0259 m, which will cause larger elastic force than case A, and then an undesired longitude oscillation tags along. Simultaneously, the minimum distance is smaller than case A. The smaller distance will increase the likelihood of collision between MUs and decrease the opening area of the connecting net, which does harm to the capture mission. A final shape of the TSNR is presented in Fig. 5.17.

References

1. Sabatini M, Gasbarri P, Palmerini GB (2016) Elastic issues and vibration reduction in a tethered deorbiting mission. Adv Space Res 57(9):1951–1964
2. Kuriki Y, Namerikawa T (2014) Consensus-based cooperative formation control with collision avoidance for a multi-UAV system. In: Proceedings of American control conference, pp 2077–2082

3. Chen T, Wen H, Hu HY, Jin DP (2016) Output consensus and collision avoidance of a team of flexible spacecraft for on-orbit autonomous assembly. Acta Astronaut 121:271–281
4. Zhang X, Liu L, Feng G (2015) Leader–follower consensus of time-varying nonlinear multi-agent systems. Automatica 52:8–14
5. Shi P, Shen Q (2015) Cooperative control of multi-agent systems with unknown state-dependent controlling effects. IEEE Trans Autom Sci Eng 12(3):827–834
6. Li B, Hu Q, Yu Y et al (2017) Observer-based fault-tolerant attitude control for rigid spacecraft. IEEE Trans Aerosp Electron Syst 53(5):2572–2582
7. Wu J, Shi Y (2011) Consensus in multi-agent systems with random delays governed by a Markov chain. Syst Control Lett 60(10):863–870
8. Qu Z (2009) Cooperative control of dynamical systems: applications to autonomous vehicles. Springer, Berlin
9. Polycarpou MM, Mears MJ (1998) Stable adaptive tracking of uncertain systems using nonlinearly parametrized on-line approximators. Int J Control 70(3):363–384
10. Ge SS, Wang C (2004) Adaptive neural control of uncertain MIMO nonlinear systems. IEEE Trans Neural Netw 15(3):674–692
11. Zhang F, Huang P (2017) Releasing dynamics and stability control of maneuverable tethered space net. IEEE/ASME Trans Mechatron 22(2):983–993
12. Ge SS, Liu XM, Goh CH, Xu LG (2016) Formation tracking control of multiagents in constrained space. IEEE Trans Control Syst Technol 24(3):992–1003
13. Merheb AR, Gazi V, Sezer-Uzol N (2016) Implementation studies of robot swarm navigation using potential functions and panel methods. IEEE-ASME Trans Mechatron 21(5):2556–2567
14. Yao B, Tomizuka M (1997) Adaptive robust control of SISO nonlinear systems in a semi-strict feedback form. Automatica 33(5):893–900
15. Krstic M, Kanellakopoulos I, Kokotovic PV (1995) Nonlinear and adaptive control design. Wiley, New York
16. Tao G (1997) A simple alternative to the Barbalat lemma. IEEE Trans Autom Control 42 (5):698

Part II
Space Tethered Formation

Chapter 6
Dynamics Modeling

Research on space robotics [1, 2] has drawn much attention in recent years and tethered space system [3–5] is one of the current hot spots in the field of space robotics. Space Tethered Formation System is a kind of tethered space system, which presents many attractive and potential advantages in space applications. These advantages can be summarized as follows: A variable configuration baseline for interferometric observations achieved by deploying or retracting the tethers [6, 7]; Coverage of the entire plane carried out continuously by rotating the whole formation system; Reduction of propellant consumption in control system of each spacecraft [8]. The Tethered Formation System can have various configurations, but all of the configurations can be separated into three basic configurations according to the shape they look like, which are Opened-String configuration, Closed-String configuration, and Hub–Spoke configuration (demonstrated in Fig. 6.1). The Opened-String configuration (Fig. 6.1a) consists of several spacecraft, and each of them is connected to another by tether, forming as an Opened-String configuration. The Closed-String configuration (Fig. 6.1b) is similar to the Opened-String configuration except that all spacecraft are connected to each other end to end, forming as a Closed-String configuration. The Hub–Spoke configuration (Fig. 6.1c) is a radial configuration with a master spacecraft in the center of the formation system. The radial tethers are released from the master spacecraft, and a sub-spacecraft is connected at the end of each radial tether.

A lot of work has been carried out on the dynamics of the Tethered Formation System, whereas fewer studies have been presented on the dynamics of the rotating HS-TFS. In reference [9], the dynamics of a tethered-connected three-body system is studied. The dynamics of certain multi-tethered satellite formations containing a parent (or central) body are examined in reference [8]. In this study, the satellites are regarded as point masses; the tethers in the formation system are massless and straight, and the motion of the parent body of the formation is prescribed. The Lagrangian formulation is used to derive the equations of motion of the tethered

P. Huang and F. Zhang, *Theory and Applications of Multi-Tethers in Space*,
Springer Tracts in Mechanical Engineering,
https://doi.org/10.1007/978-981-15-0387-0_6

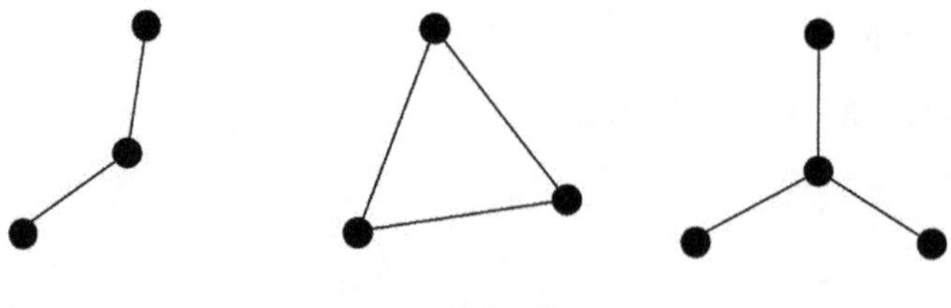

Fig. 6.1 Basic configurations of the tethered formation

system. Some researchers have focused on chains of N-bodies to obtain a general formulation of the system's dynamics [10]. Others have more specifically looked at three or four tethered bodies as a means of modeling a micro- or variable-gravity laboratory attached to the space station [11, 12]. The construction of a ring of satellites around the Earth might be advantageous for future space colonization and will be the solution to problems associated with the high volume of space traffic [13, 14]. Some researchers have examined the attitude control of spinning tethered formations of spacecraft modeled as rigid bodies [15]. Finally, research has been aimed at the problem of two bodies connected by multiple tethers [16]. However, most researchers introduce the Lagrangian formulation to obtain the dynamics model of tethered system. In addition, the tethers are usually considered as rigid body and the elasticity are always neglected.

6.1 Single-Tethered System

In the research of traditional satellite formations, the virtual structure method [17] is widely used. This method regards the entire formation as a single structure, which facilitates the modeling of the dynamics and the design of the controller. In order to obtain the equilibrium conditions of the space multi-tether system, this section will use the virtual structure method to simplify the tethered system to a single rigid body. From the geometric characteristics of the space multi-tether system of several basic configurations, we can see that in the process of spinning the tether system, the virtual structure closest to its configuration is an axisymmetric flat-cylindrical rigid body. Assume that the center of mass of the rigid body is in the Earth's Kepler circular orbit. Figure 6.2 shows a simplified schematic diagram of the spatial multiline system. In this section, we will establish the spin dynamics model of the single rigid body simplified model, and study its equilibrium state and give the equilibrium conditions.

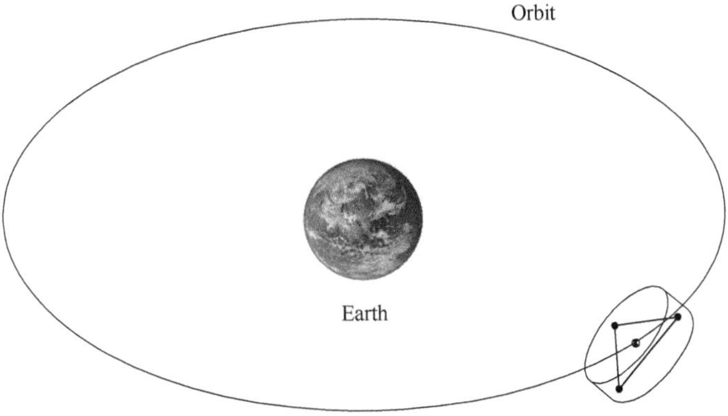

Fig. 6.2 The schematic figure of space multi-tether system

6.1.1 Coordinate Description

Three reference coordinate systems are defined as follows:

(1) Inertial coordinate system $\hat{\imath}$: the origin is located at the center of mass of the earth, and $\hat{\imath} - \hat{\jmath}$ plane is the orbital surface;
(2) Body coordinate system $\hat{b}$: The inertial coordinate system is transformed from the 3-1-3 coordinate, and is the main inertia coordinate system of the single rigid body. Each coordinate axis is consistent with the principal inertia axis passing through the centroid, ϕ, θ, ψ is Euler angle, $\hat{b}_1$, $\hat{b}_2$ is transverse axis, $\hat{b}_3$ is the spin axis;
(3) Transition coordinate system $\hat{a}$.

The detailed coordinate systems definition are shown in Fig. 6.3.

(a) We know that in Eulerian rotation, each rotation axis is a coordinate axis of the rotated coordinate system, and each rotation angle is Euler angles. Therefore, the attitude matrix determined by Euler angles is the product of the cubic

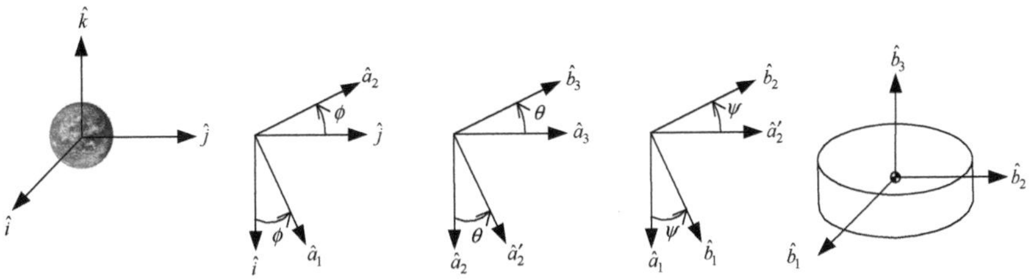

Fig. 6.3 Definition of coordinate system

coordinate transformation matrix. The standard formula for these coordinate transformation matrices is

$$R_1(\alpha_1) = \begin{bmatrix} 1 & 0 & 0 \\ 0 & \cos(\alpha_1) & \sin(\alpha_1) \\ 0 & -\sin(\alpha_1) & \cos(\alpha_1) \end{bmatrix} \tag{6.1}$$

$$R_2(\alpha_2) = \begin{bmatrix} \cos(\alpha_2) & 0 & -\sin(\alpha_2) \\ 0 & 1 & 0 \\ \sin(\alpha_2) & 0 & \cos(\alpha_2) \end{bmatrix} \tag{6.2}$$

$$R_3(\alpha_3) = \begin{bmatrix} \cos(\alpha_3) & \sin(\alpha_3) & 0 \\ -\sin(\alpha_3) & \cos(\alpha_3) & 0 \\ 0 & 0 & 1 \end{bmatrix} \tag{6.3}$$

Therefore, the attitude matrix from the system $\hat{i}$ to the system $\hat{b}$ is

$$C^{bi} = R_3(\psi)R_1(\theta)R_3(\phi) \tag{6.4}$$

namely,

$$\begin{Bmatrix} \hat{b}_1 \\ \hat{b}_2 \\ \hat{b}_3 \end{Bmatrix} = C^{bi} \begin{Bmatrix} \hat{i}_1 \\ \hat{i}_2 \\ \hat{i}_3 \end{Bmatrix} \tag{6.5}$$

$$C^{bi} = \begin{bmatrix} \cos\psi\cos\phi - \cos\theta\sin\psi\sin\phi & \cos\psi\sin\phi + \cos\theta\sin\psi\cos\phi & \sin\psi\sin\phi \\ -\sin\psi\cos\phi - \cos\theta\cos\psi\sin\phi & -\sin\psi\sin\phi + \cos\theta\cos\psi\cos\phi & \cos\psi\sin\theta \\ \sin\theta\sin\phi & -\sin\theta\cos\phi & \cos\theta \end{bmatrix} \tag{6.6}$$

Instead, the attitude matrix from the system $\hat{b}$ to system $\hat{i}$ is

$$\begin{aligned} C^{ib} &= (C^{ib})^{-1} \\ &= \begin{bmatrix} \cos\psi\cos\phi - \cos\theta\sin\psi\sin\phi & -\sin\psi\cos\phi - \cos\theta\cos\psi\sin\phi & \sin\theta\sin\phi \\ \cos\psi\sin\phi + \cos\theta\sin\psi\cos\phi & -\sin\psi\sin\phi + \cos\theta\cos\psi\cos\phi & -\sin\theta\cos\phi \\ \sin\psi\sin\phi & \cos\psi\sin\theta & \cos\theta \end{bmatrix} \end{aligned} \tag{6.7}$$

namely,

$$\begin{Bmatrix} \hat{i}_1 \\ \hat{i}_2 \\ \hat{i}_3 \end{Bmatrix} = C^{ib} \begin{Bmatrix} \hat{b}_1 \\ \hat{b}_2 \\ \hat{b}_3 \end{Bmatrix} = R_3(-\phi)R_1(-\theta)R_3(-\psi) \begin{Bmatrix} \hat{b}_1 \\ \hat{b}_2 \\ \hat{b}_3 \end{Bmatrix} \tag{6.8}$$

In addition, the attitude matrix from the system $\hat{b}$ to the system $\hat{a}$ is

$$C^{ab} = \begin{bmatrix} \cos\psi & -\sin\psi & 0 \\ \cos\theta\sin\psi & \cos\theta\cos\psi & -\sin\theta \\ \sin\theta\sin\psi & \sin\theta\cos\psi & \cos\theta \end{bmatrix} \tag{6.9}$$

namely,

$$\begin{Bmatrix} \hat{a}_1 \\ \hat{a}_2 \\ \hat{a}_3 \end{Bmatrix} = C^{ab} \begin{Bmatrix} \hat{b}_1 \\ \hat{b}_2 \\ \hat{b}_3 \end{Bmatrix} = R_1(-\theta)R_3(-\psi) \begin{Bmatrix} \hat{b}_1 \\ \hat{b}_2 \\ \hat{b}_3 \end{Bmatrix} \tag{6.10}$$

6.1.2 Dynamics Modeling

The spin dynamics of a cylindrically symmetric, roughly cylindrical rigid body is a problem of the rigid body rotating around a fixed point in classical mechanics. The attitude dynamics equation can be derived from the rigid body momentum moment formula and theorem, i.e., Newton–Euler method: rigid body pair. The rate of change of the angular momentum at a fixed point in the inertial space is equal to the sum of all external forces acting on the rigid body at this point, i.e.:

$$\vec{M} = \vec{H} \tag{6.11}$$

where $\vec{M}$ is the sum of the external moments that the rigid body receives and $\vec{H}$ is the angular momentum of the rigid body, and it can be expressed as

$$\vec{H} = {}^{\hat{b}}I^{b}\vec{\omega}^{bi} \tag{6.12}$$

where ${}^{\hat{b}}\vec{\omega}^{bi}$ is the representation of the angular velocity of the rigid body relative to the inertial coordinate system. For the rigid body's moment of inertia matrix, set it as

$${}^{\hat{b}}I = \begin{bmatrix} A & 0 & 0 \\ 0 & A & 0 \\ 0 & 0 & C \end{bmatrix} \tag{6.13}$$

where, A, C are the rotational inertia of the rigid body's transverse axis and spin axis, respectively, and the left-upper label, $\hat{b}$, is represented in the body coordinate system.

According to [18], there is a relationship between the derivatives of any vector in the two coordinate systems as follows:

$$ {}^{\hat{\imath}}\frac{d}{dt}\vec{Z} = {}^{\hat{b}}\frac{d}{dt}\vec{Z} + \vec{\omega}^{bi} \times \vec{Z} \tag{6.14}$$

Therefore, Eq. (6.11) can be converted to

$$\vec{M} = {}^{\hat{\imath}}\frac{d}{dt}(I\vec{\omega}^{bi}) = {}^{\hat{b}}\frac{d}{dt}(I\vec{\omega}^{bi}) + \vec{\omega}^{bi} \times (I\vec{\omega}^{bi}) = I\dot{\vec{\omega}}^{bi} + \vec{\omega}^{bi} \times I\vec{\omega}^{bi} \tag{6.15}$$

where $\dot{\vec{\omega}}^{bi}$ is the derivative of $\vec{\omega}^{bi}$ in body coordinate system.

Assume that $\vec{M} = [M_1 \quad M_2 \quad M_3]^T$, $\vec{\omega} = [\omega_1 \quad \omega_2 \quad \omega_3]^T$, Eq. (6.15) can be written as

$$M_1 = A\dot{\omega}_1 + (C - A)\omega_2\omega_3 \tag{6.16}$$

$$M_2 = A\dot{\omega}_2 + (A - C)\omega_1\omega_3 \tag{6.17}$$

$$M_3 = C\dot{\omega}_3 \tag{6.18}$$

It can be further simplified as

$$\dot{\omega}_1 = \frac{M_1}{A} + \frac{(A - C)}{A}\omega_2\omega_3 \tag{6.19}$$

$$\dot{\omega}_2 = \frac{M_2}{A} - \frac{(A - C)}{A}\omega_1\omega_3 \tag{6.20}$$

$$\dot{\omega}_3 = \frac{M_3}{C} \tag{6.21}$$

It can also be written in the following matrix form:

$$\dot{\vec{\omega}} = I^{-1}\vec{M} - I^{-1}\vec{\omega}^{x}I\vec{\omega} \tag{6.22}$$

where

$$\vec{\omega}^{x} = \begin{bmatrix} 0 & -\omega_3 & \omega_2 \\ \omega_3 & 0 & -\omega_1 \\ -\omega_2 & \omega_1 & 0 \end{bmatrix} \tag{6.23}$$

Due to

$$ {}^{\hat{*}}\vec{\omega}^{bi} = \dot{\psi}\hat{b}_3 + \dot{\phi}\hat{a}_3 + \dot{\theta}\hat{a}_1 \tag{6.24} $$

It can be obtained by Eq. (6.9):

$$ \hat{a}_3 = \sin\psi\sin\theta\hat{b}_1 + \cos\psi\sin\theta\hat{b}_2 + \cos\theta\hat{b}_3 \tag{6.25} $$

$$ \hat{a}_1 = \cos\psi\hat{b}_1 - \sin\psi\hat{b}_2 \tag{6.26} $$

Substituting Eqs. (6.25) and (6.26) into Eq. (6.24):

$$ {}^{\hat{b}}\vec{\omega}^{bi} = \omega_1\hat{b}_1 + \omega_2\hat{b}_2 + \omega_3\hat{b}_3 \tag{6.27} $$

where

$$ \begin{aligned} \omega_1 &= \dot{\phi}\sin\psi\sin\theta + \dot{\theta}\cos\psi \\ \omega_2 &= \dot{\phi}\cos\psi\sin\theta - \dot{\theta}\sin\psi \\ \omega_3 &= \dot{\phi}\cos\theta + \dot{\psi} \end{aligned} \tag{6.28} $$

The expression of $\vec{M}$ also needs to be obtained. Assuming that the rigid body is only subjected to the gravitational force of the earth, so the external moment is only the gravity gradient moment. In order to calculate the external torque, another coordinate system, the orbital coordinate system, is introduced here. As shown in Fig. 6.4, the system always rotates within the orbital plane, and the axis is always directed to the rigid body's center of mass.

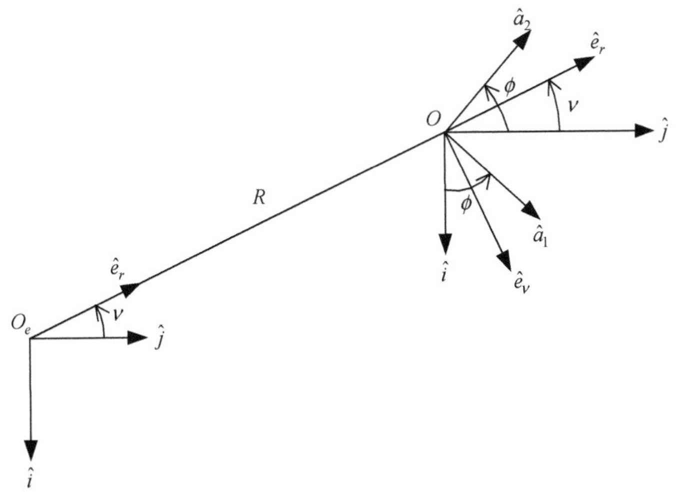

Fig. 6.4 Definition of orbital coordinate system

Assume that the height of the circular orbit of the rigid body is R, then, ${}^{\hat{e}}\vec{R} = R\hat{e}_r$, where $\hat{e}_r = (-\sin v)\hat{i} + (\cos v)\hat{j}$, v is the angle between the $\hat{j}$ axis and the $\hat{e}_r$ axis. Also in the $\hat{i}$ system, $\vec{R}$ can be rewritten as

$$^{\hat{i}}\vec{R} = (-R\sin v)\hat{i} + (R\cos v)\hat{j} \tag{6.29}$$

Then transit it into body system:

$$^{\hat{b}}\vec{R} = C^{bi}\vec{R} = X\hat{b}_1 + Y\hat{b}_2 + Z\hat{b}_3 \tag{6.30}$$

where

$$\begin{aligned} X &= -R(\sin v\cos\psi\cos\phi - \sin v\cos\theta\sin\psi\sin\phi \\ &\quad - \cos v\cos\psi\sin\phi - \cos v\cos\theta\sin\psi\cos\phi) \\ Y &= R(\sin v\sin\psi\cos\phi + \sin v\cos\theta\cos\psi\sin\phi \\ &\quad - \cos v\sin\psi\sin\phi + \cos v\cos\theta\cos\psi\cos\phi) \\ Z &= -R(\sin v\sin\theta\sin\phi + \cos v\sin\theta\cos\phi) \end{aligned} \tag{6.31}$$

Therefore, the three components of $\vec{M}$ can be expressed as

$$\begin{aligned} M_1 &= 3\mu_{\oplus}R^{-5}YZ(C - A) \\ M_2 &= 3\mu_{\oplus}R^{-5}XZ(A - C) \\ M_3 &= 3\mu_{\oplus}R^{-5}XY(A - A) = 0 \end{aligned} \tag{6.32}$$

where $\mu_{\oplus} = GM_{\oplus}$ is the gravitational constant; G is the gravitational constant; $M_{\oplus}$ is the earth's mass; X, Y, Z are the three components of the position vector of the rigid body's centroid. Substituting Eqs. (6.28) and (6.23) into Eq. (6.22), we can get the angular velocity differential equation:

$$\begin{aligned} \begin{Bmatrix} \dot{\omega}_1 \\ \dot{\omega}_2 \\ \dot{\omega}_3 \end{Bmatrix} &= \begin{bmatrix} \frac{1}{A} & 0 & 0 \\ 0 & \frac{1}{A} & 0 \\ 0 & 0 & \frac{1}{C} \end{bmatrix} \begin{bmatrix} M_1 \\ M_2 \\ M_3 \end{bmatrix} \\ &\quad - \begin{bmatrix} \frac{1}{A} & 0 & 0 \\ 0 & \frac{1}{A} & 0 \\ 0 & 0 & \frac{1}{C} \end{bmatrix} \begin{bmatrix} 0 & -\omega_3 & \omega_2 \\ \omega_3 & 0 & -\omega_1 \\ -\omega_2 & \omega_1 & 0 \end{bmatrix} \begin{bmatrix} A & 0 & 0 \\ 0 & A & 0 \\ 0 & 0 & C \end{bmatrix} \begin{bmatrix} \omega_1 \\ \omega_2 \\ \omega_3 \end{bmatrix} \end{aligned} \tag{6.33}$$

then, we have

$$\dot{\omega}_1 = \frac{M_1}{A} + \frac{(A - C)\omega_2\omega_3}{A} \tag{6.34}$$

$$\dot{\omega}_2 = \frac{M_2}{A} + \frac{(C-A)\omega_1\omega_3}{A} \tag{6.35}$$

$$\dot{\omega}_3 = \frac{M_3}{C} \tag{6.36}$$

where

$$\begin{aligned} M_1 &= 3\mu_{\oplus}R^{-5}YZ(C-A) \\ M_2 &= 3\mu_{\oplus}R^{-5}XZ(A-C) \\ M_3 &= 3\mu_{\oplus}R^{-5}XY(A-A) = 0 \end{aligned} \tag{6.37}$$

$$\begin{aligned} \omega_1 &= \dot{\phi}\sin\psi\sin\theta + \dot{\theta}\cos\psi \\ \omega_2 &= \dot{\phi}\cos\psi\sin\theta - \dot{\theta}\sin\psi \\ \omega_3 &= \dot{\phi}\cos\theta + \dot{\psi} \end{aligned} \tag{6.38}$$

In addition, Eq. (6.28) can be translated into

$$\vec{\omega}^{bi} = \begin{bmatrix} \dot{\phi}\sin\psi\sin\theta + \dot{\theta}\cos\psi \\ \dot{\phi}\cos\psi\sin\theta - \dot{\theta}\sin\psi \\ \dot{\phi}\cos\theta + \dot{\psi} \end{bmatrix} = \begin{bmatrix} \sin\psi\sin\theta & \cos\psi & 0 \\ \cos\psi\sin\theta & -\sin\psi & 0 \\ \cos\theta & 0 & 1 \end{bmatrix} \begin{Bmatrix} \dot{\phi} \\ \dot{\theta} \\ \dot{\psi} \end{Bmatrix} \tag{6.39}$$

Set that

$$K = \begin{bmatrix} \sin\psi\sin\theta & \cos\psi & 0 \\ \cos\psi\sin\theta & -\sin\psi & 0 \\ \cos\theta & 0 & 1 \end{bmatrix} \tag{6.40}$$

Define the Euler angle vector as $\vec{\varepsilon} = [\phi \quad \theta \quad \psi]^T$, then we have

$$\vec{\omega}^{bi} = K\dot{\vec{\varepsilon}} \tag{6.41}$$

or

$$\dot{\vec{\varepsilon}} = K^{-1}\vec{\omega}^{bi} \tag{6.42}$$

Therefore, the Euler angle differential equation is

$$\dot{\vec{\varepsilon}} = \begin{Bmatrix} \dot{\phi} \\ \dot{\theta} \\ \dot{\psi} \end{Bmatrix} = \frac{1}{\sin\theta} \begin{bmatrix} \sin\psi & \cos\psi & 0 \\ \cos\psi\sin\theta & -\sin\psi\sin\theta & 0 \\ -\sin\psi\cos\theta & -\cos\psi\cos\theta & \sin\theta \end{bmatrix} \begin{Bmatrix} \omega_1 \\ \omega_2 \\ \omega_3 \end{Bmatrix} \tag{6.43}$$

So far, Eqs. (6.43) and (6.33) constitute six attitude kinematics and dynamic equations of an axisymmetric flat-cylindrical rigid body in a circular orbit, and three Euler angles and three angular velocity components constitute the state vector of the kinetic equation, i.e.:

$$\vec{X} = \{\phi \quad \theta \quad \psi \quad \omega_1 \quad \omega_2 \quad \omega_3\}^T \tag{6.44}$$

6.1.3 LP Stability

For the spin-rigid body in the circular orbit, there are two equilibrium states, namely, Thomson equilibrium and Likins-Pringle (LP) equilibrium, the latter includes conic and hyperbolic two situations, in which the axis of rotation of Thomson equilibrium is perpendicular to the plane of the track, the hyperbolic axis of rotation is perpendicular to the local vertical, and the conical axis of rotation is perpendicular to the track tangent. The common point of these two kinds of equilibrium states is that the rotation axis of the rigid body remains fixed relative to the orbital coordinate system, and they can be mutually converted. When the cone angle is zero, the LP balance is changed to Thomson equilibrium, therefore, only LP balance conditions need to be found in this section.

When a spin-rigid body is in orbit, if there is a deviation between the angular momentum and the angular velocity, it will cause the rigid body to move freely, and the gravity gradient moment will also cause the forced precession of the rigid body. In the LP balance state, the rigid body must maintain the state of the ground, so it is necessary to balance the free precession of the rigid body, the forced precession and the rotational motion around the earth, so that the rigid body precesses at the orbital angular velocity. The LP balance condition of the above simplified rigid body model is deduced in the following.

The rigid body dynamic equation can be described as

$$\begin{aligned} A\dot{\omega}_1 - (A - C)\omega_2\omega_3 &= 3\mu_{\oplus}R^{-5}YZ(C - A) \\ A\dot{\omega}_2 - (C - A)\omega_1\omega_3 &= 3\mu_{\oplus}R^{-5}XZ(A - C) \\ C\dot{\omega}_3 &= 0 \end{aligned} \tag{6.45}$$

Equation (6.38) can be derived as

$$\begin{aligned} \dot{\omega}_1 &= \ddot{\phi}\sin\psi\sin\theta + \dot{\phi}\dot{\psi}\cos\psi\sin\theta + \dot{\phi}\dot{\theta}\sin\psi\cos\theta + \ddot{\theta}\cos\psi - \dot{\theta}\dot{\psi}\sin\psi \\ \dot{\omega}_2 &= \ddot{\phi}\cos\psi\sin\theta - \dot{\phi}\dot{\psi}\sin\psi\sin\theta + \dot{\phi}\dot{\theta}\cos\psi\cos\theta - \ddot{\theta}\sin\psi - \dot{\theta}\dot{\psi}\cos\psi \\ \dot{\omega}_3 &= \ddot{\theta}\cos\theta - \dot{\phi}\dot{\theta}\sin\theta + \ddot{\psi} \end{aligned} \tag{6.46}$$

Based on Eq. (6.31), we have

$$\begin{aligned} X &= -R(svc\psi c\phi - svc\theta s\psi s\phi - cvc\psi s\phi - cvc\theta s\psi c\phi) \\ Y &= R(svs\psi c\phi + svc\theta c\psi s\phi - cvs\psi s\phi + cvc\theta c\psi c\phi) \\ Z &= -R(svs\theta s\phi + cvs\theta c\phi) \end{aligned} \tag{6.47}$$

where $s^* = \sin^*$, $c^* = \cos^*$.

In the LP equilibrium state, the orientation of the rigid body rotation axis $\hat{e}$ in the rotation orbital coordinate system should remain unchanged, and the cone angle θ should remain constant in the case of a circular orbit; in addition, the rate of ϕ should also remain unchanged, and equals to the orbital angular rate $\dot{v}$; rigid body spin angular rate $\dot{\psi}$ should be constant. When $\phi = v$, $S = \dot{\psi}/\dot{v}$, LP balance conditions can be expressed as

$$\begin{aligned} &\theta = const &\quad \phi &= v \\ &\dot{\theta} = 0 &\quad \dot{\phi} &= \dot{v} = const \\ &\ddot{\theta} = 0 &\quad \ddot{\phi} &= \ddot{v} = 0 \end{aligned} \tag{6.48}$$

$$\begin{aligned} \dot{\psi} &= S\dot{v} = const \\ \ddot{\psi} &= 0 \\ \psi &= S\dot{v}t + \psi_0 \end{aligned} \tag{6.49}$$

where t is time, ψ_0 is the initial value of ψ.

Therefore, the expression of X, Y, Z can be simplified as

$$\begin{aligned} X &= -R(svc\psi cv - s^2vc\theta s\psi - cvc\psi sv - c^2vc\theta s\psi) \\ &= R(s^2vc\theta s\psi + c^2vc\theta s\psi) = R(c\theta s\psi) \end{aligned} \tag{6.50}$$

$$\begin{aligned} Y &= R\left(svs\psi cv + s^2vc\theta c\psi - cvs\psi sv + c^2vc\theta c\psi\right) \\ &= R(s^2vc\theta c\psi + c^2vc\theta c\psi) = R(c\theta c\psi) \end{aligned} \tag{6.51}$$

$$Z = -R(s^2vs\theta + c^2vs\theta) = -Rs\theta \tag{6.52}$$

Substituting Eqs. (6.48) and (6.49) into Eqs. (6.38) and (6.46) yield:

$$\begin{aligned} \omega_1 &= \dot{v}\sin(S\dot{v}t)\sin\theta \\ \omega_2 &= \dot{v}\cos(S\dot{v}t)\sin\theta \\ \omega_3 &= \dot{v}\cos\theta + S\dot{v} \end{aligned} \tag{6.53}$$

$$\begin{aligned}\dot{\omega}_1 &= S\dot{v}^2\cos(S\dot{v}t)\sin\theta\\ \dot{\omega}_2 &= -S\dot{v}^2\sin(S\dot{v}t)\sin\theta\\ \dot{\omega}_3 &= 0\end{aligned} \tag{6.54}$$

Substituting the newly obtained Eqs. (6.48)–(6.54) into the first equation in (6.45), there are

$$\dot{v}^2\left(\frac{\mathrm{CS}}{c\theta(C-A)}+4\right)=0 \tag{6.55}$$

Solving Eq. (6.55), there are

$$\dot{v}=0 \tag{6.56}$$

or

$$\frac{\mathrm{CS}}{c\theta(C-A)}+4=0 \tag{6.57}$$

The first solution does not meet the actual situation and is ignored. Solving the second case further obtains the LP balance condition as

$$\mathrm{S}=\frac{\dot{\psi}}{\dot{v}}=\frac{4\cos\theta(A-C)}{\mathrm{C}} \tag{6.58}$$

Similarly, for the second formula in (6.45), there are

$$\dot{v}^2\left(\frac{\mathrm{CS}}{c\theta(C-A)}+4\right)=0 \tag{6.59}$$

The result obtained by solving the above equation is the same as Eq. (6.57). Therefore, simplifying the rigid body's LP equilibrium conditions is

$$\mathrm{S}=\frac{\dot{\psi}}{\dot{v}}=\frac{4\cos\theta(A-C)}{\mathrm{C}} \tag{6.60}$$

In order to analyze the stability of the abovementioned equilibrium state, the linear stability analysis was carried out with reference to the method of [19].

The following form of the fourth-order characteristic polynomial is given in [19]:

$$\lambda^4+2b\lambda^2+c=0 \tag{6.61}$$

Its eigenvalue is

$$\lambda^2 = \left[-b \pm (b^2 - c)^{\frac{1}{2}}\right] \tag{6.62}$$

where

$$b = \frac{7 + 3\frac{C}{A}\left[3\left(\frac{C}{A} - 1\right)\cos^2\theta - 2\right]}{2} \tag{6.63}$$

$$c = 3\left(1 - \frac{C}{A}\right)\left(4 - 3\frac{C}{A}\right)\sin^2\theta \tag{6.64}$$

If λ^2 is a complex number or a positive real number, the real part is positive and the equilibrium is unstable. So there are three situations that can lead to instability:

(1) when $c<0$, λ^2 is a positive real number;
(2) when $(b^2 - c)<0$, λ^2 is plural;
(3) when $b<0$, the value of c will result in λ^2 a complex number or a positive real number.

Only when the above three conditions do not appear, the equilibrium will be stable.

Let $K = \frac{C}{A}$, $c<0$ in case (1) can be expressed as

$$3(1 - K)(4 - 3K)\sin^2\theta<0 \tag{6.65}$$

Assuming there are $1<K<2$ for cylindrical cylinders, then $(1 - K)<0$; when $\theta \neq 0, \pi$, then, $\sin^2\theta > 0$. Therefore, the unstable conditions in case (1) are

$$K<\frac{4}{3} \tag{6.66}$$

For case (3), there is

$$b = \frac{7 + 3\frac{C}{A}\left[3\left(\frac{C}{A} - 1\right)\cos^2\theta - 2\right]}{2}<0 \tag{6.67}$$

It can be simplified as

$$|\cos\theta| < \sqrt{\frac{6K - 7}{9K(K - 1)}} \tag{6.68}$$

namely,

$$|\theta| > \arccos\left(\sqrt{\frac{6K-7}{9K(K-1)}}\right) \tag{6.69}$$

For case (2), there are

$$\frac{(7+3K[3(K-1)\cos^2\theta - 2])^2}{4} - 3(1-K)(4-3K)(1-\cos^2\theta) < 0 \tag{6.70}$$

The above equation can be rewritten as

$$x\cos^4\theta + y\cos^2\theta + z < 0 \tag{6.71}$$

where

$$\begin{aligned} x &= \frac{(9K(K-1))^2}{4} \\ y &= \frac{(7-6K)9K(K-1)}{2} - 3(K-1)(4-3K) \\ z &= \frac{(7-6K)^2}{4} + 3(K-1)(4-3K) \end{aligned} \tag{6.72}$$

Equation (6.71) can be rewritten as

$$x(\cos^2 - \gamma)(\cos^2 - \delta) < 0 \tag{6.73}$$

where

$$\begin{aligned} \delta &= \frac{-y \pm \sqrt{y^2 - 4xz}}{2x} \\ \gamma &= \frac{z}{x\delta} \end{aligned} \tag{6.74}$$

Because of $x > 0$, there are

$$|\cos\theta| < \sqrt{\gamma}, |\cos\theta| > \sqrt{\delta} \tag{6.75}$$

or

$$|\cos\theta| > \sqrt{\gamma}, |\cos\theta| < \sqrt{\delta} \tag{6.76}$$

Without loss of generality, assume that $\gamma \leq \delta$, the unstable conditions under case (2) are

$$|\theta| < \arccos(\sqrt{\gamma}), |\theta| > \arccos(\sqrt{\delta}) \tag{6.77}$$

Assume that the axially symmetric oblong cylinder has $K = 2$, and the stability condition for the equilibrium state obtained by substituting three kinds of instability conditions is

$$|\theta| < 46.434^\circ \tag{6.78}$$

The LP equilibrium conditions for the on-orbit flight of a rigid body are deduced from the LP equilibrium conditions of the cylindrically symmetrical oblong-cylindrical rigid body deduced above.

From Eq. (6.60), we can see that ψ and ϕ do not have significant sums. Therefore, their initial values are relatively arbitrary, but they need to take into account the influence of v. Because the value of ϕ determines the value of v; θ is the cone angle. However, there is a singularity in the kinetic equation when it is zero, which does not apply to the LP equilibrium but corresponds to the Thomson equilibrium.

From Eq. (6.39), we can see that $\vec{\omega}^{bi}$ is a function of Euler angles and Euler angle velocities. Therefore, the initial value of Euler angles can be calculated simply by obtaining the initial value of Euler angles.

Since θ should remain unchanged in the LP balance condition and the orbital coordinate system should have the same rotation speed as the orbital angular velocity, there are

$$\dot{\theta}_0 = 0 \tag{6.79}$$

$$\dot{\phi}_0 = \dot{v}_0 \tag{6.80}$$

Finally, the initial values obtained from LP balance conditions are

$$\dot{\psi}_0 = \dot{v}_0 \cos\theta_0 \frac{4(A - C)}{C} \tag{6.81}$$

We have obtained a simplified model of the space multi-rope system—the LP equilibrium condition of an axisymmetric rectangular cylindrical rigid body and the LP equilibrium initial conditions of on-orbit flight. The accuracy of the model can be verified by numerical simulation below. And whether the simplified model can obtain the LP equilibrium state under the initial condition of LP balance.

6.1.4 Numerical Simulation

Some simulation parameters are shown in Table 6.1. Other parameters will be given in each simulation scenario. The simulation environment is MATLAB and ODE45 integrator is used.

In order to be able to analyze and verify intuitively and reliably, we expect to get the following simulation graphics:

(1) The variation of $\vec{H}$ and $\vec{\omega}$ in the body coordinate system $\hat{b}$;
(2) The variation of $\vec{H}$, $\vec{\omega}$ and $\hat{b}_3$ axis in the inertial coordinate system $\hat{\imath}$;
(3) The variation of $\hat{b}_3$ axis in the orbital coordinate system $\hat{e}$;
(4) The variation of the angle change between $\vec{H}$, $\vec{\omega}$ and $\hat{b}_3$ axis;
(5) The variation of state variable $\vec{X}$.

Simulation scenario 1: Gravity gradient moment ignored
In addition to the simulation parameters in Table 6.1, there are $\dot{\phi}_0 = 0.2\,\text{rad}/s$, $\dot{\theta}_0 = 0°$, $\dot{\psi}_0 = 0.2\,\text{rad}/s$. The simulation time is 50 s, and the simulation step length is 0.1. The simulation results are shown in Figs. 6.5, 6.6, 6.7.

Figure 6.5 shows the variation of $\vec{H}$, $\vec{\omega}$ and $\hat{b}_3$ axis in the inertial coordinate system. It can be seen that the nondimensionalized angular momentum $\vec{H}$ is always fixed in the inertial coordinate system and the orientation is always the same. $\vec{H}$, $\vec{\omega}$ and $\hat{b}_3$ are in the same plane. $\vec{\omega}$ and $\hat{b}_3$ axis is moving in a circular motion around $\vec{H}$. This is because the rigid body is affected by the gravity gradient moment and there is free precession. Similar to the axisymmetric spin satellite, $\vec{\omega}$ will make two conical motions simultaneously. Due to $\vec{H}$ is always fixed, the space cone also stays in the space keep fixed, the rigid body's attitude movement is that the rigid body rotates around the spin axis, and the body cone rolls on the space cone. The tangents of these two cones are the direction of the angular velocity.

Figure 6.6 shows the angle change between $\vec{H}$, $\vec{\omega}$ and $\hat{b}_3$ axis. It can be seen from the figure that the angle between them is always the same, verifying that in

Table 6.1 Simulation parameter

Parameters	Value
Orbit height	3000 km
ν_0	0°
ϕ_0	0°
θ_0	28°
ψ_0	0°
$\dot{\nu}$	0.0006952 rad/s
$\mu_\oplus$	$3.986004415 \times 10^8\ \text{m}^3/\text{s}^2$
A, C	25, 50

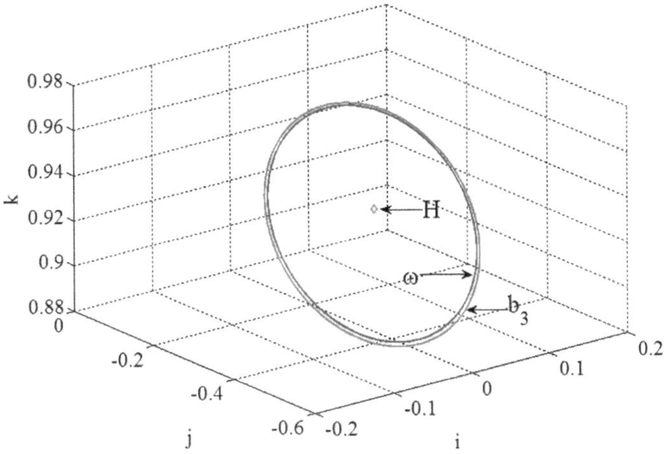

Fig. 6.5 The variation of $\vec{H}$, $\vec{\omega}$ and $\hat{b}_3$ in inertial coordinate system

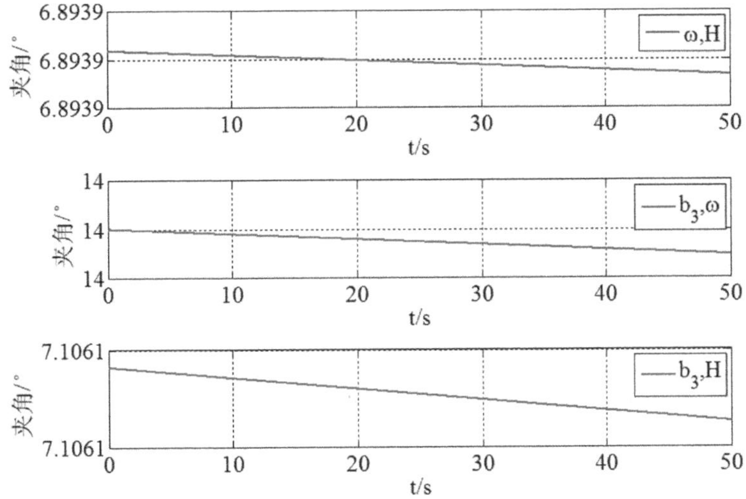

Fig. 6.6 The variation of angles between $\vec{H}$, $\vec{\omega}$ and $\hat{b}_3$ axis

Fig. 6.5. Figure 6.7 shows the changes of six state variables. It can be seen from the figure that the angular velocity changes periodically.

Therefore, we can judge that the simulation results in this simulation situation are in line with the actual situation, and the model of the cylindrically symmetrical cylindrical body with rough symmetry is available.

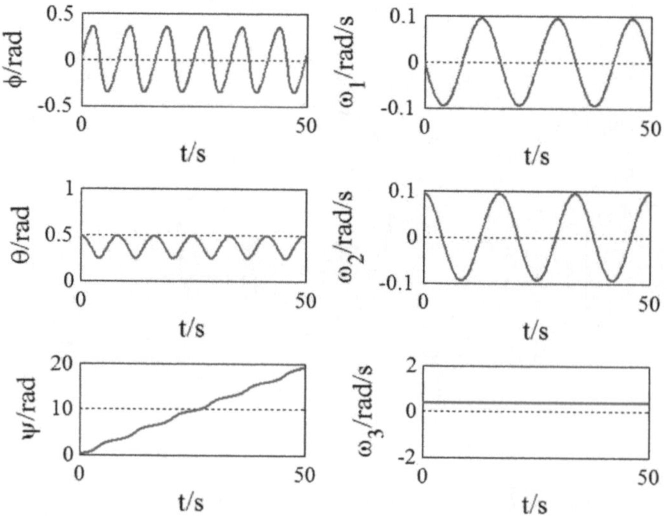

Fig. 6.7 The variation of state variables

Simulation scenario 2: Adding a gravity gradient moment effect
Simulation scenario 1 verifies the correctness of the model. In this simulation scenario, we will continue to verify whether the initial condition of the rigid body equilibrium deduced in the previous section is correct.

Assume that the initial condition of rigid body motion is the initial LP balance condition deduced in the above section. When the effect of gravity gradient moment is added to the single rigid body dynamic model, we want to know if the rigid body can achieve LP equilibrium. The simulation parameters still use the data in Table 6.1. The other parameters are $\dot{\phi}_0 = \dot{v}_0$, $\dot{\theta}_0 = 0\,\text{rad}/s$, $\dot{\psi}_0$ calculated from the LP balance condition. The simulation time is 1 track period and the simulation step length is 10. The simulation result is shown in Fig. 6.8.

Figure 6.8a shows the variation of $\vec{H}$, $\vec{\omega}$ and $\hat{b}_3$ axis in the inertial coordinate system. It can be seen that when the gravity gradient moment is added in the single rigid body model, $\vec{H}$ will no longer remain fixed in the inertial coordinate system, and keep in a stable periodic circular motion as $\vec{\omega}$ and $\hat{b}_3$ around $\hat{k}$ axis. Therefore, the rigid body not only has a free precession, but also a forced precession. In addition, the radius of the circular trajectory of these three elements from the largest to the smallest is $\vec{H}$, $\vec{\omega}$ and $\hat{b}_3$. This is mainly related to the size of A, C and θ_0.

Figure 6.8b shows the change of the angle between $\vec{H}$, $\vec{\omega}$ and $\hat{b}_3$ axis. It can be seen that the angle between them remains constant. The angle between $\vec{\omega}$ and $\hat{b}_3$ axis is always 28°, the same as θ_0. The angle between $\vec{\omega}$ and $\vec{H}$ is maintained as 13.112°, and the angle between $\hat{b}_3$ and $\vec{H}$ is maintained as 14.888°.

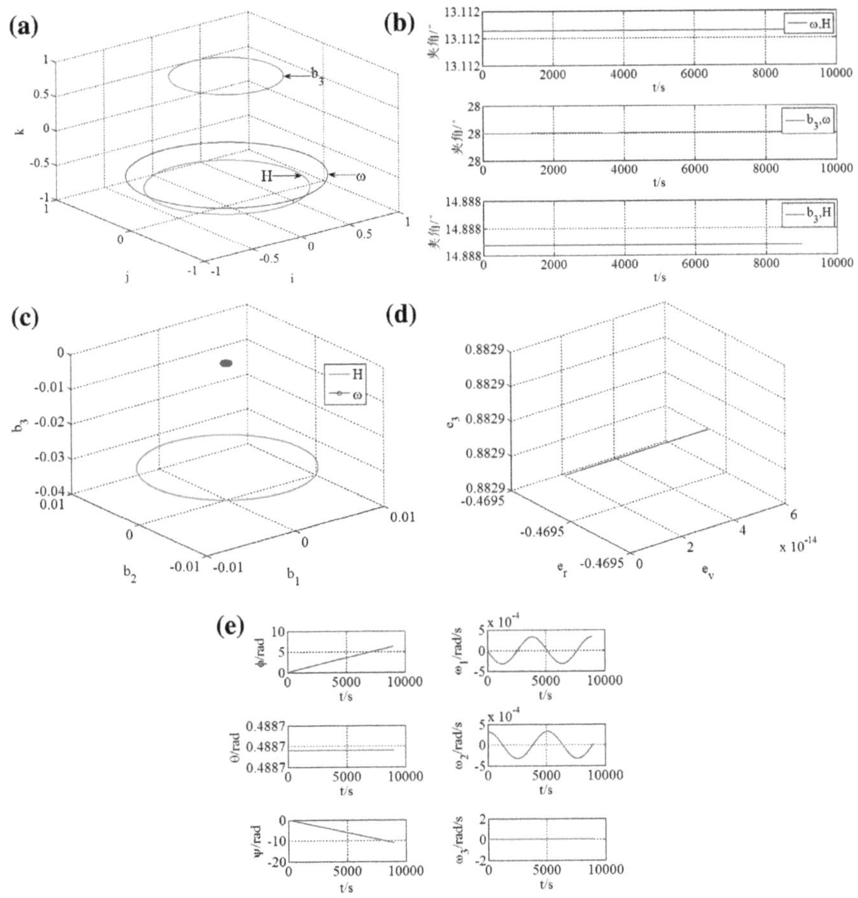

Fig. 6.8 Simulation results of scenario 2

Figure 6.8c shows that the variation of $\overrightarrow{H}$, $\vec{\omega}$ in the body coordinate system. It can be seen that $\vec{\omega}$ is fixed in the body coordinate system and $\overrightarrow{H}$ can rotate stably around the axis $\hat{b}_3$.

Figure 6.8d shows the variation of the $\hat{b}_3$ axis in the orbital coordinate system. It can be seen that the $\hat{b}_3$ axis remains fixed and the magnitude of the error in $\hat{e}_v$ axis is only 10^{-14}.

Figure 6.8e shows the variation of state variables. The value of θ remains unchanged and $\vec{\omega}$ changes periodically. Then, the existence of LP equilibrium is verified.

Based on the above analysis, it can be concluded that the LP balance initial conditions deduced in the above section can stabilize the rigid body in the LP equilibrium state.

The simulation results of the comprehensive simulation scenarios 1 and 2 show that the dynamic model of the simplified model of the space multiline system established in Sect. 6.1.2 is effective and can better reflect the basic dynamics in a Kepler's circular orbit. In addition, the LP balance initial conditions deduced in Sect. 6.1.3 allow the rigid body to achieve a stable LP equilibrium.

6.2 Closed-Chain Formation System

6.2.1 Coordinate Description

Space multi-tethered system can be divided into three basic configurations, which are ring, hub-and-spoke and pyramid configuration, depending on the geometry configuration, and the other complex ones could all be deduced from the above three basic configurations. In this section, we mainly focus on the typical three-body ring tethered system. The earth-observing mission concept is illustrated in Fig. 6.9.

The three-body ring tethered system in Fig. 6.9 consists of three spacecraft and three tethers, which are connected end to end. It is an ideal configuration is an equilateral triangle as shown in Fig. 6.10.

In Fig. 6.10, m_1, m_2 and m_3 denote the mass of the three spacecrafts. ρ_{12} is the length of the connected tether between m_1 and m_2; ρ_{23} is the length of the connected tether between m_2 and m_3; ρ_{31} is the length of the connected tether between m_3 and m_1.

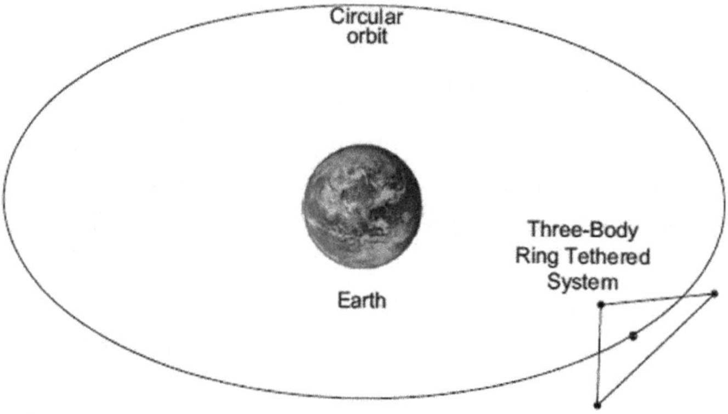

Fig. 6.9 The earth-observing mission concept

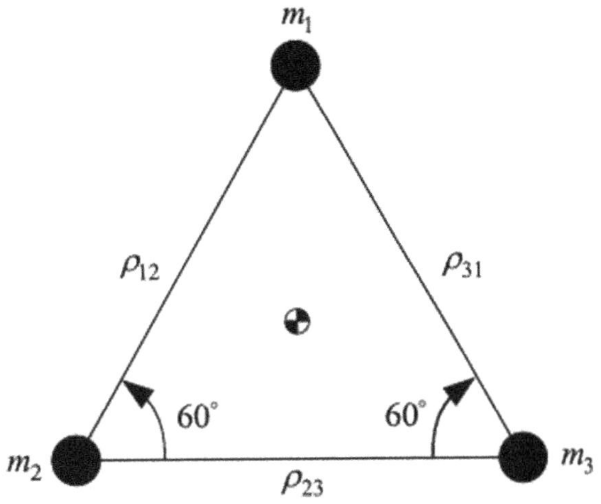

Fig. 6.10 Ideal configuration of three-body ring tethered system

6.2.2 Dynamics Modeling

Some assumptions for dynamic modeling convenience are made here.

(1) The three spacecraft are assumed as mass points with the same mass. Hence, point i means the ith spacecraft.
(2) The connected tethers are all flexible, but massless.
(3) The tethered system has been completely deployed and the initial length of the three tethers are same and the tethers are in tension and straight.
(4) The mass center of the tethered system is in a Kepler circular orbit.

The reference coordinate systems used during the modeling are given in Fig. 6.11.

As shown in Fig. 6.11, $\hat{i}$ is the inertial frame with its origin O_e located at the mass center of the earth; $\hat{i}_1 - \hat{i}_2$ plane is the orbital plane. $\hat{e}$ is the orbital reference frame with the origin O_s located at the mass center of the tethered system; $\hat{e}_1$-axis is along with $\boldsymbol{O_eO_s}$, $\hat{e}_2$-axis is along with the flying direction of the system, and $\hat{e}_3$-axis is perpendicular with the orbital plane and point upwards; ν is the angel going from $\hat{i}_1$-axis to $\boldsymbol{O_eO_s}$. $\hat{b}$ is the body frame transformed from $\hat{e}$ with $\hat{b}_1$-axis always pointing to mass point 1; α is the angel going from $\hat{e}_2$-axis to $\hat{b}_2$-axis and β is the angel going from the orbital plane to $\hat{b}_1$-axis. $\hat{s}$ frame is transformed from $\hat{e}$ by rotating around $\hat{e}_2$-axis with an angel $-\beta$; $\hat{s}_3$-axis is always pointing to a fixed point in the inertial frame and the tethered system is always spinning in $\hat{s}_1 - \hat{s}_2$ plane.

The transition matrix from $\hat{e}$ to $\hat{b}$ is

$$C^{be} = R_2(-\beta)R_3(\alpha) \tag{6.82}$$

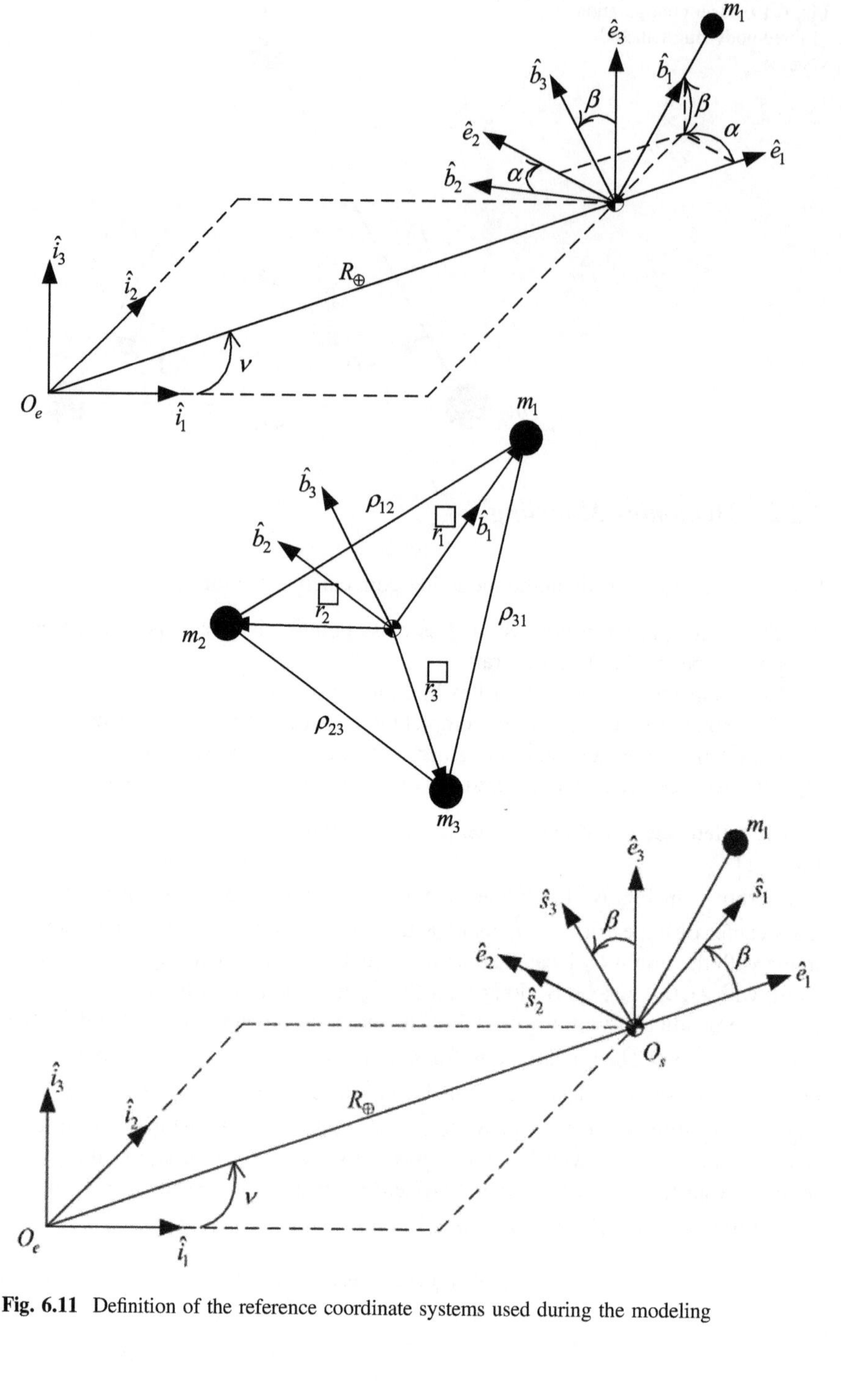

Fig. 6.11 Definition of the reference coordinate systems used during the modeling

And the one from $\hat{i}$ to $\hat{b}$ is

$$C^{bi} = R_2(-\beta)R_3(\alpha + v) \tag{6.83}$$

For convenience, we assume the state vector to be:

$$\mathbf{X} = \{\mathbf{R}_1\ \mathbf{V}_1\ \mathbf{R}_2\ \mathbf{V}_2\ \mathbf{R}_3\ \mathbf{V}_3\}^T \tag{6.84}$$

which consists of the position vectors and velocity vectors of three mass points in the inertial frame and has eighteen components. Because $\dot{\mathbf{R}}_i = \mathbf{V}_i$, we still need another nine differential equations to complete the state equations. The constraint conditions now we have are: the mass center of the tethered system is in a Kepler circular orbit; the tethered system spins in the $\hat{s}_1 - \hat{s}_2$ plane; the three tethers have the same initial length. Thus, the differential equations we needed decline to five after excluding the four constraint equations. These five differential equations also correspond to five independent general variables that are more visualized: α, β, v, $\mathbf{R}_\oplus$, ρ, where ρ is the length of the connected tether, $\mathbf{R}_\oplus$ is the position vector of O_s in the inertial frame. It is noticeable that these five general variables and the state components can be transformed into each other.

We know from Fig. 6.11 that the relative position vectors of the three mass points with respect to O_s in the body frame are

$$^{\hat{b}}\mathbf{r}_1 = \begin{bmatrix} \frac{\sqrt{3}}{3}\rho & 0 & 0 \end{bmatrix}^T \tag{6.85}$$

$$^{\hat{b}}\mathbf{r}_2 = \begin{bmatrix} \frac{-\sqrt{3}}{6}\rho & \frac{\rho}{2} & 0 \end{bmatrix}^T \tag{6.86}$$

$$^{\hat{b}}\mathbf{r}_3 = \begin{bmatrix} \frac{-\sqrt{3}}{6}\rho & \frac{-\rho}{2} & 0 \end{bmatrix}^T \tag{6.87}$$

Then, one can be written as

$$^{\hat{i}}\mathbf{R}_i = {}^{\hat{i}}\mathbf{R}_\oplus + C^{ib\hat{b}}\mathbf{r}_i \tag{6.88}$$

$$^{\hat{i}}\mathbf{V}_i = {}^{\hat{i}}\mathbf{V}_\oplus + {}^{\hat{i}}\mathbf{v}_i \tag{6.89}$$

where $^{\hat{i}}\mathbf{v}_i$ is the variation rate of $\mathbf{r}_i$ in the inertial frame, and:

$$^{\hat{i}}\mathbf{v}_i = {}^{\hat{i}}\frac{d}{dt}{}^{\hat{b}}\mathbf{r}_i = {}^{\hat{b}}\frac{d}{dt}{}^{\hat{b}}\mathbf{r}_i + {}^{\hat{b}}\boldsymbol{\omega}_{bi} \times {}^{\hat{b}}\mathbf{r}_i \tag{6.90}$$

where $\boldsymbol{\omega}_{bi}$ is the angular rate of O_s in the inertial frame and can be written in three frames:

$$\boldsymbol{\omega}_{bi} = \dot{v}\hat{i}_3 + \dot{\alpha}\hat{e}_3 + \dot{\beta}\hat{b}_2 \tag{6.91}$$

When expressed in the body frame, we have

$$^{\hat{b}}\boldsymbol{\omega}_{bi} = \mathrm{C}^{bi}\begin{bmatrix} 0 \\ 0 \\ \dot{v} \end{bmatrix} + \mathrm{C}^{be}\begin{bmatrix} 0 \\ 0 \\ \dot{\alpha} \end{bmatrix} + \begin{bmatrix} 0 \\ \dot{\beta} \\ 0 \end{bmatrix} \tag{6.92}$$

where $\dot{v}$ is the orbital angular rate. We know that the angular momentum of the tethered system is

$$h = \left| {}^{\hat{i}}\mathbf{R}_{\oplus} \times {}^{\hat{i}}\mathbf{V}_{\oplus} \right| = \dot{v}\left| {}^{\hat{i}}\mathbf{R}_{\oplus} \right|^2 \tag{6.93}$$

Thus:

$$\dot{v} = \frac{\left| {}^{\hat{i}}\mathbf{R}_{\oplus} \times {}^{\hat{i}}\mathbf{V}_{\oplus} \right|}{\left| {}^{\hat{i}}\mathbf{R}_{\oplus} \right|^2} \tag{6.94}$$

The geometric meaning of the vectors used above are shown in Fig. 6.12, where $\mathbf{R}_i$ going from O_e to point i, $\mathbf{r}_i$ going from O_s to point i and $\mathbf{r}_{ij}$ going from point i to point j.

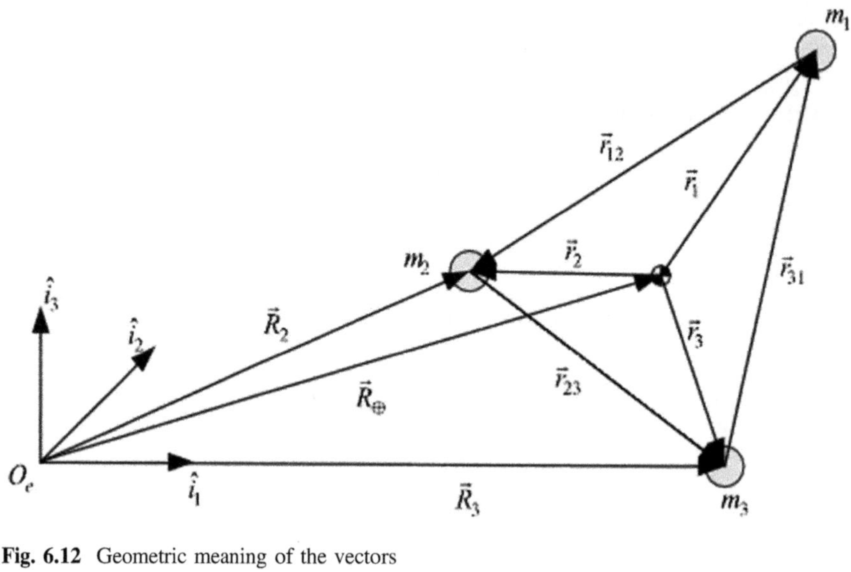

Fig. 6.12 Geometric meaning of the vectors

Assume $\dot{\mathbf{X}} = f(\mathbf{X}, t)$, we have

$$\dot{\mathbf{R}}_i = \mathbf{V}_i \tag{6.95}$$

$$\dot{\mathbf{V}}_i = \ddot{\mathbf{R}}_i \tag{6.96}$$

where $\ddot{\mathbf{R}}_i$ includes the acceleration terms induced by the gravity and the connected tethers, and the gravity acceleration is

$$\dot{\mathbf{R}}_{i_{gravity}} = \frac{-\mu_{\oplus}}{|\mathbf{R}_i|^3}\mathbf{R}_i \tag{6.97}$$

Since the tethers are assumed to be massless and flexible, they can be seemed as viscously damped springs that can only hold tension. Similar to the conclusion in single tether system, the tension in each tether is given by

$$T_{ij} = \begin{cases} K_s(\rho_{ij} - \rho_o) + \mu_d \dot{\rho}_{ij} & (\rho_{ij} - \rho_o) \geq 0 \\ 0 & (\rho_{ij} - \rho_o) \leq 0 \end{cases} \tag{6.98}$$

Then the acceleration induced by the tethers is

$$\dot{\mathbf{R}}_{i_{tether}} = \frac{\delta_{nm} T}{m_i}\left(\frac{\mathbf{r}_{ij}}{\rho_{ij}}\right) \tag{6.99}$$

where K_s is the tether stiffness, μ_d is the tether damping coefficient, ρ_o is the natural length of the tether, ρ_{ij} is the actual length of the tether, $\dot{\rho}_{ij}$ is the variation rate of ρ_{ij}, and:

$$\mathbf{r}_{ij} = {}^{\hat{i}}\mathbf{R}_j - {}^{\hat{i}}\mathbf{R}_i \tag{6.100}$$

$$\rho_{ij} = |\mathbf{r}_{ij}| \tag{6.101}$$

$$\dot{\rho}_{ij} = \frac{\mathbf{r}_{ij} \cdot \mathbf{V}_{ij}}{\rho_{ij}} \tag{6.102}$$

where $\mathbf{V}_{ij}$ is the relative velocity from point j to point i and can be written as

$$\mathbf{V}_{ij} = {}^{\hat{i}}\mathbf{V}_j - {}^{\hat{i}}\mathbf{V}_i \tag{6.103}$$

In addition, δ_{nm} denotes the direction of the tether tension acting on point i:

$$\delta_{nm} = \begin{cases} 1 & n = i \\ -1 & m = i \\ 0 & otherwise \end{cases} \tag{6.104}$$

where $n \neq m$.

Therefore, the whole state equations of the three-body ring tethered system are

$$\dot{\mathbf{R}}_1 = \mathbf{V}_1 \tag{6.105}$$

$$\dot{\mathbf{R}}_2 = \mathbf{V}_2 \tag{6.106}$$

$$\dot{\mathbf{R}}_3 = \mathbf{V}_3 \tag{6.107}$$

$$\begin{aligned}\dot{\mathbf{V}}_1 = & \frac{-\mu_\oplus}{|\mathbf{R}_1|^3}\mathbf{R}_1 + \frac{K_s(\rho_{12}-\rho_o)}{m_1}\left(\frac{\mathbf{r}_{12}}{\rho_{12}}\right) - \frac{K_s(\rho_{31}-\rho_o)}{m_1}\left(\frac{\mathbf{r}_{31}}{\rho_{31}}\right) \\ & + \frac{\mu_d\dot{\rho}_{12}}{m_1}\left(\frac{\mathbf{r}_{12}}{\rho_{12}}\right) - \frac{\mu_d\dot{\rho}_{31}}{m_1}\left(\frac{\mathbf{r}_{31}}{\rho_{31}}\right)\end{aligned} \tag{6.108}$$

$$\begin{aligned}\dot{\mathbf{V}}_2 = & \frac{-\mu_\oplus}{|\mathbf{R}_2|^3}\mathbf{R}_2 + \frac{K_s(\rho_{23}-\rho_o)}{m_2}\left(\frac{\mathbf{r}_{23}}{\rho_{23}}\right) - \frac{K_s(\rho_{12}-\rho_o)}{m_2}\left(\frac{\mathbf{r}_{12}}{\rho_{12}}\right) \\ & + \frac{\mu_d\dot{\rho}_{23}}{m_2}\left(\frac{\mathbf{r}_{23}}{\rho_{23}}\right) - \frac{\mu_d\dot{\rho}_{12}}{m_2}\left(\frac{\mathbf{r}_{12}}{\rho_{12}}\right)\end{aligned} \tag{6.109}$$

$$\begin{aligned}\dot{\mathbf{V}}_3 = & \frac{-\mu_\oplus}{|\mathbf{R}_3|^3}\mathbf{R}_3 + \frac{K_s(\rho_{31}-\rho_o)}{m_3}\left(\frac{\mathbf{r}_{31}}{\rho_{31}}\right) - \frac{K_s(\rho_{23}-\rho_o)}{m_3}\left(\frac{\mathbf{r}_{23}}{\rho_{23}}\right) \\ & + \frac{\mu_d\dot{\rho}_{31}}{m_3}\left(\frac{\mathbf{r}_{31}}{\rho_{31}}\right) - \frac{\mu_d\dot{\rho}_{23}}{m_3}\left(\frac{\mathbf{r}_{23}}{\rho_{23}}\right)\end{aligned} \tag{6.110}$$

6.2.3 LP Stability

First, we simplify the tethered system into an oblate cylinder using the virtual structure method [17] and assume that one of the three mass points are in the highest position and O_s is in the $\hat{b}_1$-axis (Fig. 6.13).

Thus, the initial values of the five general variables are

$$^{\hat{i}}\mathbf{R}_\oplus = \mathrm{X}_\oplus \hat{i}_1$$

$$^{\hat{i}}\mathbf{V}_\oplus = \upsilon = \sqrt{\frac{\mu_\oplus}{|^{\hat{i}}\mathbf{R}_\oplus|}}\hat{i}_2$$

$$\rho \text{ is arbitrary}$$

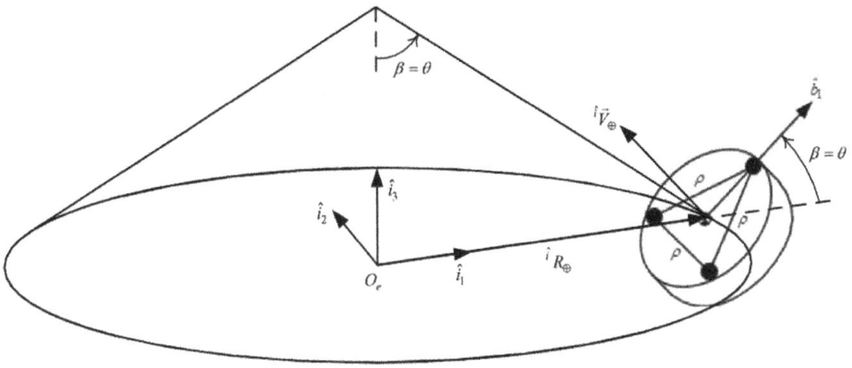

Fig. 6.13 LP equilibrium of the three-body ring tethered system

$$\alpha = 0$$
$$\beta = \theta \tag{6.111}$$

After that, we could obtain the initial values of $\mathbf{R}_i$ depending on Eq. (6.88).

The initial angular velocity of the simplified cylinder expressed in the inertial frame is

$$\boldsymbol{\omega}^{bi} = \dot{\psi}\hat{b}_3 + \dot{\upsilon}\hat{i}_3 = \dot{\psi}\hat{b}_3 + \dot{\upsilon}\hat{e}_3 \tag{6.112}$$

which can be denoted in the body frame:

$$^{b}\boldsymbol{\omega}_{bi} = \begin{bmatrix} 0 \\ 0 \\ \dot{\psi} \end{bmatrix} + \mathbf{C}^{be}\begin{bmatrix} 0 \\ 0 \\ \dot{\upsilon} \end{bmatrix} \tag{6.113}$$

From the LP Equilibrium condition of the simplified cylinder [20], we know that:

$$\dot{\psi} = \dot{\upsilon}\cos\theta\frac{4(\mathrm{A}-\mathrm{C})}{\mathrm{C}} \tag{6.114}$$

However, because of the tether's flexibility, a fixed $\dot{\psi}$ will generate a new steady-state tether length, ρ_{ss}, which will change the moment of inertia of the rigid body, A and C. Subsequently, they will induce a new $\dot{\psi}$ depending on Eq. (6.114). Therefore, to get the actual and steady initial value of $\dot{\psi}$, the iteration process in Fig. 6.14 should be executed until the value of ρ_{ss} at this time almost equals with that in the next time. Afterward, $\dot{\psi}_{ss}$, $\boldsymbol{\omega}^{bi}$ and $\mathbf{V}_i$ could be determined in turn.

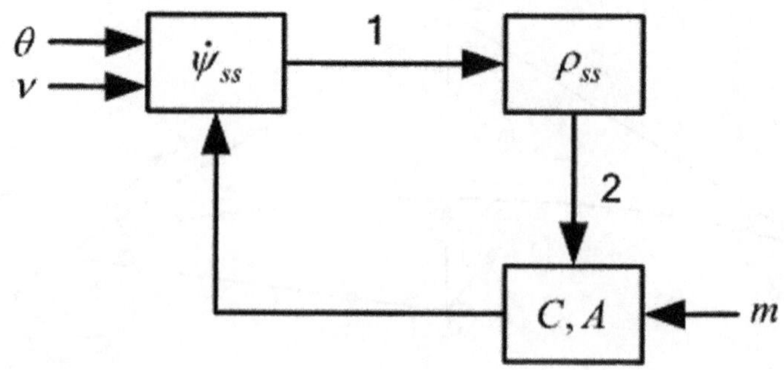

Fig. 6.14 Iteration process to compute $\dot{\psi}_{ss}$

In Fig. 6.14, process 1 is to compute ρ_{ss} from $\dot{\psi}_{ss}$ by balancing the centrifugal force and the tether tension:

$$\frac{m}{3}\rho_{ss}(\dot{\psi}_{ss} + (\dot{\upsilon}\hat{e}_3 \cdot \hat{b}_3))^2 = K_s(\rho_{ss} - \rho_o) \tag{6.115}$$

So ρ_{ss} can be obtained as

$$\rho_{ss} = \frac{K_s\rho_o}{K_s - \frac{m}{3}(\psi_{ss} + (\dot{\upsilon}\hat{e}_3 \cdot \hat{b}_3))^2} \tag{6.116}$$

Process 2 is to compute A and C from ρ_{ss}:

$$\mathrm{C}_{sys} = 3m(\mathrm{r}_{ss})^2 = m\rho_{ss}^2 \tag{6.117}$$

$$\mathrm{A}_{ss} = m(\mathrm{r}_{ss})^2\binom{n}{2} \quad n > 2 \tag{6.118}$$

where n is the number of the mass points and $n = 3$ in this section.
Hence,

$$\mathrm{A}_{ss} = m(\mathrm{r}_{ss})^2\binom{3}{2} = \frac{1}{2}m\rho_{ss}^2 \tag{6.119}$$

We should notice that:

$$\frac{\mathrm{C}_{sys}}{\mathrm{A}_{sys}} = 2 \tag{6.120}$$

As so far, we have obtained the state equations and the initial value of **X**, and then some dynamic analysis about the three-body ring tethered system could be performed by numerical simulation.

6.2.4 Numerical Simulation

To make the analysis more effective, we would like to transform the state values into some other interested variables, for example, ${}^{\hat{i}}\mathbf{R}_{\oplus}$, ${}^{\hat{i}}\mathbf{V}_{\oplus}$, ${}^{\hat{i}}\mathbf{h}$, and so on.

The formulas to compute ${}^{\hat{i}}\mathbf{R}_{\oplus}$ and ${}^{\hat{i}}\mathbf{V}_{\oplus}$ are

$$ {}^{\hat{i}}\mathbf{R}_{\oplus} = \mathbf{R}_1 \frac{m_1}{\sum_i m_i} + \mathbf{R}_2 \frac{m_2}{\sum_i m_i} + \mathbf{R}_3 \frac{m_3}{\sum_i m_i} \tag{6.121} $$

$$ {}^{\hat{i}}\mathbf{V}_{\oplus} = \mathbf{V}_1 \frac{m_1}{\sum_i m_i} + \mathbf{V}_2 \frac{m_2}{\sum_i m_i} + \mathbf{V}_3 \frac{m_3}{\sum_i m_i} \tag{6.122} $$

Then, ${}^{\hat{i}}\mathbf{r}_i$ and ${}^{\hat{i}\hat{i}_1}\mathbf{v}$ could be given by

$$ {}^{\hat{i}}\mathbf{r}_i = {}^{\hat{i}}\mathbf{R}_i - {}^{\hat{i}}\mathbf{R}_{\oplus} \tag{6.123} $$

$$ {}^{\hat{i}\hat{i}_1}\mathbf{v} = {}^{\hat{i}\hat{i}_1}\mathbf{V} - {}^{\hat{i}\hat{i}_{\oplus}}\mathbf{V} \tag{6.124} $$

In addition, the angular momentum of the tethered system expressed in the inertial frame is

$$ {}^{\hat{i}}\mathbf{h} = \sum_i {}^{\hat{i}}\mathbf{r}_i \times m_i {}^{\hat{i}\hat{i}_1}\mathbf{v} = m \sum_i {}^{\hat{i}}\mathbf{r}_i \times {}^{\hat{i}\hat{i}_1}\mathbf{v} \tag{6.125} $$

which denotes the rotation axis's direction of the tethered system and should keep constant in ideal equilibrium state.

Besides, we could also obtain ρ_{ij} and $\dot{\rho}_{ij}$ depending on Eqs. (6.100) and (6.102), respectively.

We find that these interesting variables obtained above are all expressed in the inertial frame and will be helpful to analyze the absolute dynamics of the tethered system. However, we have to transform them into the $\hat{e}$ and $\hat{s}$ frames to analyze the relative dynamics. The detailed transmission process is as follows.

When it is transformed into the $\hat{e}$ frame, we have

$$ {}^{\hat{e}}\mathbf{r}_i = \mathbf{C}^{ei}\, {}^{\hat{i}}\mathbf{r}_i $$

$$ {}^{\hat{e}}\mathbf{h}_i = \mathbf{C}^{ei}\, {}^{\hat{i}}\mathbf{h}_i $$

$$ \upsilon = \tan^{-1}\left(\frac{\mathrm{Y}_{\oplus}}{\mathrm{X}_{\oplus}}\right) \tag{6.126} $$

where

$$C^{ei} = R_3(\upsilon) \tag{6.127}$$

When transformed into the **s** frame, there have

$$\begin{aligned} {}^{\hat{s}}\mathbf{r}_i &= C^{si}\,{}^{\hat{i}}\mathbf{r}_i \\ {}^{\hat{s}}\mathbf{h}_i &= C^{si}\,{}^{\hat{i}}\mathbf{h}_i \end{aligned} \tag{6.128}$$

where

$$C^{si} = R_2(-\beta_0)R_3(\upsilon) \tag{6.129}$$

where β_0 is the initial cone angle.

The parameters used in the simulation are shown in Table 6.2.

The initial state of the three-body ring tethered system in the $\hat{e}$ frame corresponding to the initial parameters is shown in Fig. 6.15.

The simulation period is 10 orbits, the step size is 1 s, and the results are shown in Figs. 6.16, 6.17 and 6.18.

Figure 6.16a, b are, respectively, the variation trend of ${}^{\hat{e}}\mathbf{r}_i$ and ${}^{\hat{s}}\mathbf{r}_i$. We could see that the configuration of the tethered system breaks down quickly after the simulation starts and the system can not spin in the $\hat{s}_1 - \hat{s}_2$ plane steadily. Figure 6.16c presents the nondimensional ${}^{\hat{s}}\mathbf{h}_i$ and (d) shows its three components. One could find that the direction of ${}^{\hat{s}}\mathbf{h}_i$ would not keep fixed during the simulation with h_2 oscillating near zero value and h_3 near 1. On the whole, the pointing direction of the tethered system cannot be maintained. Figure 6.16e shows the variation trend of ρ_{ij}, from which we find that the three connected tethers all slacken seriously. Thus, they cannot hold up the configuration of the system. Besides, we notice that there is a fixed sequence during the slackening process. The first one to slacken is ρ_{23}, which

Table 6.2 Simulation parameters

Parameters	Value
$\mu_\oplus$	$3.986004415 \times 10^8\ \mathrm{m^3/s^2}$
m_i	200 kg
K_s	20.0 Nm
μ_d	0.05 kg/s
ρ_o	10.0 km
ρ_0	ρ_o
α_0	$0°$
β_0	$28°$
${}^{\hat{i}}\vec{R}_\oplus$	$(6378.13655 + 3000)\hat{i}_1$
${}^{\hat{i}}\vec{V}_\oplus$	$\sqrt{\frac{\mu_\oplus}{\lvert {}^{\hat{i}}\vec{R}_\oplus \rvert}}\hat{i}_2$

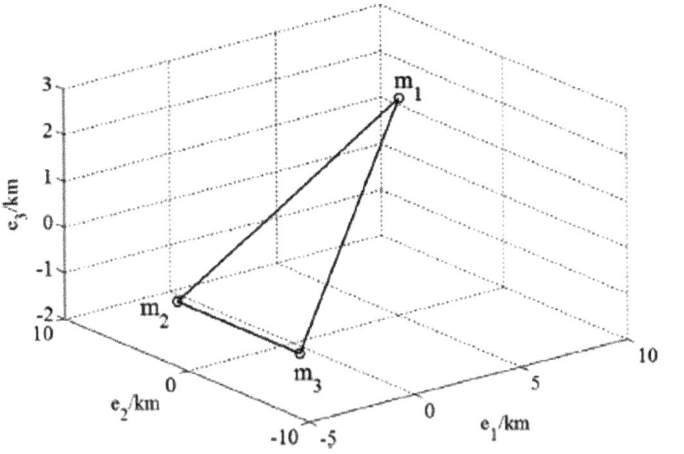

Fig. 6.15 Initial state of the tethered system

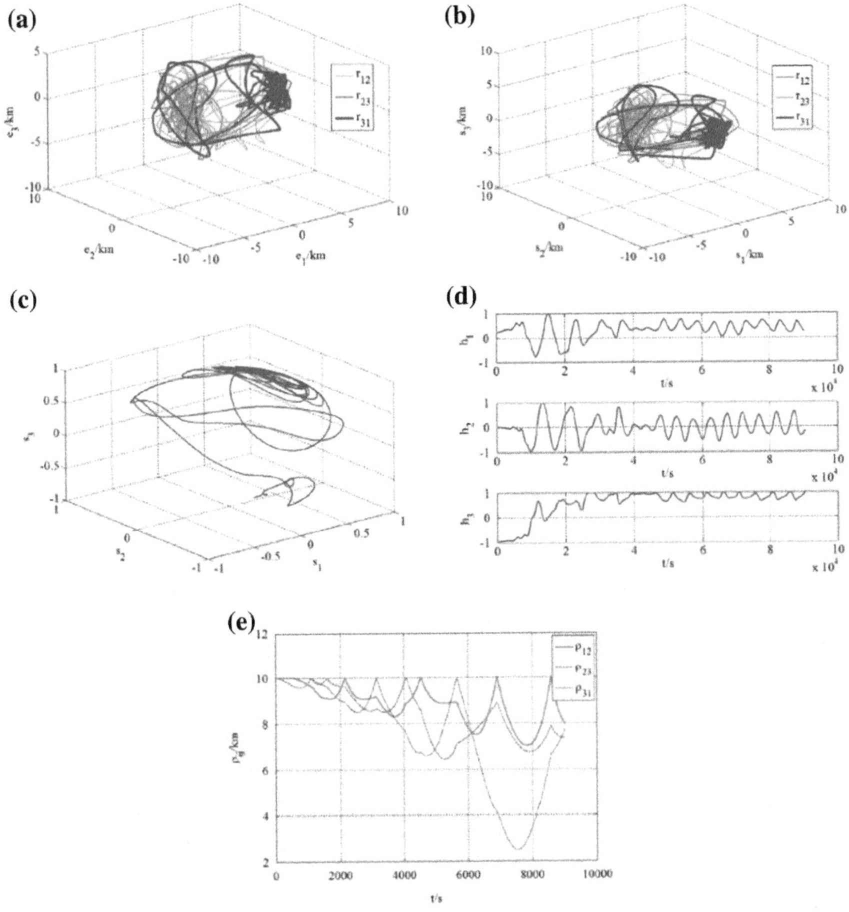

Fig. 6.16 Simulation results in uncontrolled situation

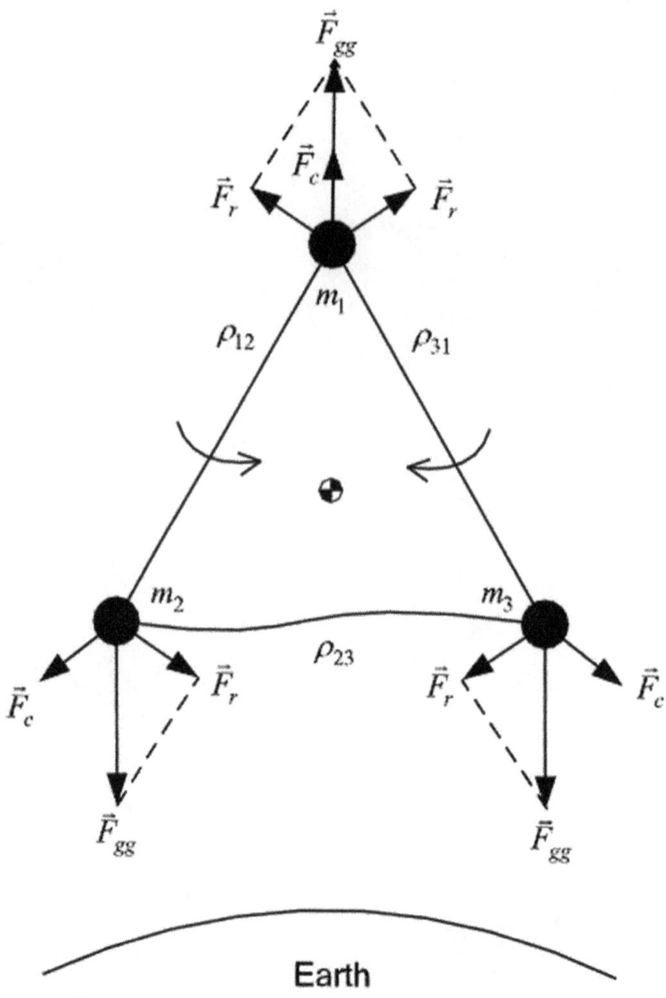

Fig. 6.17 Explanation about the tether slackening phenomenon

happens almost at the same time the system begins spinning. The later ones are ρ_{12} and ρ_{31}, which is related to the spin rate and will be explained in detail in Fig. 6.17.

We assume that $\beta = 0°$ for convenience in Fig. 6.17, which presents the initial state of the tethered system. One can find that ρ_{23} is completely in horizon when the simulation starts. F_c is the centrifugal force acting on the spacecraft arising from the tethered system's self-spinning. F_r is the component of the spacecraft's gravity in the direction perpendicular to the tether line and always exists in pairs on the two spacecraft connected by one tether. F_r will lead to the clockwise rotation of ρ_{31} and anticlockwise rotation of ρ_{12}, which will provoke ρ_{23}'s first to slacken, and the next one to slacken will also be the horizontal one. In Fig. 6.17, the horizontal component of F_c acting on spacecraft 2 or 3 must be bigger than that of F_r to avoid

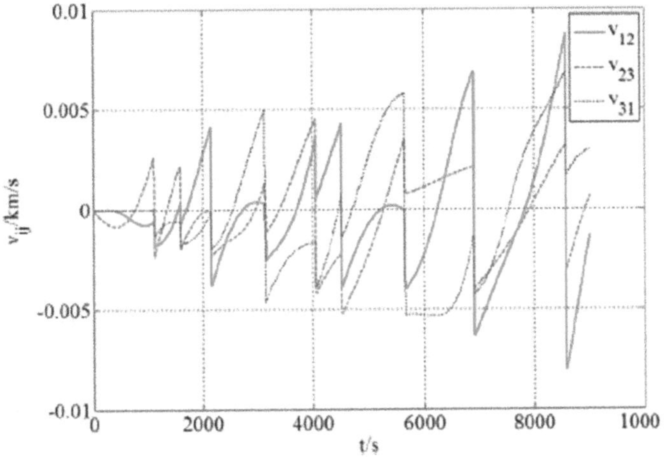

Fig. 6.18 Variation rates of the tethers' length

slackening. The increasing of F_c commonly means the increase of the system's spin rate. We just propose the lower limit of the spin rate here.

According to the analysis above, we have

$$F_c \sin(60^\circ) > F_r \sin(60^\circ) \tag{6.130}$$

After simplification:

$$F_c > F_r \tag{6.131}$$

But:

$$F_r = F_{gg} \sin(30^\circ) = 3Lm\dot{\upsilon}^2 \sin(30^\circ) \tag{6.132}$$

where $\dot{\upsilon}$ is the orbital angular rate, m is the mass of the spacecraft, and $L = \rho/2$ is the distance between the spacecraft and the tether's midpoint. When it is at the end of the tether, we have

$$F_r = 3\rho m\dot{\upsilon}^2 \sin(30^\circ) \tag{6.133}$$

F_c is

$$F_c = mr(\dot{\psi} + \dot{\upsilon})^2 = m\frac{\rho}{\sqrt{3}}(\dot{\psi} + \dot{\upsilon})^2 \tag{6.134}$$

Substitute F_c and F_r into Eq. (6.131):

$$m\frac{\rho}{\sqrt{3}}(\dot{\psi}+\dot{\upsilon})^2 > 3\rho m\dot{\upsilon}^2 \sin(30^\circ) \tag{6.135}$$

We obtain:

$$\frac{\dot{\psi}}{\dot{\upsilon}} < -2.61 \tag{6.136}$$

or

$$\frac{\dot{\psi}}{\dot{\upsilon}} > 0.61 \tag{6.137}$$

In the simulation case there are $\dot{\psi} \approx -0.00123$ rad/s and $\dot{\upsilon} \approx 6.9517 \times 10^{-4}$ rad/s, so:

$$\frac{\dot{\psi}}{\dot{\upsilon}} = -1.766 \tag{6.138}$$

which was not included in Eq. (6.136) or Eq. (6.137). Therefore, when the tethered system rotates in the direction shown in Fig. 6.17, ρ_{23} will be the first to slacken and deviate from horizon but will have a nonzero component in the vertical direction, which will lead to the redeployment of ρ_{23}. However, when the tether stretch to the normal length ρ_o, it will rebound right now and return to the slackening state. ρ_{12} and ρ_{31} will also trap in this phenomenon one after another, which can be seen in Figs. 6.18 and 6.19.

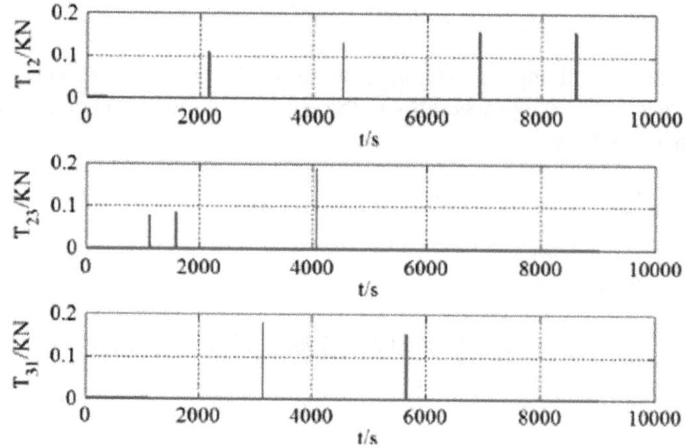

Fig. 6.19 Tensions in the tethers

The tensions in the tether are presented in Fig. 6.19. One can find that the tensions almost remain zero during the simulation except some 100 N level impulse forces at the moment when the tether rebounds suddenly. In general, the three tethers in the tethered system almost always keep in slack when $\beta \neq 0°$, and cannot help to maintain the configuration of the system.

In summary, the three-body ring tethered system cannot achieve steady equilibrium under the LP equilibrium initial conditions when $\beta \neq 0°$. Instead, the configuration of the system will break down because of the tether's slackening and rebound. Therefore, active control strategies are urgently needed to maintain the configuration, which will be discussed in the following section.

6.3 Hub–Spoke Formation System

The Hub–Spoke configuration is provided with a particular trait: many formation control operations can be achieved by the master spacecraft independently, including deployment, retraction, and rotation of the whole formation system. This trait contributes two significant advantages compared with other configurations: simplification of the formation control, and reduction of fuel consumption of the sub-spacecraft. Therefore, the Hub–Spoke configuration has been widely used in Tethered Formation Systems.

6.3.1 Coordinate Description

It is very common to employ suitable low-order models of a dynamic system for control system design. Because of the flexibility of the tether, dynamics of tethered space system is governed by complicated nonlinear equations. Therefore, it is acceptable to derive the single-tethered system dynamic model as a rigid, inextensible model [3, 21]. And in some cases, based on this simplified model, the simulation results are good enough for analysis and controller design [22]. Thus, in order to study the deployment and retraction dynamics of the rotating Hub–Spoke Tethered Formation System, a relatively simple analytical model is derived. For simplification, only two-body Hub–Spoke Tethered Formation System is considered and the analytical results can be enlarged to the multibody situation.

The following assumptions are made for the analytical model of the two-body Hub–Spoke Tethered Formation System:

(1) The master spacecraft is a rigid hub, which can provide rotating torque. The sub-spacecraft are regarded as mass points.
(2) The radial tethers are assumed to be symmetrically straight relative to the center of the master spacecraft, and the centrifugal force acting on the sub-spacecraft results in the tether tension.

(3) The masses of the radial tethers are ignored.
(4) The gravity gradient and the elasticity in the radial tethers are neglected.
(5) The rotating motion is in a planar plane, and the out-of-plane motion of the system is ignored.
(6) Energy dissipation caused by deformation, friction, and environmental effects are neglected.

The two-body Hub–Spoke Tethered Formation System model considered in this paper is demonstrated in Fig. 6.20. O_0 denotes the center of the master spacecraft, and O_2 denotes the point from which stems the tether connected to sub-spacecraft1. Three coordinate systems are introduced, which are denoted by $O_0x_0y_0z_0$, $O_1x_1y_1z_1$ and $O_2x_2y_2z_2$. $O_1x_1y_1z_1$ is defined after the rotation from the $O_0x_0y_0z_0$ about the O_0z_0-axis by an angle θ. The O_2x_2 axis points positively outward from O_2 along the tether connected to sub-spacecraft1. The O_2y_2 axis is perpendicular to O_2x_2. The angular velocity of the master spacecraft is ω, and the $O_2x_2y_2$ is rotating around the master spacecraft with the same angular velocity of $\omega + \dot{\varphi}_1$. The length of the tethers connected to sub-spacecraft1 and sub-spacecratft2 are denoted by L_1 and L_2.

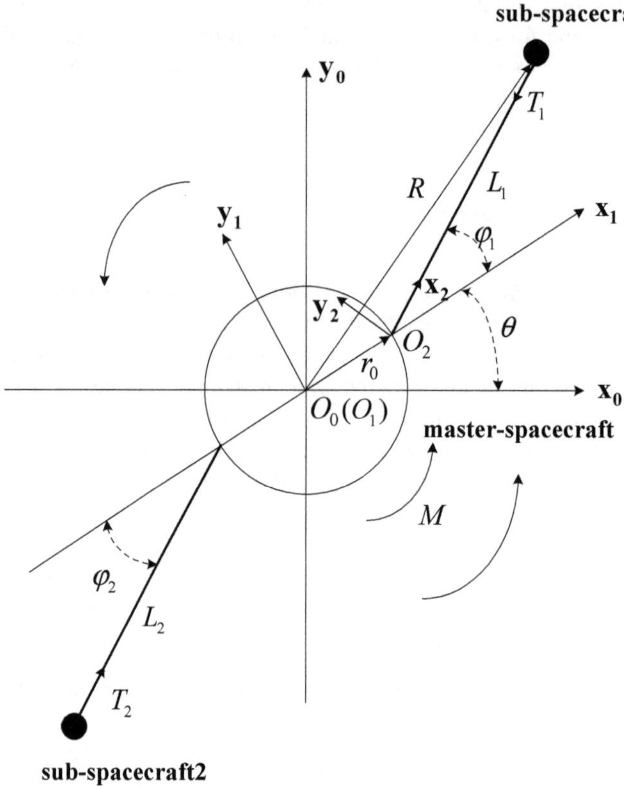

Fig. 6.20 The analytical model for the two-body HS-TFS

The tension force of the tethers connected to sub-spacecraft1 and sub-spacecratft2 are denoted by T_1 and T_2, which are generated by tether reel mechanism. The masses of sub-spacecraft1 and sub-spacecratft2 are denoted by m_1 and m_2. The r_0 denotes the rotating radius of the master spacecraft. The I denotes the moment of inertia of masterspacecraft. The torque of the master spacecraft is the control variable for the deployment and retraction control of the rotating Hub–Spoke Tethered Formation System in this paper, which is denoted by M.

6.3.2 Dynamics Modeling

The change in angular momentum for the central master spacecraft around its axis of rotation is

$$I\ddot{\theta} = I\dot{\omega} = r_0 T_1 \sin\varphi_1 + r_0 T_2 \sin\varphi_2 + M \tag{6.139}$$

where θ is the rotation angle of the master spacecraft, φ_1 and φ_2 are the angles between the straight tether's direction and radial direction. Because stiffness and damping are not included, the equation of motion for the sub-spacecraft1 can simply presented as follows:

$$\mathbf{F} = m_1\ddot{\mathbf{R}} \tag{6.140}$$

Its position $\mathbf{R}$ is presented in Fig. 6.20 and is expressed as

$$\mathbf{R} = \mathbf{r_0} + \mathbf{L_1} \tag{6.141}$$

where $\mathbf{L_1}$ is the vector of the tether connected master spacecraft and sub-spacecraft1. Then, the derivative of $\mathbf{R}$ become

$$\dot{\mathbf{R}} = \boldsymbol{\omega} \times \mathbf{r_0} + \dot{\mathbf{L}}_1 + \left(\boldsymbol{\omega} + \dot{\varphi}_1\mathbf{e}_z^{(2)}\right) \times \mathbf{L_1} \tag{6.142}$$

$$\begin{aligned}\ddot{\mathbf{R}} = {} & \dot{\boldsymbol{\omega}} \times \mathbf{r_0} + \boldsymbol{\omega} \times (\boldsymbol{\omega} \times \mathbf{r_0}) + \ddot{\mathbf{L}}_1 + 2\left(\boldsymbol{\omega} + \dot{\varphi}_1\mathbf{e}_z^{(2)}\right)\dot{\mathbf{L}}_1 \\ & + \left(\dot{\boldsymbol{\omega}} + \ddot{\varphi}_1\mathbf{e}_z^{(2)}\right) \times \mathbf{L_1} + \left(\boldsymbol{\omega} + \dot{\varphi}_1\mathbf{e}_z^{(2)}\right) \times \left(\left(\boldsymbol{\omega} + \dot{\varphi}_1\mathbf{e}_z^{(2)}\right) \times \mathbf{L_1}\right)\end{aligned} \tag{6.143}$$

where $\mathbf{L}_1'$ is the derivative of $\mathbf{L_1}$ and $\mathbf{e}_z^{(2)}$ is the unit vector of $O_2x_2y_2z_2$.

Projected in coordinate $O_2x_2y_2z_2$, the equations of motion become

$$\begin{cases} r_0\dot{\omega}\sin\varphi_1 - r_0\omega^2\cos\varphi_1 - L_1(\omega + \dot{\varphi}_1)^2 + \ddot{L}_1 = -T_1/m_1 \\ r_0\dot{\omega}\cos\varphi_1 + r_0\omega^2\sin\varphi_1 + L_1(\dot{\omega} + \ddot{\varphi}_1) + 2\dot{L}_1(\omega + \dot{\varphi}_1) = 0 \end{cases} \tag{6.144}$$

Similarly, the equations of motion for sub-spacecraft2 can be derived:

$$\begin{cases} r_0\dot{\omega}\sin\varphi_2 - r_0\omega^2\cos\varphi_2 - L_2(\omega+\dot{\varphi}_2)^2 + \ddot{L}_2 = -T_2/m_2 \\ r_0\dot{\omega}\cos\varphi_2 + r_0\omega^2\sin\varphi_2 + L_2(\dot{\omega}+\ddot{\varphi}_2) + 2\dot{L}_2(\omega+\dot{\varphi}_2) = 0 \end{cases} \tag{6.145}$$

Finally, the following equations of motion for the two-body Hub–Spoke Tethered Formation System are presented as follows:

$$\begin{cases} r_0\dot{\omega}\sin\varphi_1 - r_0\omega^2\cos\varphi_1 - L_1(\omega+\dot{\varphi}_1)^2 + \ddot{L}_1 = -T_1/m_1 \\ r_0\dot{\omega}\cos\varphi_1 + r_0\omega^2\sin\varphi_1 + L_1(\dot{\omega}+\ddot{\varphi}_1) + 2\dot{L}_1(\omega+\dot{\varphi}_1) = 0 \\ r_0\dot{\omega}\sin\varphi_2 - r_0\omega^2\cos\varphi_2 - L_2(\omega+\dot{\varphi}_2)^2 + \ddot{L}_2 = -T_2/m_2 \\ r_0\dot{\omega}\cos\varphi_2 + r_0\omega^2\sin\varphi_2 + L_2(\dot{\omega}+\ddot{\varphi}_2) + 2\dot{L}_2(\omega+\dot{\varphi}_2) = 0 \\ I\dot{\omega} = r_0T_1\sin\varphi_1 + r_0T_2\sin\varphi_2 + M \end{cases} \tag{6.146}$$

Furthermore, the equations of motion for multibody Hub–Spoke Tethered Formation System can be derived similarly:

$$\begin{cases} r_0\dot{\omega}\sin\varphi_1 - r_0\omega^2\cos\varphi_1 - L_1(\omega+\dot{\varphi}_1)^2 + \ddot{L}_1 = -T_1/m_1 \\ r_0\dot{\omega}\cos\varphi_1 + r_0\omega^2\sin\varphi_1 + L_1(\dot{\omega}+\ddot{\varphi}_1) + 2\dot{L}_1(\omega+\dot{\varphi}_1) = 0 \\ \vdots \\ \vdots \\ r_0\dot{\omega}\sin\varphi_n - r_0\omega^2\cos\varphi_n - L_n(\omega+\dot{\varphi}_n)^2 + \ddot{L}_n = -T_n/m_n \\ r_0\dot{\omega}\cos\varphi_n + r_0\omega^2\sin\varphi_n + L_n(\dot{\omega}+\ddot{\varphi}_n) + 2\dot{L}_n(\omega+\dot{\varphi}_n) = 0 \\ I\dot{\omega} = r_0T_1\sin\varphi_1 + \cdots + r_0T_n\sin\varphi_n + M \end{cases} \tag{6.147}$$

Similar model has been used for deployment control of rotating space webs [23–25]. Even though the Hub–Spoke Tethered Formation System and the space webs are different in reality, the mathematical formulations of the rotating motion are similar. The difference is that the length of the tethers in the Hub–Spoke Tethered Formation System is variable and controllable, while the length of the space webs is decided by the folding pattern and unchangeable.

6.3.3 Numerical Simulation

In this section, the rotating motion of a two-body Hub–Spoke tethered formation system is taken as an example, the effect of error of rotating angle on the formation configuration is simulated under different rotating velocities and tether lengths.

First, the state variable is

$$\mathbf{x} = [x_1\ x_2\ x_3\ x_4\ x_5]^{\mathrm{T}} = [\omega\ \varphi_1\ \varphi_2\ \dot{\varphi}_1\ \dot{\varphi}_2]^{\mathrm{T}} \tag{6.148}$$

The control amount:

$$\mathbf{u} = [u_1\ u_2\ u_3]^{\mathrm{T}} = [M\ F_{t1}\ F_{t2}]^{\mathrm{T}} \tag{6.149}$$

We can have

$$\dot{\mathbf{x}} = \begin{bmatrix} \frac{p_1 + p_2 + u_1}{I + m_1 r_0^2 \sin^2 x_2 + m_2 r_0^2 \sin^2 x_3} \\ x_4 \\ x_5 \\ -\dot{\omega} - \frac{r_0}{L_1}\left(\dot{\omega}\cos x_2 + x_1^2 \sin x_2\right) + \frac{u_2}{L_1 m_1} \\ -\dot{\omega} - \frac{r_0}{L_1}\left(\dot{\omega}\cos x_3 + x_1^2 \sin x_3\right) + \frac{u_3}{L_2 m_2} \end{bmatrix} \tag{6.150}$$

where

$$p_1 = m_1 r_0 \sin x_2 \left[r_0 x_1^2 \cos x_2 + L_1 (x_1 + x_4)^2\right] \tag{6.151}$$

$$p_2 = m_2 r_0 \sin x_3 \left[r_0 x_1^2 \cos x_3 + L_2 (x_1 + x_5)^2\right] \tag{6.152}$$

Based on Eq. (6.150), simulations are performed on the variation in formation configuration caused by different angles of rotation and different tether lengths. The values of the main parameters in the simulation are shown in Table 6.3.

Effect of rotating velocity

First of all, the effect of the error of rotating angle for formation configuration with different rotating velocity is simulated. The simulation scenario is designed as follows: In the initial state, the angular velocity of Hub–Spoke tethered formation system is ω_0. When the rotating angle error (φ and $\dot{\varphi}$) occurs, the rotating angle and angular velocity will change. Then, the formation system cannot maintain the ideal uniform spin motion and Hub–Spoke configuration. In the simulation, the initial error of the rotating angle is assumed to be:

$$[\varphi_{10}\ \varphi_{20}\ \dot{\varphi}_{10}\ \dot{\varphi}_{20}]^{\mathrm{T}} = [0.1\,\text{rad}\ 0.1\,\text{rad}\ 0.1\,\text{rad}\,\text{s}^{-1}\ 0.1\,\text{rad}\,\text{s}^{-1}]^{\mathrm{T}} \tag{6.153}$$

where φ_{10} and φ_{20} are the initial values of the rotating angle error respectively, and $\dot{\varphi}_{10}$ and $\dot{\varphi}_{20}$ are the initial values of the rotating angular velocity error, respectively.

Table 6.3 Simulation parameter

Parameters	Value
L_1	10.0 m
L_2	10.0 m
m_1	5.0 kg
m_2	5.0 kg
r_0	1.0 m
I	500.0 kg m^2

The simulation results are shown in Figs. 6.21, 6.22, 6.23, 6.24, and 6.25. From Figs. 6.21, 6.22, 6.23, 6.24, and 6.25, it can be seen that when there are errors in rotating angles, rotating velocity, the rotating angles and the angular velocities of the formation system deviate from the ideal values, and they oscillate periodically. This means that the rotating velocity cannot maintain constant, and the rotating angle cannot maintain an ideal zero value. The ideal uniform rotating motion and Hub–Spoke configuration of the formation system cannot be maintained.

In addition, it can be concluded from the simulation results that after the rotating angle error appears:

(1) The greater the initial value of the rotating velocity of the formation system, the higher the oscillating frequency of the rotating velocity;
(2) The larger the initial value of the rotating velocity of the formation system, the higher the oscillation frequency of rotating angle and the smaller the amplitude;
(3) The greater the initial value of the rotating velocity of the formation system, the higher the oscillating frequency of the angular velocity.

Effect of the tether length

The effect of the error of rotating angle for formation configuration with different tether length is simulated. The simulation scenario is designed as follows: In the initial state, the angular velocity of Hub–Spoke tethered formation system is ω_0. When the rotating angle error (φ and $\dot{\varphi}$) occurs, the rotating angle and angular velocity will change. Then, the formation system cannot maintain the ideal uniform spin motion and Hub–Spoke configuration. In the simulation, the initial error of the rotating angle is assumed to be:

$$\begin{bmatrix}\omega_0 & \varphi_{10} & \varphi_{20} & \dot{\varphi}_{10} & \dot{\varphi}_{20}\end{bmatrix}^{\mathrm{T}} = \begin{bmatrix}1.0\,\mathrm{rad\,s^{-1}} & 0.1\,\mathrm{rad} & 0.1\,\mathrm{rad} & 0.1\,\mathrm{rad\,s^{-1}} & 0.1\,\mathrm{rad\,s^{-1}}\end{bmatrix}^{\mathrm{T}} \tag{6.154}$$

Fig. 6.21 Variation of ω

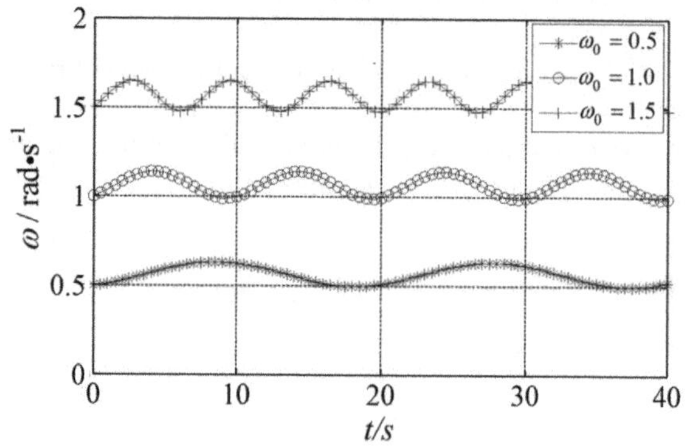

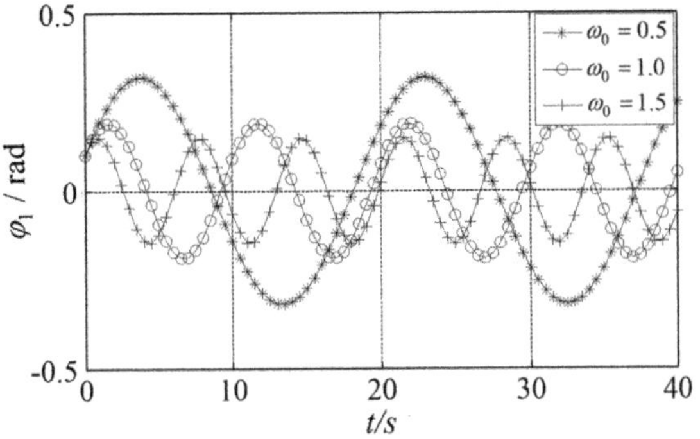

Fig. 6.22 Variation of φ_1

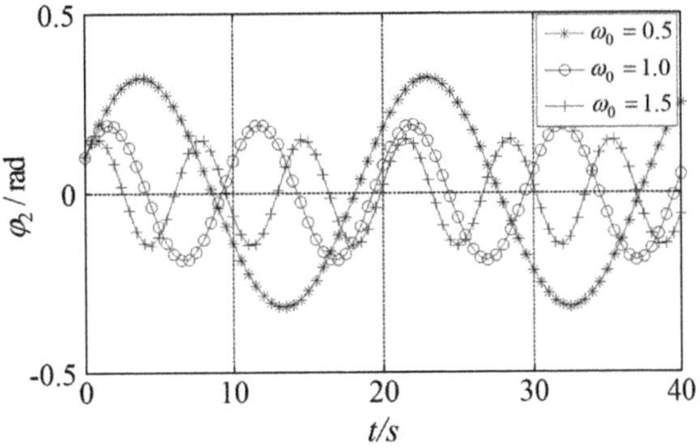

Fig. 6.23 Variation of φ_2

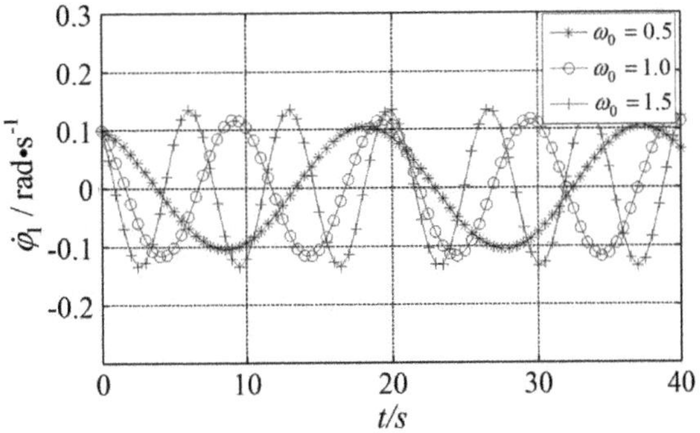

Fig. 6.24 Variation of $\dot{\varphi}_1$

Fig. 6.25 Variation of $\dot{\varphi}_2$

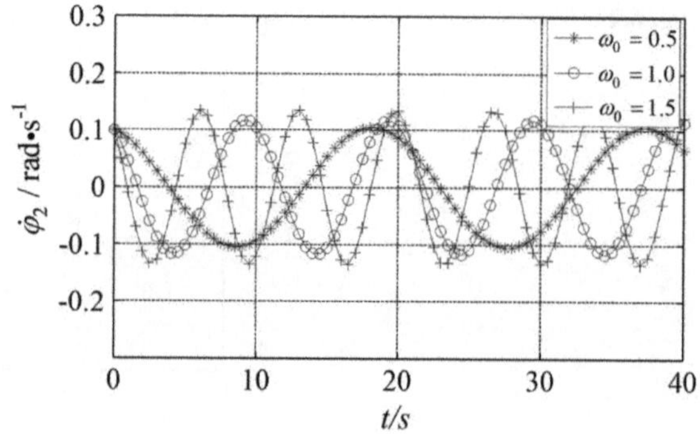

Fig. 6.26 Variation of ω

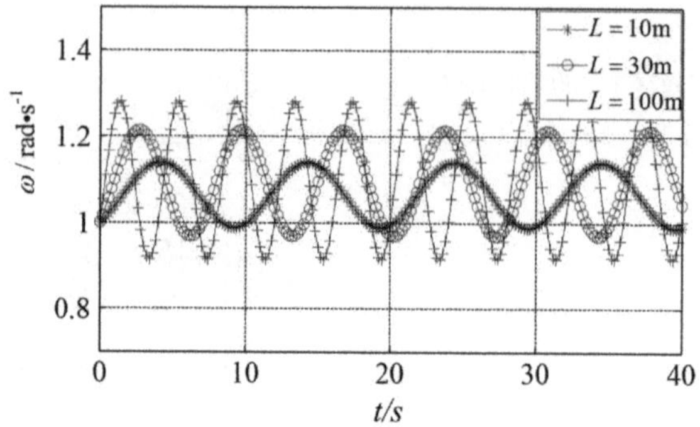

Fig. 6.27 Variation of φ_1

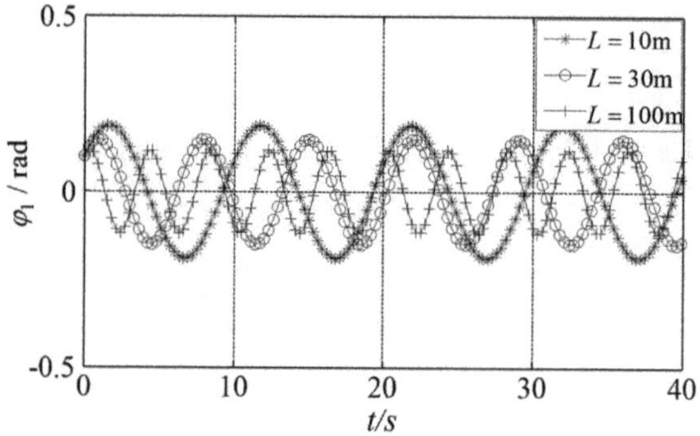

where φ_{10} and φ_{20} are the initial values of the rotating angle error, respectively, and $\dot{\varphi}_{10}$ and $\dot{\varphi}_{20}$ are the initial values of the rotating angular velocity error, respectively.

The simulation results are shown in Figs. 6.26, 6.27, 6.28, 6.29, and 6.30. From Figs. 6.26, 6.27, 6.28, 6.29, and 6.30, it can be seen that when there are errors in tether length, rotating velocity, the rotating angles and the angular velocities of the

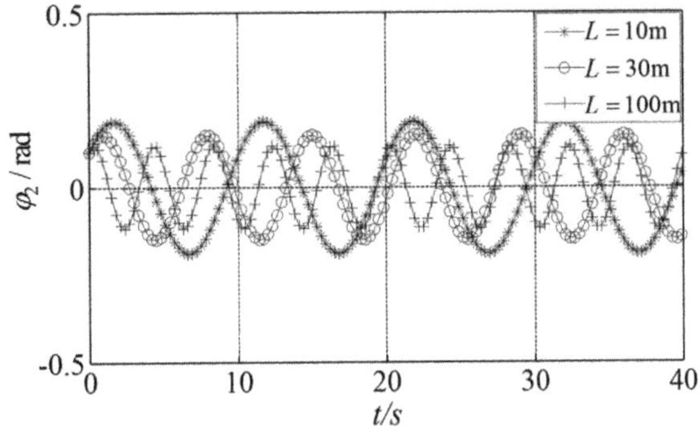

Fig. 6.28 Variation of φ_2

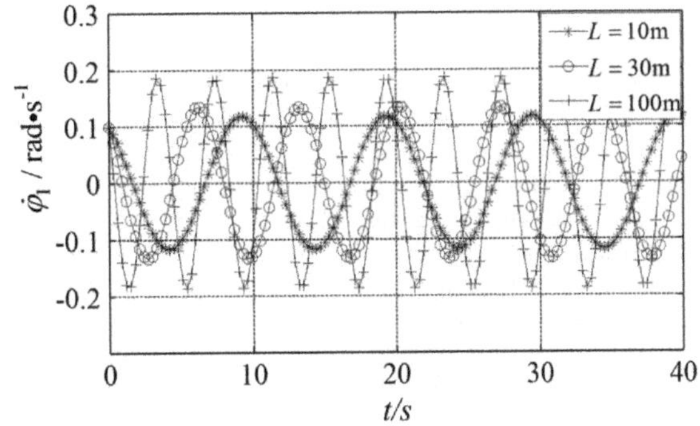

Fig. 6.29 Variation of $\dot{\varphi}_1$

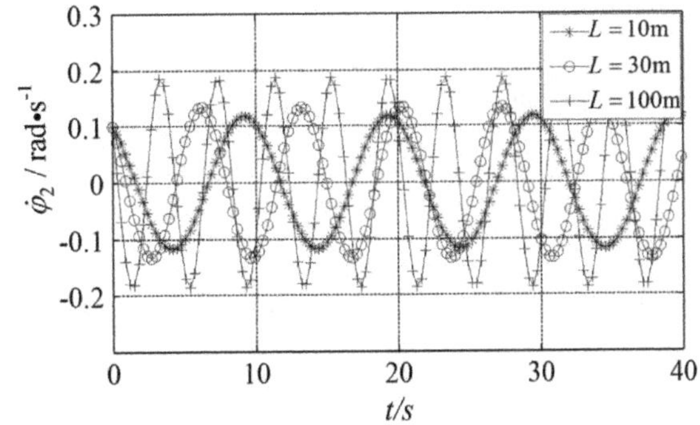

Fig. 6.30 Variation of $\dot{\varphi}_2$

formation system deviate from the ideal values, and they oscillate periodically. It can be summarized as

(1) The greater the initial value of the tether length of the formation system, the higher the oscillating frequency of the rotating velocity;

(2) The larger the initial value of the tether length of the formation system, the higher the oscillation frequency of error of rotating angle and the smaller the amplitude;
(3) The greater the initial value of the tether length of the formation system, the higher the oscillating frequency of the error of angular velocity and the great the amplitude.

References

1. Xu W, Peng J, Liang B et al (2016) Hybrid modeling and analysis method for dynamic coupling of space robots. IEEE Trans Aerosp Electron Syst 52(1):85–98
2. Peng J, Xu W, Wang Z et al (2013) Dynamic analysis of the compounded system formed by dual-arm space robot and the captured target. In: 2013 IEEE international conference on robotics and biomimetics (ROBIO). IEEE
3. Wen H, Zhu Z, Jin D et al (2015) Space tether deployment control with explicit tension constraint and saturation function. J Guid Control Dyn 39(4):916–921
4. Wen H, Zhu Z, Jin D et al (2016) Tension control of space tether via online quasi-linearization iterations. Adv Space Res 57(3):754–763
5. Wen H, Zhu Z, Jin D et al (2016) Constrained tension control of a tethered space-tug system with only length measurement. Acta Astronaut 119:110–117
6. Nakaya K, Matunaga S (2005) On attitude maneuver of spinning tethered formation flying based on virtual structure method. In: AIAA guidance, navigation, and control conference
7. Williams P (2006) Optimal deployment/retrieval of a tethered formation spinning in the orbital plane. J Spacecr Rocket 43(3):638–650
8. Pizarro-Chong A, Misra AK (2008) Dynamics of multi-tethered satellite formations containing a parent body. Acta Astronaut 63(11):1188–1202
9. Misra AK, Amier Z, Modi VJ (1988) Attitude dynamics of three-body tethered systems. Acta Astronaut 17(10):1059–1068
10. Keshmiri M, Misra AK, Modi VJ (1996) General formulation for n-body tethered satellite system dynamics. J Guid Control Dyn 19(1):75–83
11. Lorenzini EC (1987) A three-mass tethered system for micro-g/variable-g applications. J Guid Control Dyn 10(3):242–249
12. Lorenzini EC, Cosmo M, Vetrella S et al (1988) Acceleration levels on board the space station and a tethered elevator for micro and variable-gravity applications. Space Tethers Sci Space Stn Era 1:513–522
13. Breakwell JV (1981) Stability of an orbiting ring. J Guid Control Dyn 4(2):197–200
14. Beletsky VV, Levin EM (1985) Stability of a ring of connected satellites. Acta Astronaut 12 (10):765–769
15. Menon C, Bombardelli C, Bianchini G (2005) Spinning tethered formation with self-stabilising attitude control. International Astronautical Congress, Fukuoka, Japan
16. Pengelley CD (1966) Preliminary survey of dynamic stability of a cable-connected spinning space station. J Spacecr Rocket 3(10):1456–1462
17. Nakaya K, Matunaga S (2005) On attitude maneuver of spinning tethered formation flying based on virtual structure method. In: Proceedings of the AIAA guidance, navigation, and control conference and exhibit. San Francisco, California
18. Likins PW (1973) Elements of engineering mechanics. McGraw-Hill Book Company, New York
19. Likins PW (1965) Stability of a symmetrical satellite in attitudes fixed in an orbiting reference frame. J Astronaut Sci 12(1):18–24

20. Krieger G, Moreira A (2006) Spaceborne bi- and multistatic SAR: potentials and challenges. Proc Geosci Remote Sens Symp 153(3):184–198
21. Sun G, Zhu Z (2014) Fractional order tension control for stable and fast tethered satellite retrieval. Acta Astronaut 104(1):304–312
22. Sun G, Zhu Z (2014) Fractional-order tension control law for deployment of space tether system. J Guid Control Dyn 37(6):2057–2062
23. Huang P, Zhang F, Ma J et al (2015) Dynamics and configuration control of the maneuvering-net space robot system. Adv Space Res 55(4):1004–1014
24. Zhai G, Qiu Y, Liang B et al (2008) Research of attitude dynamics with time-varying inertia for space net capture robot system. J Astronaut 29(4):1131–1136
25. Zhai G, Qiu Y, Liang B et al (2007) Research of capture error and error compensate for space net capture robot. In: IEEE International conference on robotics and biomimetics. ROBIO 2007. IEEE

Chapter 7
Formation-Keeping Control of the Closed-Chain System

The three-body closed-chain system cannot achieve the ideal LP equilibrium state under uncontrolled state, the main reason are that there are slack and rebound phenomena in the tether between the three spacecraft, which is due to the low rotational rate of the tether system in the initial condition of LP equilibrium and the inability of the tether to withstand the pressure.

Many researchers have conducted extensive research on the problem of maintaining control of the formation of closed-chain systems [1]. Kumar [2] explored the feasibility of rotating formation flying of satellites using flexible tethers. The system they used was composed of three satellites connected through tethers and located at the vertices of a triangle-like configuration. Huang [3] proposed a coupling dynamics model for the tethered space robot system based on the Hamilton principle and the linear assumption. Wang [4, 5], proposed a coordinated control of tethered space robot using mobile tether attachment point during the approaching phase and post-capture for capturing a target, respectively. They [6, 7], also proposed the Maneuvering-Net Space Robot System(MNSRS), which can capture and remove the space debris dexterously and mainly focused on coupled dynamics modeling and configuration control problems. In 2008, Vogel [8] assessed the utility of tethered satellite formations for the space-based remote sensing mission. Williams [9] considered a three-spacecraft tethered formation spinning nominally in a plane inclined to the orbital plane and obtained periodic solutions for the system via direct transcription that uses tether reeling to augment the system dynamics to ensure periodicity is maintained. Mori [10] proposed a tethered satellite cluster system, which consisted of a cluster of satellites connected by tethers and which can maintain and change formation via active control of tether tension and length to save thruster fuel and improve control accuracy. By now, most of the papers about multi-tethered system focus on the dynamics modeling and analysis. Hussein [11] studied the stability and control of relative equilibria for the spinning three-craft coulomb-tether problem. This part mainly focuses on symmetric Coulomb-tether systems, where all three crafts have the same mass and nominal charge values.

P. Huang and F. Zhang, *Theory and Applications of Multi-Tethers in Space*, Springer Tracts in Mechanical Engineering,
https://doi.org/10.1007/978-981-15-0387-0_7

To use the thruster to generate extra acceleration is an effective solution for the formation-keeping problem. The control method based on the thruster is a common control strategy in satellite attitude orbit control, which has the advantages of fast response, large control torque and obvious effect. Using the thruster of the spacecraft to simulate the tether compression in the initial condition of LP equilibrium is one specific control strategy. Tether-assisted thruster control during high-speed rotation is another specific control strategy. The two control strategies are described in detail and simulated in Sects. 7.1 and 7.2, respectively.

7.1 Thruster-Based Control for Formation-Keeping

In the three-body closed-chain system, without considering the effect of the low rotational rate of the tether system in the initial condition of LP equilibrium, the main reason for its configuration failure is that the connecting tether between the spacecraft cannot withstand the pressure, and the centrifugal force provided by tether system spin is not enough to overcome the effect of gravity, the tether cannot maintain tension, so the tether is easy to relax. Under the influence of the gravity gradient, it will cause the subsequent tether rebound. Therefore, this section considers using the thrust device on both ends of the tether to simulate the compression characteristics of the tether, that is, when the distance between the two ends of the tether is less than the natural extension length of the tether, the thruster begins to work, providing a thrust contrary to the line's contraction direction, thus simulating the tensile prosperities of the tether. It can be found that in this control mode, the tether and the thruster together work like a spring, the tether is no longer the elastic rod which can only withstand the tension. This control method improves the rigidity of the tether system, makes the tether system closer to the single rigid body, weakens the negative effect of the tether, and makes the three-body closed-loop tether system easier to achieve the LP equilibrium state.

7.1.1 Controller Design

The control strategy of using thruster to simulate tether compression is described as follows:

(a) When the distance between the spacecraft ρ_0 is greater than the natural extension length of the tether, the tether is in a tight state, at the same time the tether can be used as a spring–damper, the three-body closed-chain system model unchanged;
(b) When the distance between the spacecraft is less than $\rho_o - Q_d$, the tether is already in a relaxed state, at this time the thruster device began to work to simulate the tether compression characteristics, hindering the tether to continue

to slack. Q_d is the dead zone of the length of the tether, when the length of the tether within the dead zone $[\rho_o - Q_d, \rho_o]$, although the tether is in a relaxed state, the thruster does not work. The spacecraft linked at both ends of the tether are of free movement, the existence of Q_d simulates the reaction time of the thruster.

Therefore, the acceleration term caused by the tether should be modified as follows:

$$\ddot{\boldsymbol{R}}_{i_{tether}} = \begin{cases} \frac{\delta_{nm}\left[K_s\left(\rho_{ij}-\rho_o\right)+\mu_d\dot{\rho}_{ij}\right]}{m_i}\left(\frac{\boldsymbol{r}_{ij}}{\rho_{ij}}\right) & \left(\rho_{ij}-\rho_o\right)\geq 0 \\ 0 & \left(\rho_{ij}-\rho_o\right)\leq 0 \end{cases} \tag{7.1}$$

In addition, there are:

$$\ddot{\boldsymbol{R}}_{i_{thruster}} = \begin{cases} \frac{\delta_{nm}F_t}{m_i}\left(\frac{\boldsymbol{r}_{ij}}{\rho_{ij}}\right) & \rho_{ij}<\left(\rho_o - Q_d\right) \\ 0 & \text{otherwise} \end{cases} \tag{7.2}$$

$$\delta_{nm} = \begin{cases} 1 & n = i \\ -1 & m = i \\ 0 & \text{otherwise} \end{cases} \tag{7.3}$$

where, F_t is the force produced by the thruster, which is opposite to the slack direction of the tether, and is of continuous and constant, it always appearing on the spacecraft at both ends of the tether to simulate the tether compression. The size of the length Q_d of the tether dead zone determines the dead zone width of the thruster.

7.1.2 *Case Studies*

The simulation parameters in this section are set shown in Table 7.1. In addition, there are $F_t = 2.0\,\text{kg m/s}^2 = 2.0\,\text{N}$, $Q_d = 0.1\,\text{m}$. To estimate the fuel consumption, the ignition times of the thruster N are calculated in the simulation process, so, the fuel consumption required in the three-body closed-chain system during configuration keeping control ΔV is as follows:

Table 7.1 Simulation parameters

Parameters	Value
$\mu_\oplus$	$3.986004415 \times 10^8\ \text{m}^3/\text{s}^2$
m_i	200 kg
K_s	20.0 Nm
μ_d	0.05 kg
ρ_o	10.0 km

$$\Delta V = N\frac{F_t\Delta t}{m_i} \tag{7.4}$$

where, Δt is the simulation step.

Simulation Scenario 1: without outside interference

The simulation time is 10 orbital periods, the simulation step is 1, and the simulation results are shown in Fig. 7.1.

Figure 7.1a is the position vectors of three spacecraft relative to the centroid of the tether system in the orbital coordinates. The existence of the three-body closed-loop chain tether system can achieve the desired LP equilibrium state, and the trajectories of three spacecraft coincide in the orbit coordinate.

Figure 7.1b shows the three components of the angular momentum of the dimensionless string system in the coordinate systems, it can be found that the three components have a small periodic oscillation near the constant value, which indicates that the LP equilibrium state is not completely stable, but the oscillation amplitude is also within the acceptable range compared with the uncontrolled case. Figure 7.1c, d are the lengths of three tie tethers in the 1th orbital period and in the first 500 s, respectively, which is seen in the graph, the length of the tether is periodically changed in the range of the dead zone, and the oscillation is due to the rotation of the system The gravitational gradient force of the tether is constantly changing in the direction of the body coordinate system, and when the tether is perpendicular to the local horizontal line, the gravity gradient the most remarkable effects, and when the tether is horizontal, the effect of gravity gradient disappears completely; In addition, when the length of the tether reaches the natural stretching length, the tether rebound phenomenon disappears, but it should be noted that when the length of the tether reaches the lower limit of the dead zone, the tether will stretch suddenly, which is caused by the ignition of the thruster. It is a positive phenomenon that inhibits the continued relaxation of the tether. Figure 7.1e is the length of the tether during first 500 s, it can be seen from the figure that, as the length of the tether reaches the natural stretching length, the rebound phenomenon disappears, and the interaction between the tethers disappears, and no longer occurs because the rebound of a root tether causes other tethers to rebound. Figure 7.1f, g is the tether tension and the working time of three-spacecraft thrusters, respectively, as can be seen in the figure, when the tether is slack, the thruster at the two spacecraft is always working at the same time, and if both tethers are simultaneously relaxed, the spacecraft will produce thrust in two directions, but because of the uncertainty of the direction, it is not superimposed on the Fig. 7.1g. In addition, in the 10-track period simulation time, the thruster ignition number is 98,334 times.

To observe the influence of the parameters F_t and Q_d on the control effect, several other typical simulation parameters are designed, as shown in Table 7.2. Compared to case 2, the values of F_t and Q_d in scenario 1 increases by one and two orders of magnitude. Q_d remains unchanged in case 3, while F_t lowers ore order of magnitude. In case 4, F_t remains unchanged, and the value of Q_d increases 5 times,

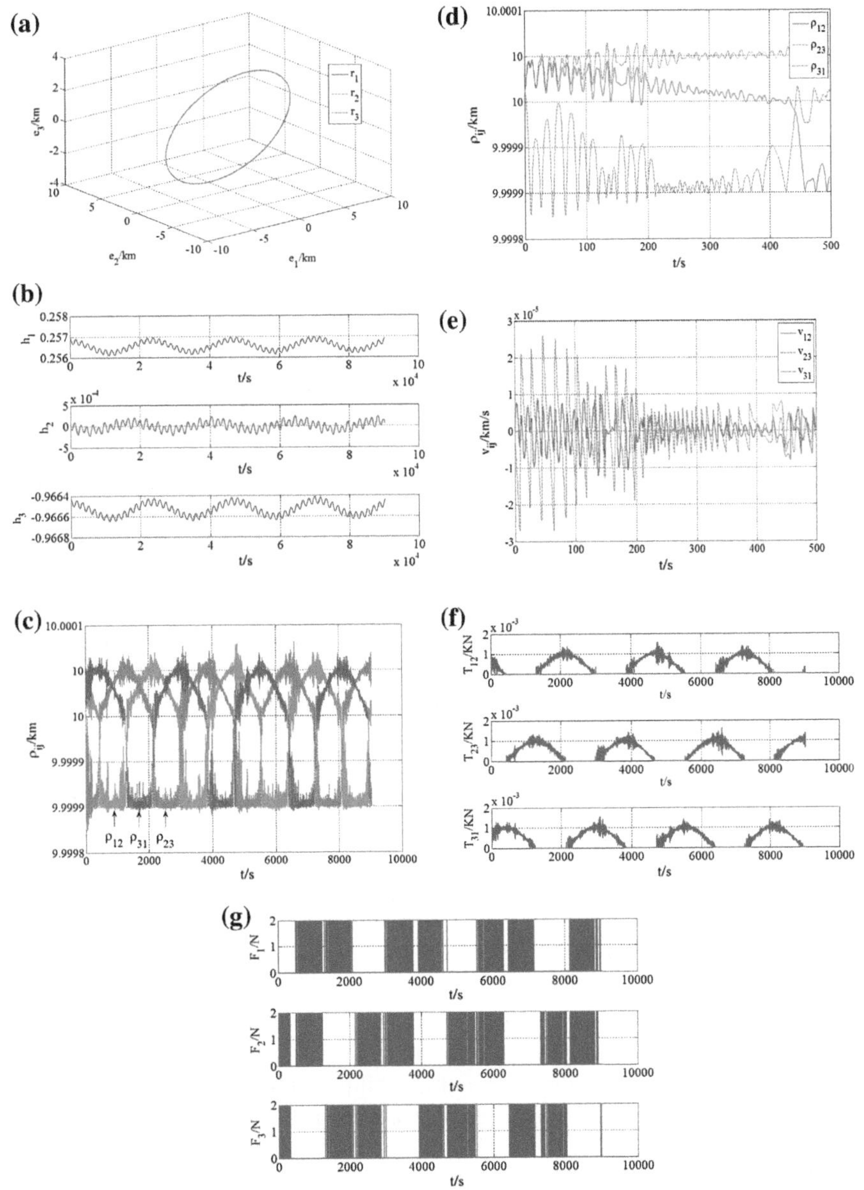

Fig. 7.1 Simulation results without interference

basically covering all cases of changes for two parameters, and the simulation results of four scenarios are shown in Figs. 7.2, 7.3, 7.4, 7.5, 7.6 and 7.7.

Figure 7.8a is the position vectors of the three spacecraft relative to the centroid of the tether system in the orbital coordinate systems, Fig. 7.3 is the three

Table 7.2 Values of F_t and Q_d energy consumption in various simulation cases

	F_t (kg m/s^2)	Q_d (m)	Ignition times	ΔV (m/s)
Original situation	2.0	0.1	98334	983.34
Case 1	20.0	1.0	20872	2087.2
Case 2	200.0	10	57662	57662
Case 3	0.2	0.1	2079352	2079.352
Case 4	2.0	0.5	98866	988.66

components of the angular momentum of the dimensionless string system in $\hat{s}$ coordinate systems, Fig. 7.4 is the length of three connected tethers in the first 500 s, Fig. 7.5 is the length of the tether during the first 500 s, Figs. 7.6 and 7.7 are the tension size in the tether and the working time of the three-spacecraft thruster. The ignition times and energy consumption in several cases are shown in Table 7.2. It can be found that in case 1, the three-body closed-chain system in case 2 and case 4 can approximate the LP equilibrium state, but the structure of the tether system in case 3 collapses after about two orbital cycles; In case 1 and case 2, the length of the tether are larger than the length of the natural extension of the tether, and improved about one or two orders of magnitude compared with the original case. Also, in case 1, because Q_d increases by one order of magnitude, the number of ignition of the thruster dropped to 20,872 times, but since the F_t value also increased by one order of magnitude, but instead ΔV increased by twice times, it is strange that in case 2, although Q_d value in the case of 1 increased by one order of magnitude, but the number of ignition increase to 57,662 times, This is due to the excessive size of F_t, which increases the oscillation frequency of the length of the tether in the dead zone, the number of thruster work increases; In case 3, because F_t value is too small, even if the thruster ignition, long time in the working state cannot inhibit the tether becoming slack, and ultimately lead to the system configuration of the tether being collapse; The number of ignition times in scenario 4 is fewer than the other three cases, is about 98,866, but is still greater than the original case, so the values of F_t and Q_d need to be carefully selected.

The results of these simulations show that both the Q_d and F_t have an effect on the ignition frequency and energy consumption ΔV, and determine the value of ΔV, in addition, the value of F_t determines the stability of the tether system. The value of the Q_d and F_t needs to be chosen in terms of tether system parameters, fuel consumption limits, thruster life, and other aspects to select. It is possible to find a suitable parameter by numerical simulation or other effective methods to achieve the stability of the tether system, and the number of ignition and ΔV is not too large. Therefore, in the absence of interference, this kind of control method which uses the thruster to simulate the tether compression can make the three-body closed-loop chain tether system achieve stable LP equilibrium state.

Simulation Scenario 2: with outside interference

Simulation scenario 1 verifies that the control method of using thruster to simulate tether compression can achieve stable LP equilibrium state in three-body

Fig. 7.2 The position vectors of three spacecraft relative to the centroid of the tether system in the orbit coordinates

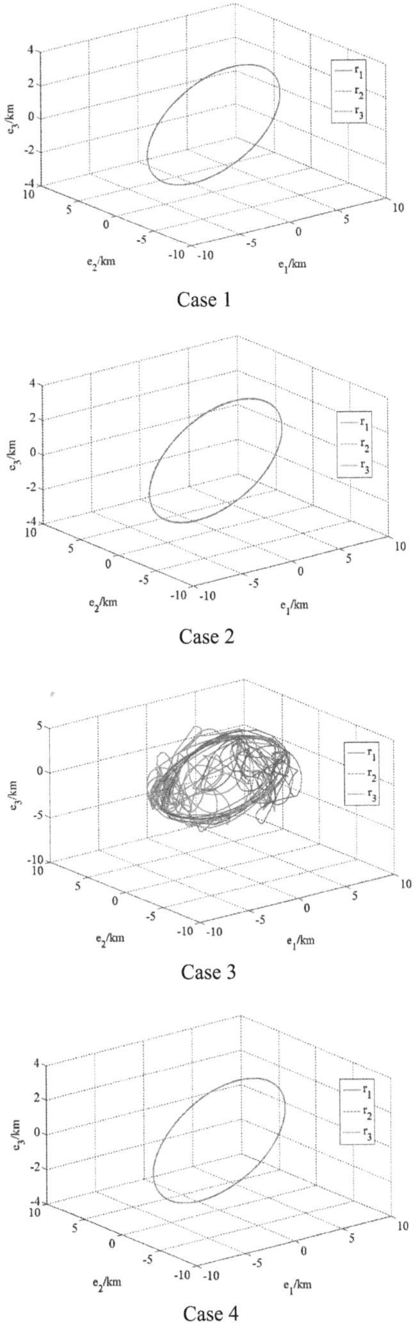

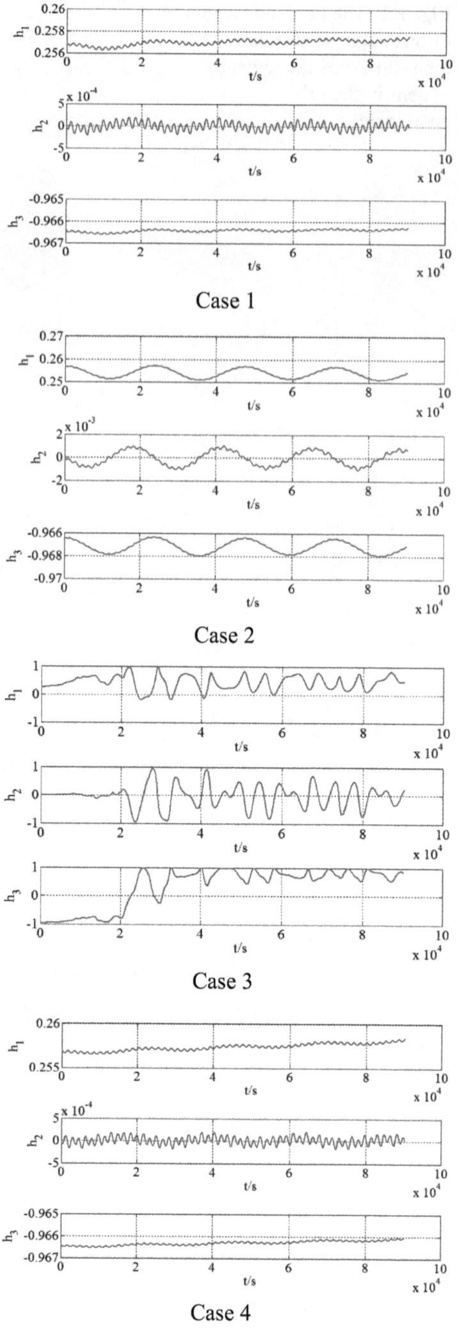

Fig. 7.3 The three components of angular momentum in the system of the dimensionless tether in the $\hat{s}$ coordinate systems

Fig. 7.4 The length of three connected tethers during the first 500 s

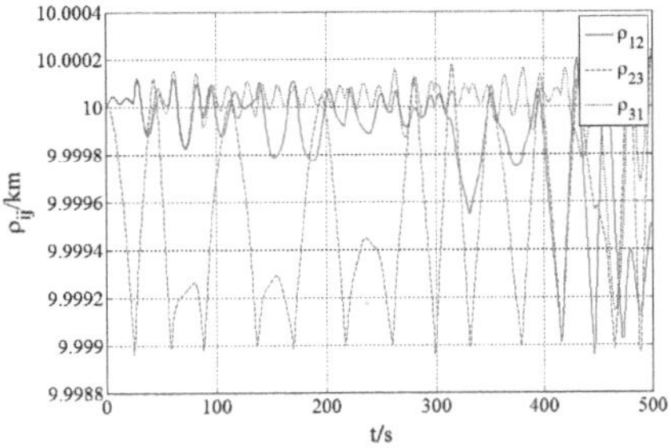

Case 1

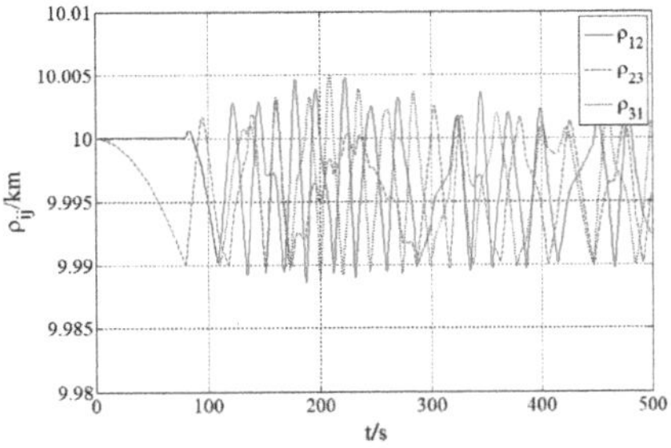

Case 2

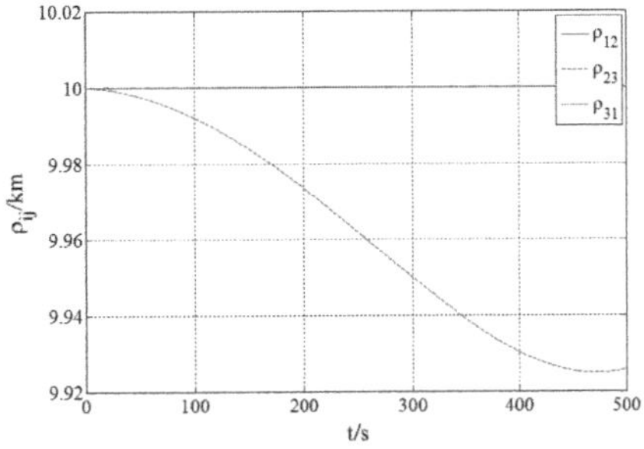

Case 3

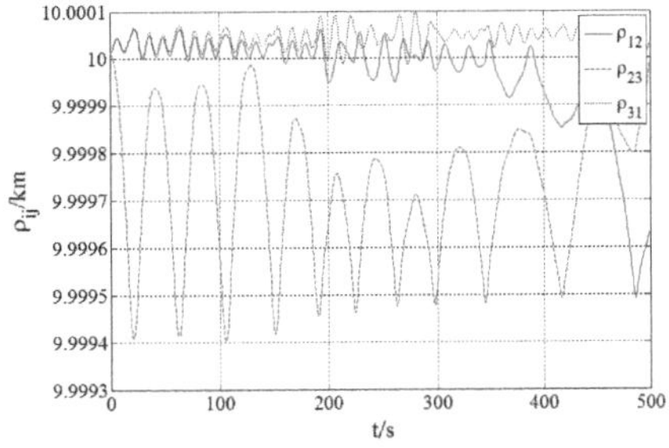

Case 4

Fig. 7.5 The length change rate of the tether during the first 500 s

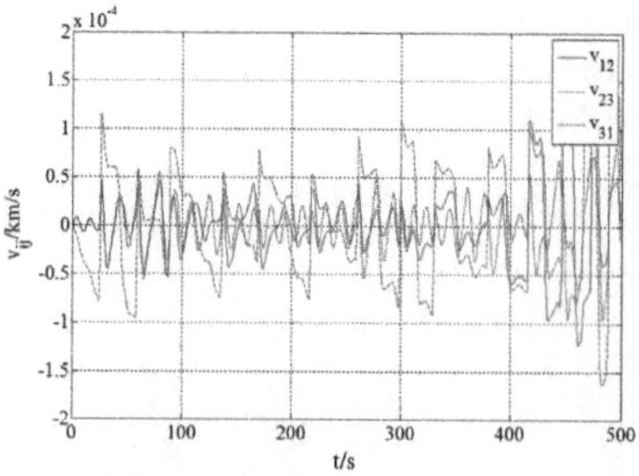

Case 1

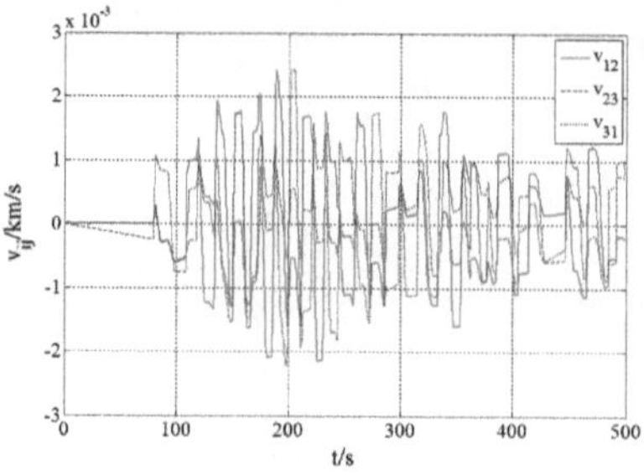

Case 2

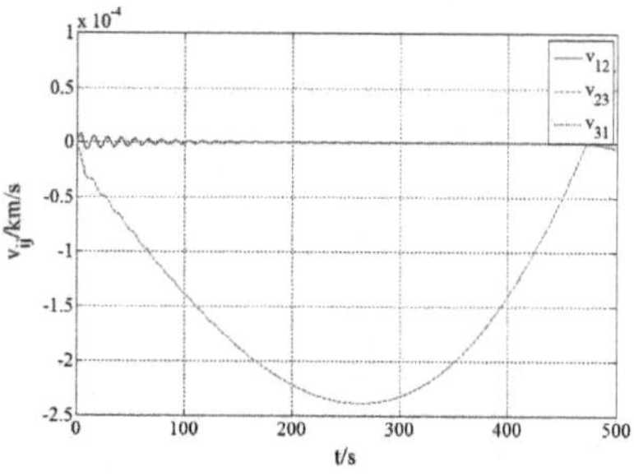

Case 3

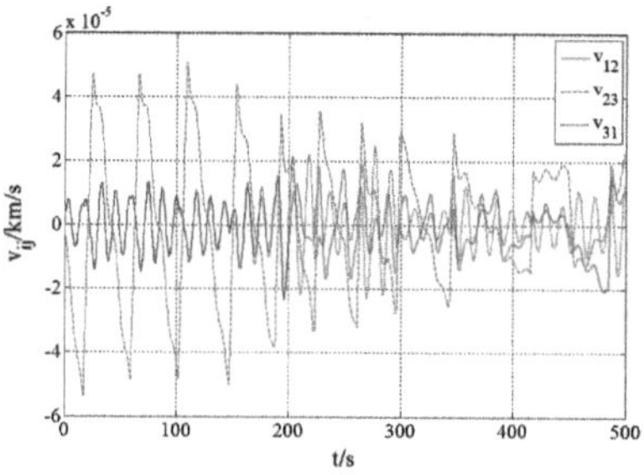

Case 4

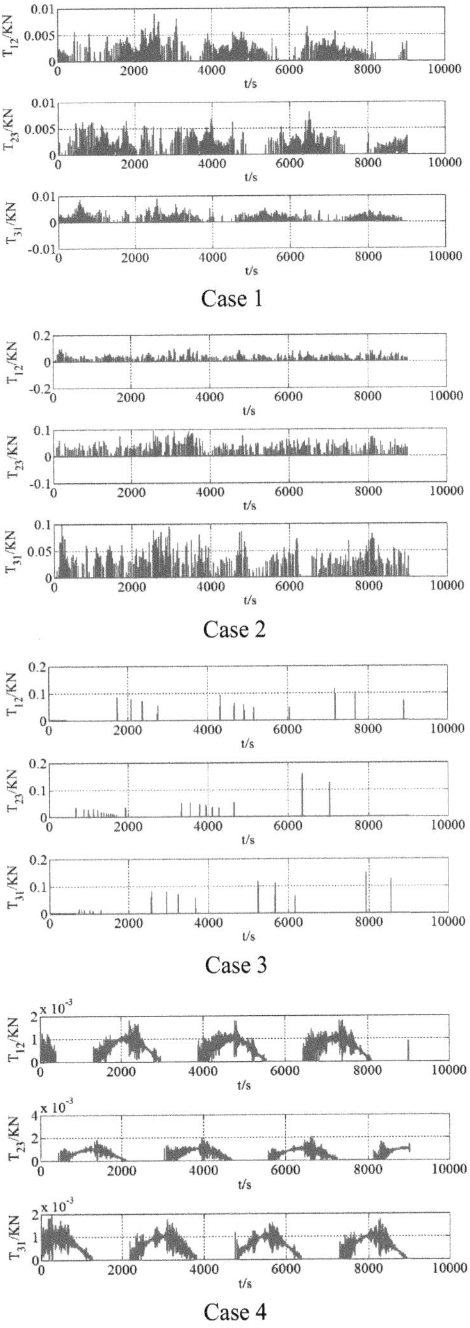

Fig. 7.6 The tension along three tethers

Fig. 7.7 The working time of three thrusters

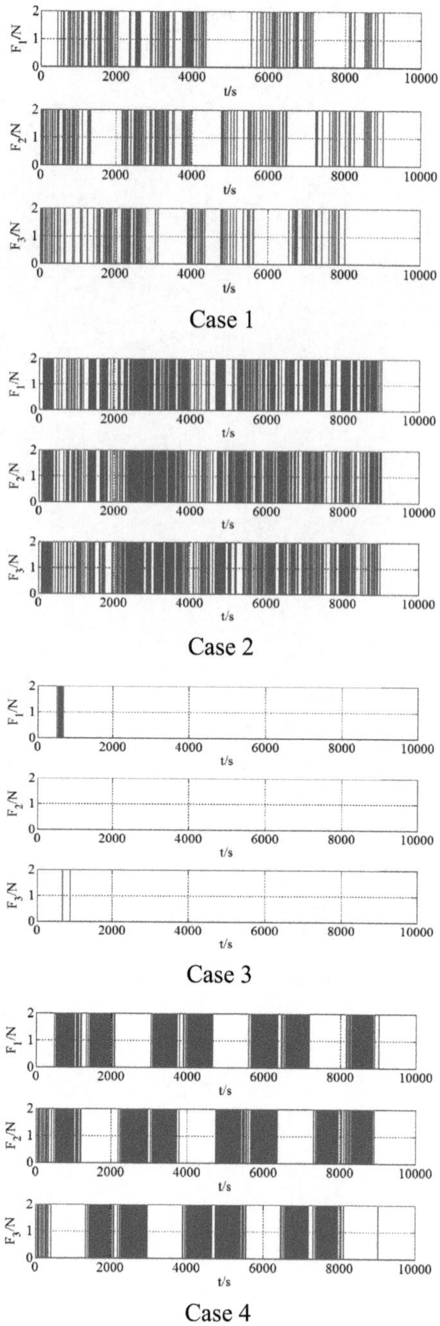

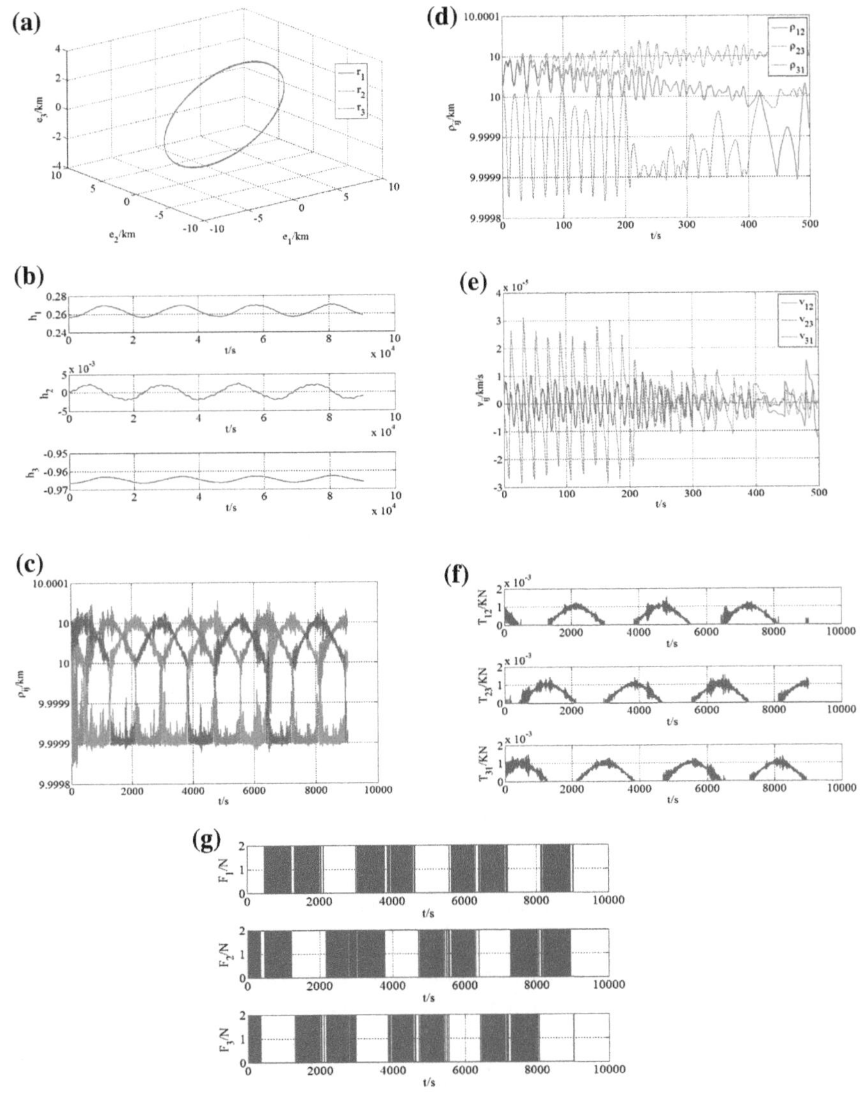

Fig. 7.8 Simulation results with interference situations

closed-chain system without interference, and this simulation will continue to verify that when there is external interference in the system model of the tether, whether the control strategy of using thruster to simulate tether compression can still make the three-body closed-loop chain system achieve stable LP equilibrium state.

The spacecraft is always subjected to various perturbation forces in the space environment when flying in orbit. These perturbations are: the Earth's shape is not spherical and uneven quality of the additional pressure, air force, the sun's gravity, the moon's gravity, as well as the pressure of sunlight pressure. Although these perturbations are about one out of 10,000 of the gravity of the Earth's center, the longtime effect will cause the spacecraft's orbit to deviate from the requirements of the application task. The most important interfering term in these perturbations is the J2 perturbation, so that only the J2 perturbation is considered here.

The acceleration caused by the J2 perturbation on the spacecraft is:

$$\ddot{\boldsymbol{R}}_{i_{J2}} = \begin{bmatrix} \frac{-3J_2\mu_{\oplus}R_{\oplus}^2(\boldsymbol{R}_i \cdot \hat{i}_1)}{2|\boldsymbol{R}_i|^5}\left(1 - \frac{(\boldsymbol{R}_i \cdot \hat{i}_3)}{|\vec{\boldsymbol{R}}_i|^2}\right) \\ \frac{-3J_2\mu_{\oplus}R_{\oplus}^2(\boldsymbol{R}_i \cdot \hat{i}_2)}{2|\boldsymbol{R}_i|^5}\left(1 - \frac{(\boldsymbol{R}_i \cdot \hat{i}_3)}{|\boldsymbol{R}_i|^2}\right) \\ \frac{-3J_2\mu_{\oplus}R_{\oplus}^2(\boldsymbol{R}_i \cdot \hat{i}_3)}{2|\boldsymbol{R}_i|^5}\left(1 - \frac{(\boldsymbol{R}_i \cdot \hat{i}_3)}{|\boldsymbol{R}_i|^2}\right) \end{bmatrix} \tag{7.5}$$

After integrating the upper equation into the motion equation of the three-body closed-loop chain system, simulation results are carried out with the original case parameters in simulation case 1, and the simulation result is shown in Fig. 7.8. Figure 7.8a is the position vectors of three spacecrafts relative to the centroid of the tether system in the orbital coordinate system, Fig. 7.8b is the three components of the angular momentum of the dimensionless string systems in the $\hat{s}$ coordinates. Figure 7.8c, d are the length of the three-link tether in the first orbital period and the first 500 s respectively. Figure 7.8e is the length of the 500 s inner tether. Figure 7.8f, g are the tension along the tether and the working moment of the three-spacecraft thruster, respectively, as can be seen from the figure, In the case of J2 interference, the three-body closed-chain system can still achieve stable LP equilibrium state. But the strange thing is that the ignition times are 97,852 times, ΔV is about 978.52 m/s, even less than no interference, this phenomenon shows that outside interference does not always play a negative role in the system, just as the gravitational gradient force can sometimes be used to assist system balance, J2 interference can sometimes produce unexpected results.

The simulation results show that even in the case of with disturbance, the control strategy of using thruster to simulate the compression characteristics of the tether can still make the three-body closed-loop chain tether system obtain stable LP equilibrium state, and the control strategy is effective.

7.2 Coordinated Formation-Keeping Control Based on Thruster and Tether Tension

7.2.1 Controller Design

Because the three-body closed-loop chain tether system is always in a high-speed rotation state and there is no slack and rebound phenomenon, the rigidity of the whole system is increased greatly compared with the low speed. It can be equivalent to a flat coarse cylindrical rigid body. First, we derive the control torque and energy consumption to maintain the LP equilibrium state at high rotational speed for the single rigid body case.

Figure 7.9 is a precession figure of a flat thick cylindrical single rigid body, H_r and H_3 are the components of the rigid angular momentum $\boldsymbol{H}$ on the $\hat{\boldsymbol{e}}_r$ axis and the $\hat{\boldsymbol{e}}_3$ axis, respectively.

According to reference [1], the relationship between the torque $\boldsymbol{M}$ and the precession angular velocity $\boldsymbol{\Omega}_p$ of the rigid body around $\hat{\boldsymbol{e}}_3$ axis and the angular momentum $\boldsymbol{H}$ is

$$\boldsymbol{M} = \boldsymbol{\Omega}_p \times \boldsymbol{H} \tag{7.6}$$

In the desired LP equilibrium state, the precession angular velocity $\boldsymbol{\Omega}_p$ of a rigid body and the orbital angular velocity $\dot{\boldsymbol{v}}\hat{\boldsymbol{e}}_3$ should be the same, namely:

$$\boldsymbol{\Omega}_p = \dot{\boldsymbol{v}}\hat{\boldsymbol{e}}_3 \tag{7.7}$$

Therefore, from the Eqs. (7.7) and (7.8) we can obtain the external force–torque $\boldsymbol{M}_{req}$ to maintain the single rigid body orientation is

$$\boldsymbol{M}_{req} = \begin{vmatrix} \hat{e}_v & \hat{e}_r & \hat{e}_3 \\ 0 & 0 & \dot{v} \\ 0 & H_r & H_3 \end{vmatrix} = -\dot{\boldsymbol{v}}\boldsymbol{H}_r\hat{\boldsymbol{e}}_v \tag{7.8}$$

It is known from the above formula that to counteract the effect of gravity gradient torque, the $\boldsymbol{M}_{req}$ must be parallel to the $-\hat{\boldsymbol{e}}_v$ axis. In addition, to obtain $\boldsymbol{M}_{req}$, we must first obtain H_r.

The angular momentum of a rigid body in the body coordinate system is

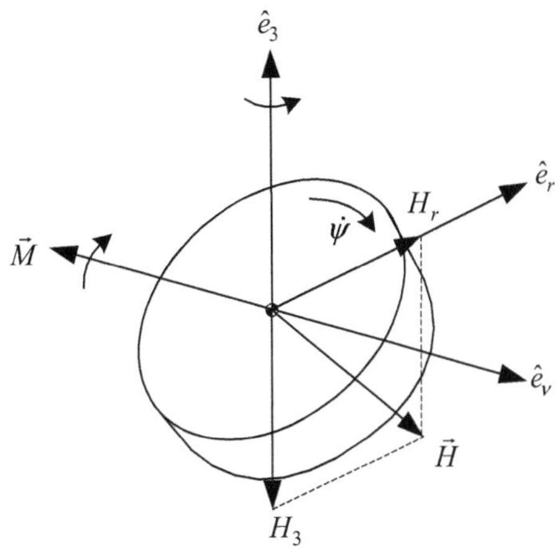

Fig. 7.9 The precession of a flat course cylindrical single rigid body

$$ {}^{\hat{b}}\boldsymbol{H} = {}^{\hat{b}}\boldsymbol{I}^{\hat{b}}\omega^{bi} = \begin{bmatrix} A & 0 & 0 \\ 0 & A & 0 \\ 0 & 0 & C \end{bmatrix} {}^{\hat{b}}\omega^{bi} \tag{7.9} $$

And:

$$ {}^{\hat{b}}\boldsymbol{H} = \begin{bmatrix} A & 0 & 0 \\ 0 & A & 0 \\ 0 & 0 & C \end{bmatrix} \begin{bmatrix} \sin\psi\sin\theta & \cos\psi & 0 \\ \cos\psi\sin\theta & -\sin\psi & 0 \\ \cos\theta & 0 & 1 \end{bmatrix} \begin{Bmatrix} \dot{f} \\ \dot{\theta} \\ \dot{\psi} \end{Bmatrix} \tag{7.10} $$

In the desired LP equilibrium state, there is $\dot{f} = \dot{v}$, $\dot{\theta} = 0$, therefore

$$ {}^{\hat{b}}\boldsymbol{H} = \begin{Bmatrix} A\sin\psi\sin\theta\dot{v} \\ A\cos\psi\sin\theta\dot{v} \\ C\dot{\psi} + C\cos\theta\dot{v} \end{Bmatrix} \tag{7.11} $$

To convert it to $\hat{\boldsymbol{e}}$ system, we have

$$ {}^{\hat{e}}\boldsymbol{H} = \boldsymbol{C}^{e b \hat{b}}\boldsymbol{H} = \begin{bmatrix} \cos\psi & -\sin\psi & 0 \\ \cos\theta\sin\psi & \cos\psi\cos\theta & -\sin\theta \\ \sin\psi\sin\theta & \cos\psi\sin\theta & \cos\theta \end{bmatrix} \begin{bmatrix} A\sin\psi\sin\theta\dot{v} \\ A\cos\psi\sin\theta\dot{v} \\ C\dot{\psi} + C\cos\theta\dot{v} \end{bmatrix} = \begin{bmatrix} H_y \\ H_r \\ H_3 \end{bmatrix} \tag{7.12} $$

H_r can be obtained from the upper equation, so

$$ {}^{\hat{e}}\boldsymbol{M}_{req} = \begin{bmatrix} -\dot{v}H_r \\ 0 \\ 0 \end{bmatrix} = \begin{bmatrix} -\dot{v}\left\{-\sin\theta\left(C\dot{\psi} + C\cos\theta\dot{v}\right) + A\cos\theta\sin\theta\dot{v}\right\} \\ 0 \\ 0 \end{bmatrix} \tag{7.13} $$

To convert it to $\hat{\boldsymbol{b}}$ system, we have

$$ {}^{\hat{b}}\boldsymbol{M}_{req} = \boldsymbol{C}^{be\hat{e}}\boldsymbol{M}_{req} = \begin{bmatrix} -\cos\psi\dot{v}\left\{-\sin\theta\left(C\dot{\psi} + C\cos\theta\dot{v}\right) + A\cos\theta\sin\theta\dot{v}\right\} \\ \sin\psi\dot{v}\left\{-\sin\theta\left(C\dot{\psi} + C\cos\theta\dot{v}\right) + A\cos\theta\sin\theta\dot{v}\right\} \\ 0 \end{bmatrix} \tag{7.14} $$

Therefore, the flat thick cylindrical body at high rotational speed to keep the LP equilibrium conditions requires external control torque:

$$\boldsymbol{M}_{out} = \boldsymbol{M}_{req} - \boldsymbol{M}_{gg} \tag{7.15}$$

The fuel consumption ΔV required to provide the above control torque is calculated below.

The assumed control moment $\boldsymbol{M}_{out}$ is provided by the thruster device, and its resulting thrust is a continuous constant. If the mass of the rigid body is m and the radius is r, then

$$\boldsymbol{F}_{thrusters} = \frac{\boldsymbol{M}_{out}}{r} = m\Delta \boldsymbol{a} = m\frac{\Delta V}{\Delta t} \tag{7.16}$$

where, the Δt is the simulation step, and we have

$$\Delta \boldsymbol{V} = \frac{\boldsymbol{M}_{out}\Delta t}{rm} \tag{7.17}$$

7.2.2 *Case Studies*

First, the flat thick cylindrical single rigid body model is validated to verify whether the LP equilibrium state can be maintained after the speed is improved. The value of the SR will no longer be determined by the LP equilibrium initial condition, but given a high rotational speed SR = −5.0, the simulation time is 10 orbital periods and the simulation step is 10.

Here, we can consider the state is controllable when the ${}^{\hat{b}}\boldsymbol{M}_{req}$ replaced by the original $\boldsymbol{M}_{gg}$, and only when $\boldsymbol{M}_{gg}$ exists, the state is uncontrolled. The simulation results of controlled state and uncontrolled state under the above simulation parameters are shown in Figs. 7.10, 7.11, 7.12, 7.13 and 7.14. Figure 7.10 is the change curve of $\boldsymbol{H}$, ω and $\hat{\boldsymbol{b}}_3$ in the inertial coordinate system, Fig. 7.11 is the angle between $\boldsymbol{H}$, ω and $\hat{\boldsymbol{b}}_3$ axes, Fig. 7.12 is $\boldsymbol{H}$, ω in the body coordinate system, Fig. 7.13 is the $\hat{\boldsymbol{b}}_3$ axis in the orbit coordinate system, Fig. 7.14 is the components of the state vector $\boldsymbol{X}$. Under uncontrolled state, we can find that $\boldsymbol{H}$, ω and $\hat{\boldsymbol{b}}_3$ axes cannot be in stable rotation state around the $\hat{\boldsymbol{k}}$ axis, but there is a strange irregular oscillation, and the angle between them is no longer maintain a fixed value in the vicinity of the constant value of about 2° of the upper and lower oscillations, and θ also no longer remain fixed, but $\vec{\omega}$ is similar to the controlled state in the body coordinate system, the basic fixed, and the amplitude is on the level of 10^{-4} rad/s, but the oscillation is more regular and orderly when the state is controlled; Unlike the state is uncontrolled, $\boldsymbol{H}$, $\vec{\omega}$ and $\hat{\boldsymbol{b}}_3$ axes can rotate stably around the $\hat{\boldsymbol{k}}$ axis, the rigid body achieves the desired LP equilibrium state, $\vec{\omega}$ in the body coordinate system, and the angular momentum $\boldsymbol{H}$ can rotate around the $\hat{\boldsymbol{b}}_3$ axis stably, In the orbit coordinate system the $\hat{\boldsymbol{b}}_3$ axis also basically maintains the fixed, the θ value

Fig. 7.10 H, ω and $\hat{b}_3$ axes in inertial systems

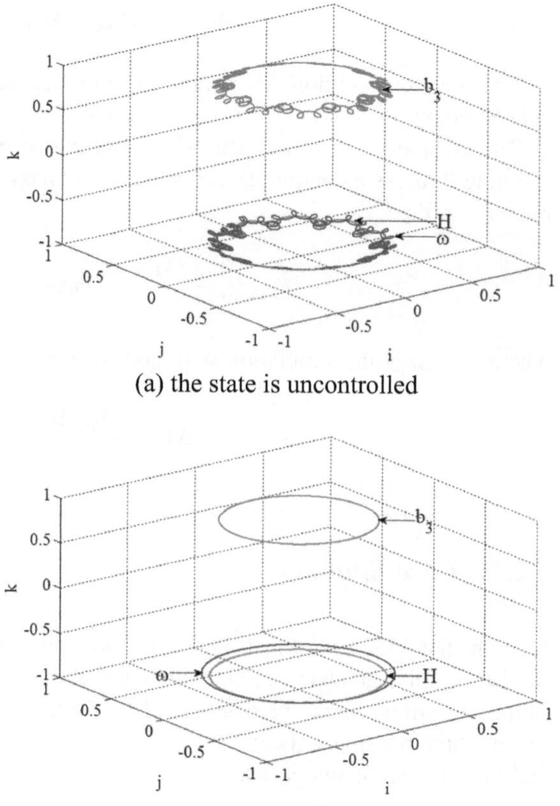

basically remains unchanged, $\vec{\omega}$ changes in a small periodic, therefore may explain uses the ${}^{b}\hat{\boldsymbol{M}}_{req}$ to replace $\boldsymbol{M}_{gg}$, the rigid body can attain the LP equilibrium State, the control strategy is feasible to the single rigid body, the above ${}^{b}\hat{\boldsymbol{M}}_{req}$ calculation process is correct. Formulas are available.

The above control strategy is applied to the three-body closed-loop chain tether system, and the equivalent is simplified to a flat coarse cylindrical rigid body. Assuming that the natural extension length ρ_o of the tether is 10 km between three spacecraft and the mass of each spacecraft is 200 kg, the equivalent rigid body radius $r = \sqrt{3}\rho_o/3 = 5.774\,\text{km}$ and the equivalent rigid total mass m is 600 kg. It is known from the theoretical mechanics knowledge that the moment of inertia of the flat coarse cylindrical homogeneous rigid body is

$$A = \frac{1}{4}mr^2 \tag{7.18}$$

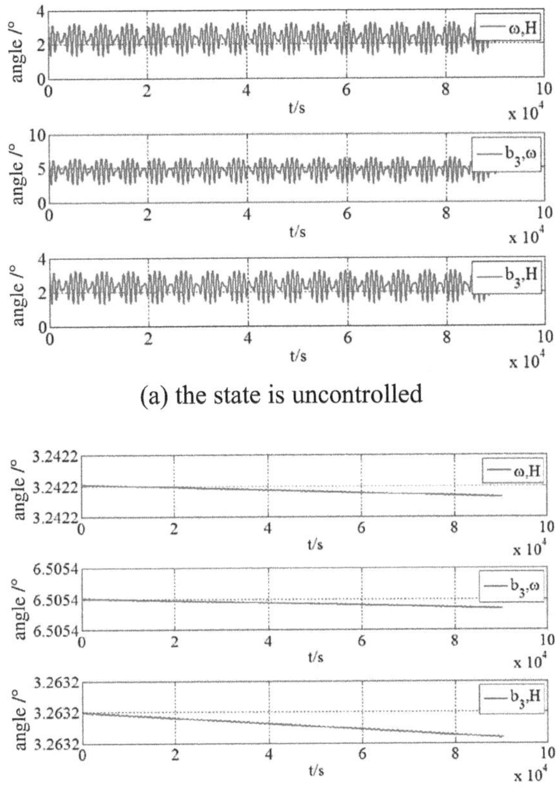

(a) the state is uncontrolled

(b) the state is controlled

Fig. 7.11 The angle between the $\boldsymbol{H}$, $\boldsymbol{\omega}$ and $\hat{\boldsymbol{b}}_3$ axes

$$C = \frac{1}{2}mr^2 \tag{7.19}$$

The simulation time is 10 orbital period, the simulation step is 10 s, and SR = −5.0, then the simulation results are shown in Fig. 7.15.

Figure 7.15a is the $\boldsymbol{H}$, $\boldsymbol{\omega}$ and $\hat{\boldsymbol{b}}_3$ in the inertial coordinate system, and (b) is the angle between the $\boldsymbol{H}$, $\boldsymbol{\omega}$ and $\hat{\boldsymbol{b}}_3$ axes, (c) is $\boldsymbol{H}$, $\boldsymbol{\omega}$ in the body coordinate system, (d) the $\hat{\boldsymbol{b}}_3$ axis in the orbital coordinate system, and (e) the components of the state vector $\boldsymbol{X}$. It can be seen from the figure that the equivalent simplified model of the three-body closed-loop chain tether system is able to achieve the LP equilibrium State, $\boldsymbol{H}$, $\boldsymbol{\omega}$ and $\hat{\boldsymbol{b}}_3$ axes can rotate stably and periodically around the $\hat{\boldsymbol{k}}$ axis, and the angle between them is also maintained as a fixed value; $\boldsymbol{\omega}$ is fixed in the body coordinate system and $\boldsymbol{H}$ can rotate around the $\hat{\boldsymbol{b}}_3$ axis stably, the $\hat{\boldsymbol{b}}_3$ axis in the orbital coordinate system is basically maintain fixed, the value of θ remains unchanged basically, $\boldsymbol{\omega}$ is with a small periodic oscillation, $\Delta V \approx 346.61$ m/s, and far less than the fuel consumption when use the thruster to simulate the tether compression characteristics, but it should be explained that the results here are only

Fig. 7.12 The ***H***, *ω* in body coordinate system

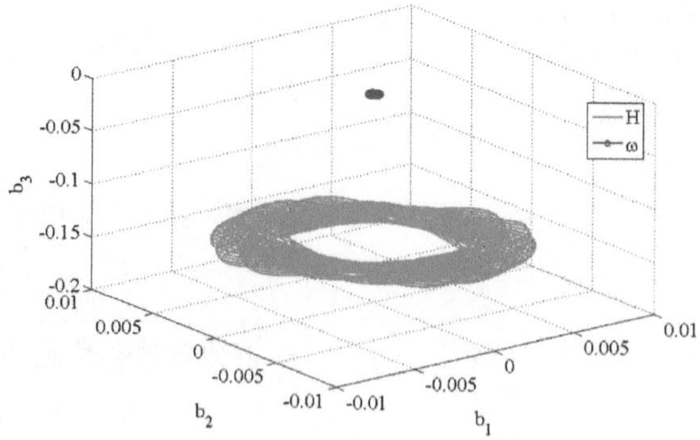

(a) the state is uncontrolled

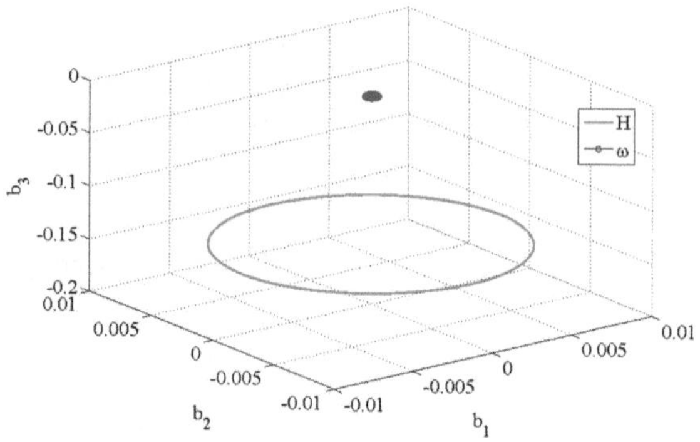

(b) the state is controlled

for the equivalent simplified model of the three-body closed-loop chain tether system, and the distribution and combination of fuel consumption on three spacecraft are not taken into account. It is only when the problem is studied thoroughly that the control strategy is better than the control strategy which is using the thruster to simulate the tether compression characteristics, but it is certain that the two control strategies based on thruster can achieve the LP equilibrium state of the three-body closed-chain system.

Therefore, in order to improve the spin rate of space multi-tether system and use the thruster to assist the control strategy of the tether system orientation, it is feasible to add extra control torque to the tether system by the thruster device of the spacecraft, and overcome the influence of gravity gradient moment. It solves the problem that the three-body closed-loop chain tether system cannot achieve LP equilibrium state under uncontrolled state.

Fig. 7.13 The $\hat{b}_3$ axis in the orbit coordinate system

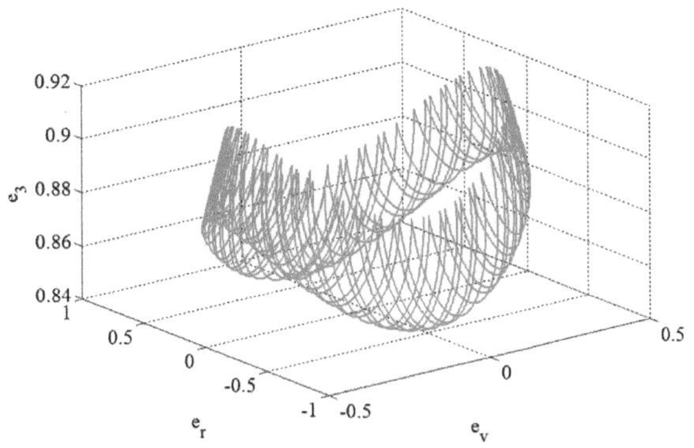

(a) the state is uncontrolled

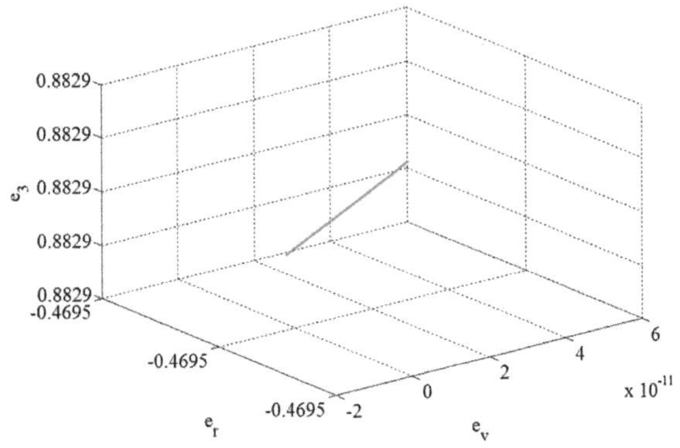

(b) the state is controlled

Fig. 7.14 The components of the state vector $\boldsymbol{X}$

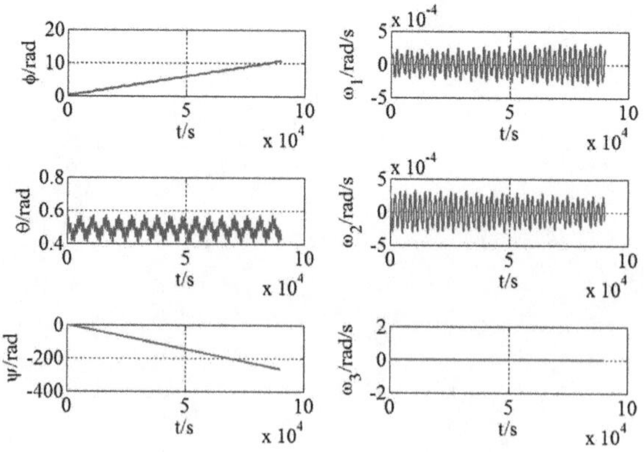

(a) the state is uncontrolled

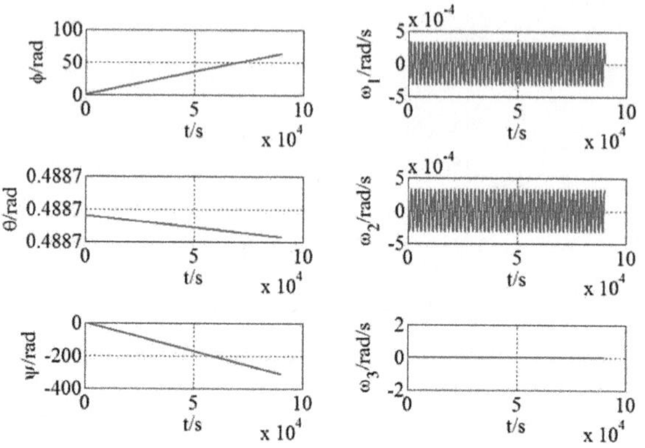

(b) the state is controlled

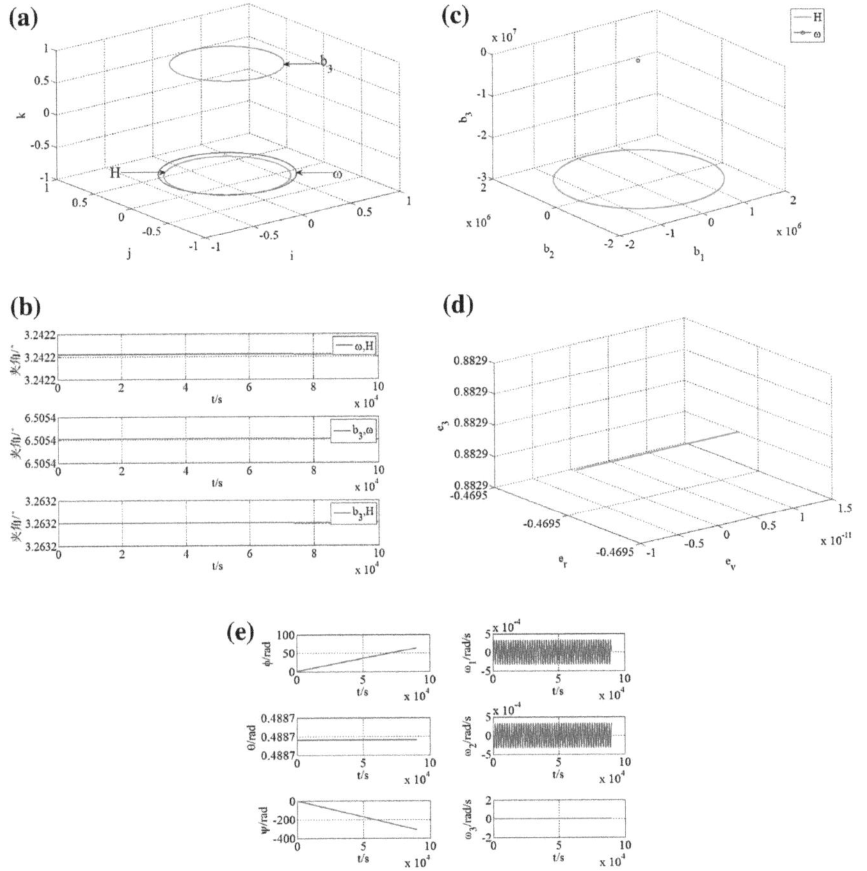

Fig. 7.15 Simulation results of thruster auxiliary tether system

References

1. Wertz JR, Larson WJ (1999) Space mission analysis and design. California Microcosm Press, El Segundo
2. Kumar KD, Yasaka T (2004) Rotating formation flying of three satellites using tethers. J Spacecr Rocket 41(6):973–985
3. Huang PF, Hu ZH, Meng ZJ et al (2015) Coupling dynamics modelling and optimal coordinated control of tethered spacerobot. Aerosp Sci Technol 41(2):36–46
4. Wang DK, Huang PF et al (2014) Coordinated control of tethered space robot using mobile tether attachment point in approaching phase. Adv Space Res 54(6):1077–1091
5. Wang DK, Huang PF et al (2015) Coordinated stabilization of tumbling targets using tethered space manipulators. IEEE Trans Aerosp Electron Syst 51(3):2420–2431
6. Huang PF, Zhang F, Ma J (2015) Dynamics and configuration control of the maneuvering-net space robot system. Adv Space Res 55(4):1004–1014
7. Huang PF, Hu ZH, Zhang F (2016) Dynamic modelling and coordinated controller designing for the manoeuvrable tether-net space robot system. Multibody Syst Dyn 36(2):115–141

8. Vogel KA (2006) Dynamics and control of tethered satellite formations for the purpose of space-based remote sensing. Air Force Institute of Technology, Ohio
9. Williams P (2006) Periodic optimal control of a spinning earth-pointing tethered satellite formation. In: Proceedings of the AIAA/AAS astrodynamics specialist conference and exhibit, Keystone, Colorado, America
10. Mori O (2007) Formation and attitude control for rotational tethered satellite clusters. J Spacecr Rocket 44(1):211–220
11. Hussein I, Schaub H (2009) Stability and control of relative equilibria for the three-spacecraft coulomb tether problem. Acta Astronaut 65(5):738–754

Chapter 8
Deployment and Retrieval Control of the Hub-Spoke System

The Hub-Spoke Tethered Formation System (HS-TFS) is now a hot issue in many space applications, such as multi-point measurements, providing a flexible frame for solar sail and other membrane or net structures. To achieve the valuable advantages such as reduction of fuel consumption, promotion of the formation stability, the HS-TFS is usually in the rotating state. In addition, it is necessary to change the length of the tethers to obtain a variable coverage of the entire plane of the rotating HS-TFS in some applications, that is, the deployment and retraction problems of the rotating HS-TFS. However, the rotating motion will increase the complexity of the deployment and retraction of the rotating HS-TFS. In this chapter, the deployment and retrieval of a rotating HS-TFS is investigated. First, a mathematical model is derived to describe the dynamics of the rotating HS-TFS. Then, the Gauss pseudospectral method is employed to solve the optimal deployment and retraction problems of the rotating HS-TFS. Finally, numerical simulations for deployment and retraction of the rotating HS-TFS are performed. Numerical simulation results reveal that it is necessary to apply active control on the deployment and retraction phases of the rotating HS-TFS, and after employing the optimal control (Gauss pseudospectral method), the HS-TFS can reach the desired configuration.

8.1 Natural Deployment/Retrieval Without Control

Based on the dynamic model introduced in Sect. 6.3.2, the uncontrolled deployment/retrieval are simulated in this section. The initial angular velocity of the two-body HS-TFS is $\pi/3$ rad/s and no active control is employed. Although $\pi/3$ rad/s is almost large for a spacecraft, that is, it is not appropriate for the normal mission, it is the required initial angular velocity for generating enough tension in tethers. The deployment operation begins with a releasing velocity of 0.1 m/s and the retraction operation begins with a retrieving velocity of −0.1 m/s. The process

P. Huang and F. Zhang, *Theory and Applications of Multi-Tethers in Space*, Springer Tracts in Mechanical Engineering,
https://doi.org/10.1007/978-981-15-0387-0_8

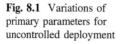

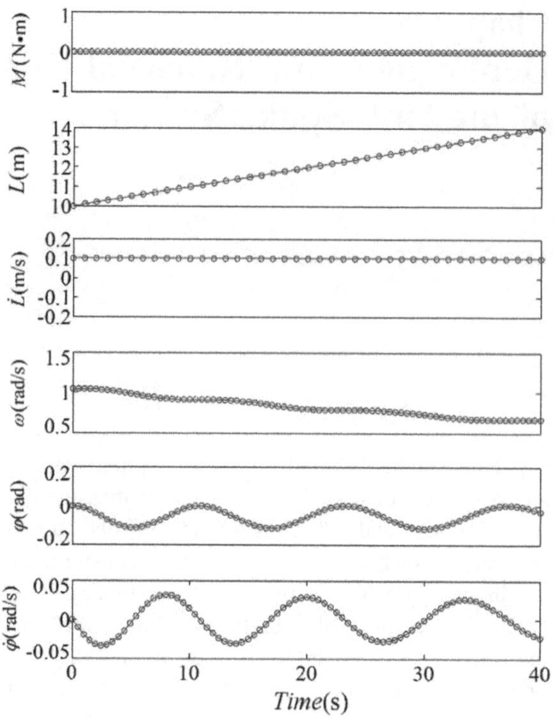

Fig. 8.1 Variations of primary parameters for uncontrolled deployment

of deployment and retraction last 40 s. The simulation results are demonstrated in Figs. 8.1 and 8.2. The graphs show the variations of primary parameters, including torque of the master-spacecraft, M; length of the tethers connected to sub-spacecraft1 and sub-spacecratft2, L; releasing or retrieving velocity of tethers connected to sub-spacecraft1 and sub-spacecratft2, $\dot{L}$; angular velocity of the master-spacecraft ω; angles between the straight tether's direction and radial direction, φ; angular velocity of φ, $\dot{\varphi}$.

Figure 8.1 demonstrates the variations of primary parameters for uncontrolled deployment. From $t = 0\,\text{s}$ to $t = 40\,\text{s}$, the $\dot{L}$ keeps as a constant value to assure a uniform deployment. The L increases with time uniformly, and the final value of L is 14 m. The M keeps zero all the time. The ω, φ, $\dot{\varphi}$ are varying continuously during the deployment operation. Compared to the initial states, the final values of ω, φ, $\dot{\varphi}$ are all changed. Particularly, the ω is decreasing with time, and the final value of ω is smaller than its initial value.

Figure 8.2 demonstrates the variations of primary parameters for uncontrolled retraction. From t = 0 s to t = 40 s, the $\dot{L}$ keeps as a constant value to maintain a uniform retraction. The L decreases with time uniformly, and the final value of L is 6 m. The M keeps zero all the time. The ω, φ, $\dot{\varphi}$ are varying continuously during the deployment operation. Compared to the initial states, the final values of ω, φ, $\dot{\varphi}$

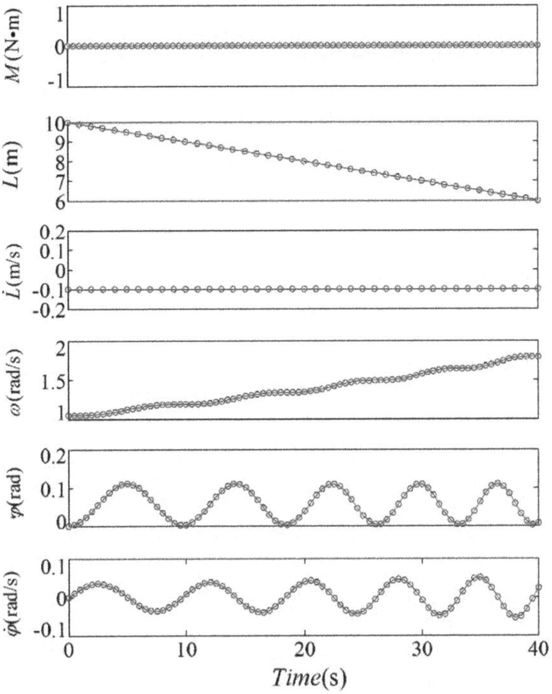

Fig. 8.2 Variations of primary parameters for uncontrolled retraction

are all changed. Particularly, the ω is increasing with time, and the final value of ω is larger than its initial value.

Simulation results in Figs. 8.1 and 8.2 revealed that the uncontrolled deployment and retraction of the two-body HS-TFS will result in undesired variations of ω, φ, $\dot{\varphi}$, that is, the initial uniform rotating motion of the formation system will be destroyed after the deployment or retraction operation. Therefore, the active control must be employed to reach the desired value of ω, φ and $\dot{\varphi}$. In addition, although no active control is employed during the uncontrolled deployment and retraction operations, the whole rotating formation system is governed and limited by a fundamental physical law: conservation of the angular momentum. The decrease of ω in uncontrolled deployment and increase of ω in uncontrolled retraction can be explained by this fundamental physical law.

8.2 Optimal Control for Deployment/Retrieval

Maintaining the uniform rotating motion is the essential condition in many space applications and at the same time, forming a variable coverage of the entire plane during deployment and retraction are necessary for the task. However, the deployment and retraction of Hub-Spoke Tethered Formation System need to change the length of the tether, namely, will destroy the uniform configuration. Thus, the major goal of this section is to employ effective control laws to achieve the deployment and retraction operations of the rotating Hub-Spoke Tethered Formation System. The control laws can bring the rotating Hub-Spoke Tethered Formation System from the initial configuration to a desired stable configuration without exceeding certain limits for the control and state variables. But it is difficult to find an appropriate controller for the nonlinear dynamics of the Hub-Spoke Tethered Formation System mentioned in previous section. In addition, it is important to ensure that the deployment and retraction operations of the Hub-Spoke Tethered Formation System end up in the correct configuration. Hence, the boundary conditions should be satisfied. Therefore, the deployment and retraction control problem of the Hub-Spoke Tethered Formation System can be classified as a kind of nonlinear optimal control problem.

The solution of optimal control problems is well known in [1–3]. Direct transcription of optimal control problems requires approximations of the integration in the cost function, the differential equations of the state-control system, the state-control constraint equations, and ideally the same collocation points can be used for all. In principle, any set of unique collocation points and any discretization methodology can be used. A better option is to use pseudospectral (PS) methods because they converge with spectral accuracy for smooth problems. The three most commonly used set of orthogonal collocation points in the pseudospectral method are Legendre–Gauss (LG), Legendre–Gauss–Radau (LGR), and Legendre–Gauss–Lobatto (LGL) points. The following three mathematical methods have been developed: the Legendre pseudospectral method (LPM) [4, 5], the Radau pseudospectral method (RPM) [6, 7], and the Gauss pseudospectral method (GPM) [8–10]. There are little differences among three methods about the time consumption when solving the same problem. RPM and GPM are much better than LPM in accuracy of approximation, accuracy of control variables and coordinate variables, and rate of convergence. However, in terms of estimation accuracy of the boundary of coordinate variables, GPM is much better than RPM and it has many advantages in dealing with the problem with initial and terminal constraints. So in this paper, the GPM is employed to solve the optimal deployment and retraction control problem of the Hub-Spoke Tethered Formation System.

The GPM can be applied to the deployment and retraction of the rotating two-body Hub-Spoke Tethered Formation System, and in this section, relatively simple uniform configuration during deployment and retraction scenarios are considered. Initially, the two-body HS-TFS is maintaining in the uniform rotating motion with a perfect Hub-Spoke configuration. Then, in order to obtain a changeable coverage of the entire plane in some space applications, the deployment and retraction operations of the formation system are performed. Finally, optimal control is employed to make the HS-TFS back to the uniform rotating configuration. For the rotating two-body HS-TFS, an optimal deployment and retraction problem can be described as follows: (1) Require minimum values of desired rotating angular velocity and an ideal Hub-Spoke configuration $x^*(t)$, which can achieve the minimization of the torque of the master-spacecraft and make the

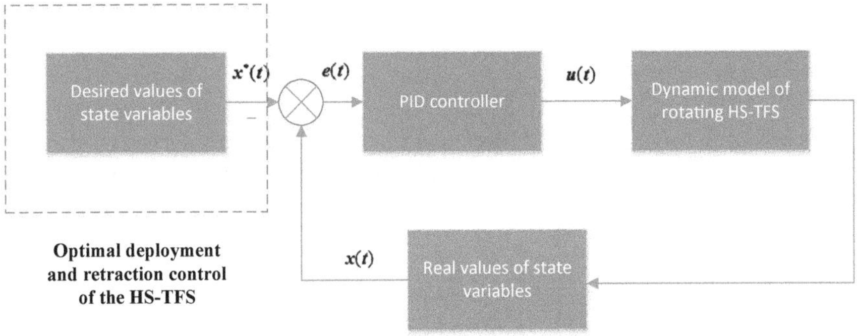

Fig. 8.3 Optimal control scheme for deployment/retraction of the rotating HS-TFS

HS-TFS maintain a uniform rotating motion after the deployment and retraction operation. (2) Employ the PID controller $(u(t))$ to eliminate the error $(e(t))$ of real state variables $x(t)$ and ideal state variables $x^*(t)$. The optimal control scheme for deployment/retraction of the rotating Hub-Spoke Tethered Formation System is demonstrated in Fig. 8.3.

8.2.1 Problem Statement

The optimal control problems considered in this section can be stated as follows:

Find the optimal control that minimizes the objective function:

$$\mathbf{J} = \theta\left[\mathbf{x}\left(t_f\right), t_f\right] + \int_0^{t_f} F[\mathbf{x}(t), \mathbf{u}(t), t]dt \tag{8.1}$$

which is subject to the nonlinear state equations:

$$\dot{\mathbf{x}}(t) = f[\mathbf{x}(t), \mathbf{u}(t), t] \tag{8.2}$$

The boundary condition of the above function is expressed as

$$g\left(\mathbf{x}(t_0), t_0, \mathbf{x}\left(t_f\right), t_f\right) = 0 \tag{8.3}$$

with the path constraints are

$$h_{\min} \leq h(\mathbf{x}(t), \mathbf{u}(t), t) = 0 \leq h_{\max} \tag{8.4}$$

where $\mathbf{x}(t)$ are the state variables, $\mathbf{u}(t)$ are the control variables and t is the time, $h_{\min}$ and $h_{\max}$ are the allowable boundary values of the path constraints.

8.2.2 Assumptions

Some assumptions are made as follows:

(1) A uniform deployment and retraction of the HS-TFS are necessary. In other words, the velocity of deployment and retraction are constant values, which are defined as

$$\begin{cases} L_1 = L_2 = L \\ \dot{L}_1 = \dot{L}_2 = \dot{L} \\ \ddot{L}_1 = \ddot{L}_2 = \ddot{L} = 0 \end{cases} \tag{8.5}$$

(2) The mass of sub-spacecraft1 is the same as sub-spacecraft2, which are expressed as

$$m_1 = m_2 = m \tag{8.6}$$

In consideration of the assumptions, the Eq. (6.146) can be simplified as

$$\begin{cases} r_0\dot{\omega}\sin\varphi_1 - r_0\omega^2\cos\varphi_1 - L(\omega+\dot{\varphi}_1)^2 = -T_1/m \\ r_0\dot{\omega}\cos\varphi_1 + r_0\omega^2\sin\varphi_1 + L(\dot{\omega}+\ddot{\varphi}_1) + 2\dot{L}(\omega+\dot{\varphi}_1) = 0 \\ r_0\dot{\omega}\sin\varphi_2 - r_0\omega^2\cos\varphi_2 - L(\omega+\dot{\varphi}_2)^2 = -T_2/m \\ r_0\dot{\omega}\cos\varphi_2 + r_0\omega^2\sin\varphi_2 + L(\dot{\omega}+\ddot{\varphi}_2) + 2L(\omega+\dot{\varphi}_2) = 0 \\ I\dot{\omega} = r_0T_1\sin\varphi_1 + r_0T_2\sin\varphi_2 + M \\ L = L_0 + \dot{L}t \end{cases} \tag{8.7}$$

where L_0 is the initial value of the length of the tethers connected to sub-spacecraft1 and sub-spacecratft2. Note that $\dot{L}$ is positive value in the uniform deployment operation, a contrarily negative value in the uniform retraction operation.

8.2.3 Constraints

According to "User's Manual for GPOPS" ([11]), the cost function, phase number, dynamics constraints, boundary conditions, inequality path constraints, and the phase continuity (linkage) constraints should be ascertained first. Especially, the phase number is chosen to be one, which is not only sufficient for us to solve the optimal problem, but also reduces the trouble of the phase continuity constraints. The constraints are demonstrated as follows:

(1) Dynamics constraints

The vector of state variables is

$$\mathbf{x} = \begin{bmatrix} x_1 & x_2 & x_3 \end{bmatrix}^{\mathrm{T}} = \begin{bmatrix} \omega & \varphi & \dot{\varphi} \end{bmatrix}^{\mathrm{T}} \tag{8.8}$$

and the vector of control variables is

$$\mathbf{u} = [u_1]^{\mathrm{T}} = [M]^{\mathrm{T}} \tag{8.9}$$

and the dynamics constraints can be written as a system of nonlinear ordinary differential equations:

$$\dot{\mathbf{x}} = \begin{bmatrix} \frac{2mr_0 \sin x_2 \left[r_0 x_1^2 \cos x_2 + L(x_1 + x_3)^2 \right] + u_1}{I + 2mr_0^2 \sin^2 x_2} \\ x_3 \\ -\dot{\omega} - \frac{r_0}{L}\left(\dot{\omega} \cos x_2 + x_1^2 \sin x_2 \right) - \frac{2\dot{L}}{L}(x_1 + x_3) \end{bmatrix} \tag{8.10}$$

(2) Boundary conditions

The boundary conditions can be written as

$$\begin{cases} \omega(t_0) = \omega_0 \\ \omega(t_f) = \omega_f \\ \varphi(t_0) = \varphi_0 \\ \varphi(t_f) = \varphi_f \\ \dot{\varphi}(t_0) = \dot{\varphi}_0 \\ \dot{\varphi}(t_f) = \dot{\varphi}_f \end{cases} \tag{8.11}$$

where $\omega_0, \varphi_0, \dot{\varphi}_0$ are the initial conditions of the state variables, $\omega_f, \varphi_f, \dot{\varphi}_f$ are the final conditions of the state variables. Besides, the conditions $\varphi_0 = \varphi_f$ and $\dot{\varphi}_0 = \dot{\varphi}_f$ should be satisfied.

(3) Path constraints

The path constraints can be written as

$$\begin{cases} \omega_{\min} \le \omega \le \omega_{\max} \\ \varphi_{\min} \le \varphi \le \varphi_{\max} \\ \dot{\varphi}_{\min} \le \dot{\varphi} \le \dot{\varphi}_{\max} \\ M_{\min} \le M \le M_{\max} \end{cases} \tag{8.12}$$

where $\omega_{\min}$ and $\omega_{\max}$ are the allowable boundary values of the rotating angular velocity. $\varphi_{\min}$ and $\varphi_{\max}$ are the allowable boundary values of φ. $\dot{\varphi}_{\min}$ and $\dot{\varphi}_{\max}$ are the allowable boundary values of $\dot{\varphi}$. $M_{\min}$ and $M_{\max}$ are the available boundary values of the torque of the master-spacecraft.

8.2.4 *Objective Function*

Because the deployment and retraction operations are based on the controllable tethers of HS-TFS, the time consumption must be lay on the releasing and retracting times of the tethers. Therefore, the objective function can be demonstrated by minimizing the torque of the master-spacecraft:

$$\mathbf{J} = \int_0^{t_f} u_1^2 dt = \int_0^{t_f} M^2 dt \tag{8.13}$$

8.3 Numerical Simulation

Several numerical simulations of the optimal deployment and retrieval of the rotating two-body HS-TFS are performed. The Gauss Pseudospectral Optimization Solver (GPOPS) is used to solving the optimal deployment and retraction problem of the HS-TFS. This is an open-source and freely available software package for solving optimal control problems, in which the GPM is employed. The parameters are assumed as follows: the mass of the sub-spacecraft m = 5 kg, the rotating radius of the master-spacecraft $r_0 = 1\,\text{m}$, the moment of inertia of master-spacecraft $I = 500.0\,\text{kg}\,\text{m}^2$, the initial value of the length of the tethers connected to sub-spacecraft 1 and sub-spacecratft2 $L_0 = 10\,\text{m}$.

8.3.1 *Torque Control of the Master Satellite*

Controlled deployment

For the controlled deployment of the two-body HS-TFS, the torque of the master-spacecraft, M, is applied during the deployment operation, and the optimal control trajectory is planned by the GPOPS. The dynamics constraints can be presented by Eq. (8.10). The boundary conditions are

$$\begin{cases} \omega(t_0) = \pi/3 & \omega(t_f) = \pi/3 \\ \varphi(t_0) = 0 & \varphi(t_f) = 0 \\ \dot{\varphi}(t_0) = 0 & \dot{\varphi}(t_f) = 0 \end{cases} \tag{8.14}$$

The path constraints are

$$\begin{cases} +\pi/12\,\text{rad/s} \le \omega \le +\pi/2\,\text{rad/s} \\ -\pi/2\,\text{rad} \le \varphi \le +\pi/2\,\text{rad} \\ -\pi/2\,\text{rad/s} \le \dot{\varphi} \le +\pi/2\,\text{rad/s} \\ -30\,\text{N}\,\text{m} \le M \le +30\,\text{N}\,\text{m} \end{cases} \tag{8.15}$$

The PID parameters are

$$\begin{cases} k_p = [1 \quad 1 \quad 1] \\ k_i = [0.05 \quad 0.05 \quad 0.05] \\ k_d = [0.5 \quad 0.5 \quad 0.5] \end{cases} \tag{8.16}$$

Note that if φ is $|\varphi| \geq \pi/2\,\text{rad}$, the tethers will be coiled around the master-spacecraft, thus the path constraint of φ should be less than $\pi/2\,\text{rad}$, that is, $-\pi/2\,\text{rad} \leq \varphi \leq +\pi/2\,\text{rad}$. The objective function is equal to Eq. (8.13). The deployment operation begins with a releasing velocity of 0.1 m/s, and the operation time persists for 40 s. The optimal control solutions based on the above constraints and conditions are demonstrated in Figs. 8.4 and 8.5.

The variations of primary parameters for controlled deployment while $\dot{L} = 0.1$ is demonstrated in Fig. 8.4. Similar to the uncontrolled deployment, the $\dot{L}$ keeps as a constant value, and the uniform deployment operation lasts for 40 s. The final value of L is 14 m. The ω, φ, $\dot{\varphi}$ are varying continuously during the deployment operation. Different from the uncontrolled deployment, the final values of ω, φ, $\dot{\varphi}$ are all equal to the desired final values in Eq. (8.15) and the active control, M, operated during the deployment phase. Figure 8.5 demonstrates the configuration variations of the two-body HS-TFS during the deployment phase. The two-body HS-TFS has

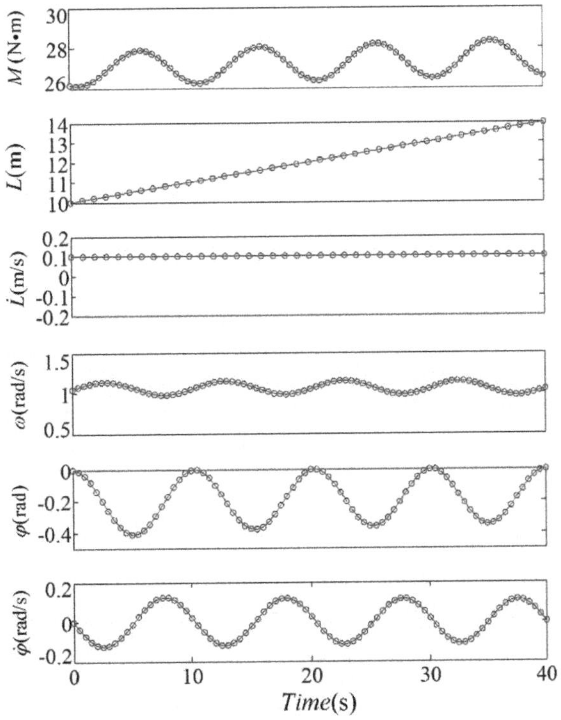

Fig. 8.4 Variations of primary parameters for controlled deployment ($\dot{L} = 0.1$ m/s)

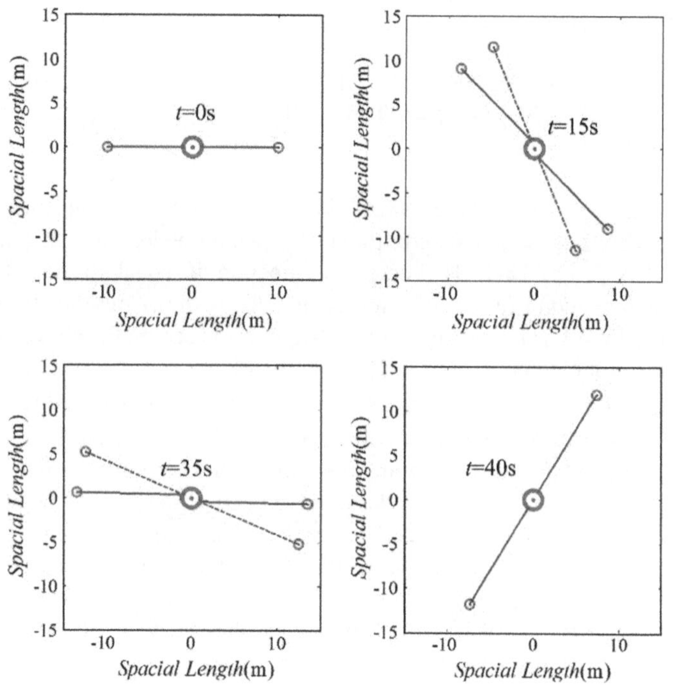

Fig. 8.5 Configuration variations for controlled deployment ($\dot{L} = 0.1$ m/s)

the same configurations in t = 0 s and t = 40 s, but the length of tethers connected sub-spacecraft1 and sub-spacecraft2 is increased.

An alternative option is to deploy the two-body HS-TFS with a relatively small releasing velocity of 0.02 m/s. The simulation results of the solutions are demonstrated in Figs. 8.6 and 8.7. Though the boundary conditions and path constraints are all satisfied, there exist some disparities compared to the simulation results in Figs. 8.4, 8.5, 8.6 and 8.7. These disparities can be summarized as: The quicker of the deployment velocity, the bigger of the variation of magnitude of the control variable (M) will be, which results in much fuel consumption and bigger magnitude of ω and φ. So it would be better to employ relative small releasing velocity for the deployment of Hub-Spoke Tethered Formation System. In this way, the fuel consumption will lower and formation configuration can be maintained more stably.

Controlled retrieval

The simulation results of the controlled retraction of two-body HS-TFS are demonstrated in Figs. 8.8, 8.9, 8.10, and 8.11. Two retractions with different retracting velocity are presented. The dynamics constraints can be presented by Eq. (8.10). The objective function is equal to Eq. (8.13). The boundary conditions, path constraints, and PID parameters are the same as Eqs. (8.14), (8.15), and (8.16). Figure 8.8 presents the variations of primary parameters for controlled retraction

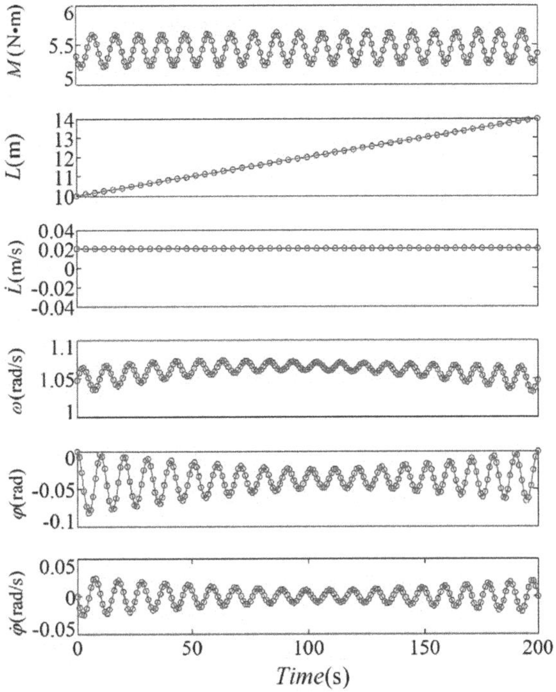

Fig. 8.6 Variations of primary parameters for controlled deployment ($\dot{L} = 0.02\,\mathrm{m/s}$)

while $\dot{L} = -0.1\,\mathrm{m/s}$, and Fig. 8.9 presents the configuration variations. The process of retraction operation lasts 40 s, and the final value of L is 6 m. The ω, φ, $\dot{\varphi}$ are varying continuously during the retraction phase. The final values of ω, φ, $\dot{\varphi}$ are all equal to the desired final values in Eq. (8.14). The active control, M, operated during the deployment phase. Figure 8.9 demonstrates the configuration variations of the controlled retraction while $\dot{L} = -0.1\,\mathrm{m/s}$. The two-body HS-TFS has the same configurations in $t = 0$ s and $t = 40$ s, but the length of tethers connected sub-spacecraft1 and sub-spacecraft2 is decreased.

Figures 8.10 and 8.11 show the optimal solutions for the controlled retraction with a relative small retracing velocity of $\dot{L} = -0.02\,\mathrm{m/s}$. Compared to the simulation results in Figs. 8.8 and 8.9, the variation amplitudes of ω, φ, $\dot{\varphi}$ are relatively small, and the variation frequency of ω, φ, $\dot{\varphi}$ are relatively high. These are similar to the simulation results of controlled deployment operation. The retraction operation costs 200 s to reach the same final value of L as the faster retracting velocity of $\dot{L} = -0.1\,\mathrm{m/s}$. Thus, it can be concluded that smaller retrieving velocity can result in relative stable retraction phase, and a reduction of the magnitude of the active control, M.

All simulation results in Sect. 8.3.1 suggest that the GPM is a feasible method for the optimal deployment and retraction control problems. Some useful conclusions are drawn: the deployment/retraction operations with small releasing/

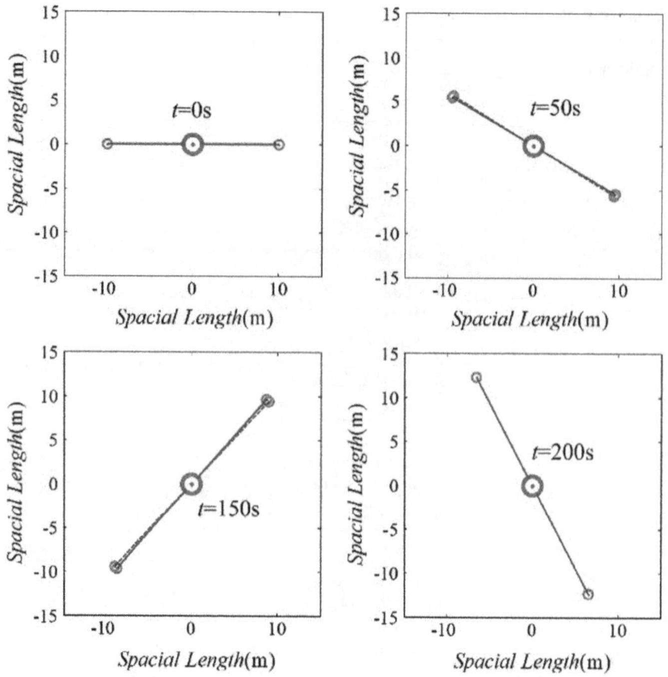

Fig. 8.7 Configuration variations for controlled deployment ($\dot{L} = 0.02$ m/s)

retracting velocity are more stable than faster situations and the magnitude of the active control in the deployment/retraction operations, the torque of the master-spacecraft, are also reduced. However, the time cost by the deployment and retraction operations inevitably increases.

8.3.2 Tension Control of the Sub-Satellite

Controlled deployment

For the controlled deployment of the two-body HS-TFS, the tension of the sub-spacecraft, F_t, is applied during the deployment operation, and the optimal control trajectory is planned by the GPOPS. The dynamics model can be presented by Eq. (8.10). The boundary conditions are

$$\begin{cases} \omega(t_0) = \pi/3 & \omega(t_f) = \pi/3 \\ \varphi(t_0) = 0 & \varphi(t_f) = 0 \\ \dot{\varphi}(t_0) = 0 & \dot{\varphi}(t_f) = 0 \end{cases} \tag{8.17}$$

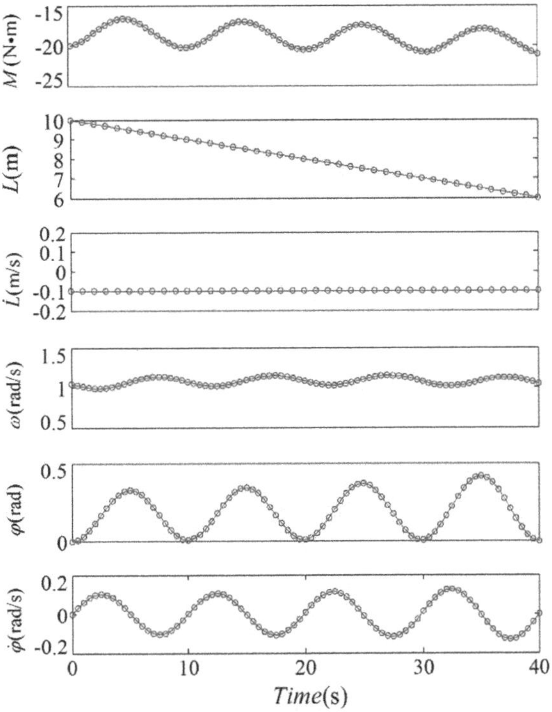

Fig. 8.8 Variations of primary parameters for controlled retraction ($\dot{L} = -0.1$ m/s)

The path constraints are

$$\begin{cases} +\pi/12\,\text{rad/s} \leq \omega \leq +\pi/2\,\text{rad/s} \\ -\pi/2\,\text{rad} \leq \varphi \leq +\pi/2\,\text{rad} \\ -\pi/2\,\text{rad/s} \leq \dot{\varphi} \leq +\pi/2\,\text{rad/s} \\ -1\text{N} \leq F_t \leq +1\text{N} \end{cases} \tag{8.18}$$

The objective function is

$$\mathbf{J} = \int_0^{t_f} F[\mathbf{x}(t), \mathbf{u}(t), t]dt = \int_0^{t_f} F_t^2 dt \tag{8.19}$$

The optimal control solutions based on the above constraints and conditions are demonstrated in Figs. 8.4 and 8.5.

The variations of parameters for controlled deployment with the tension of the sub-spacecraft while $\dot{L} = 0.1$ is demonstrated in Figs. 8.12 and 8.13. Similar to the uncontrolled deployment, the uniform deployment operation lasts for 40 s. The

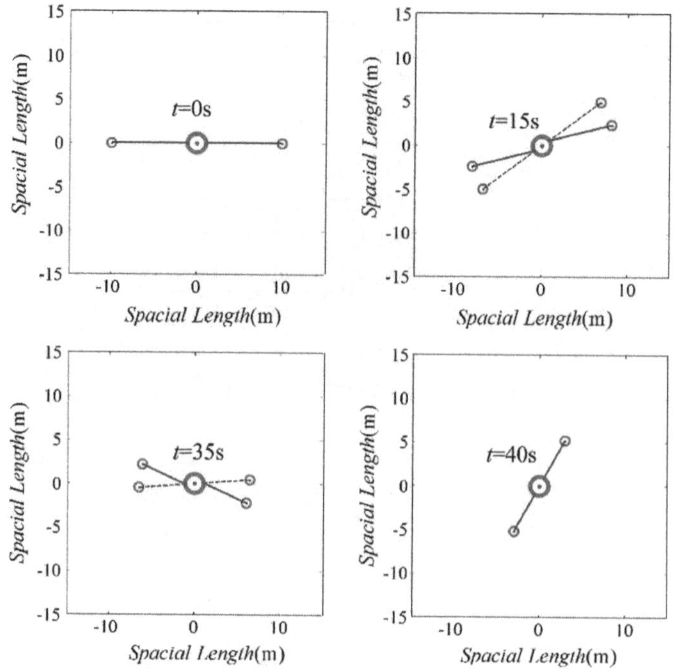

Fig. 8.9 Configuration variations for controlled retraction ($\dot{L} = -0.1$ m/s)

value of L increases from 10 to 14 m. The tension provided by sub-spacecraft is working on during all the deployment and stops when the deployment ends. Figure 8.13 shows that the initial value of ω is $\pi/3$ rad, it vibrates along the initial value during deployment, and finally maintains at $\pi/3$ rad. φ, $\dot{\varphi}$ vibrate during deployment and maintain at zero. After 40 s, ω, φ, $\dot{\varphi}$ vibrate little along zero in uncontrolled situation.

An alternative option is to deploy the two-body HS-TFS with a relatively small releasing velocity of 0.02 m/s with tension of the sub-spacecraft. The simulation results of the solutions are demonstrated in Figs. 8.14 and 8.15. The uniform deployment operation lasts for 200 s, and the value of L increases from 10 to 14 m. The tension provided by sub-spacecraft is working on during all the deployment and stops when the deployment ends. Figure 8.15 demonstrates that ω, φ, $\dot{\varphi}$ vibrate all the time during deployment. When the deployment phase ends, ω maintains at $\pi/3$ rad, φ, $\dot{\varphi}$ maintain at zero. Compared with the results of $\dot{L} = 0.1$ m/s, the

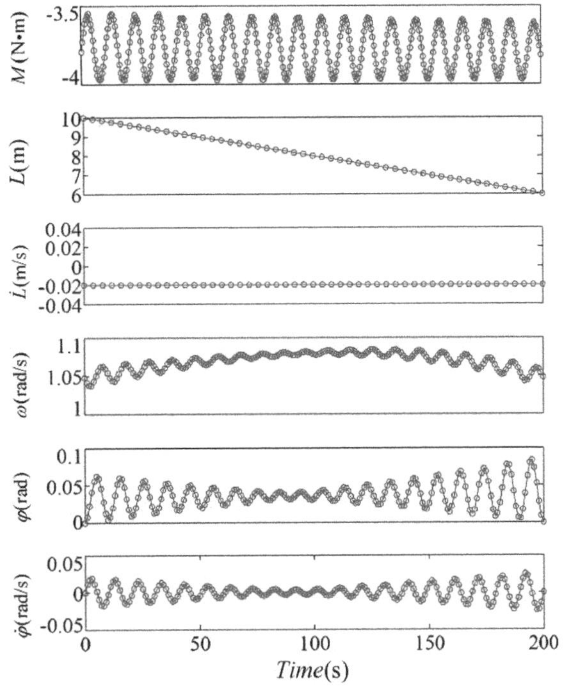

Fig. 8.10 Variations of primary parameters for controlled retraction ($\dot{L} = -0.02\,\mathrm{m/s}$)

tension of sub-spacecraft is much smaller when the deployment velocity is $\dot{L} = 0.02\,\mathrm{m/s}$, and the amplitudes of the vibration of ω, φ, $\dot{\varphi}$ are also smaller. However, the frequencies of the vibration of ω, φ, $\dot{\varphi}$ are increased.

Generally, compared with the simulation results in Sect. 8.3.1, the amplitudes of the vibration of ω, φ, $\dot{\varphi}$ are much smaller and the control consumption is also much smaller in both $\dot{L} = 0.1\,\mathrm{m/s}$ and $\dot{L} = 0.02\,\mathrm{m/s}$ cases. It can be concluded that the tension of sub-spacecraft is more appropriate for controlling the deployment of HS-TFS than the torque of master-spacecraft.

Controlled retrieval

The constraints of controlled retrieval are the same as the constraints of controlled deployment with tension of sub-spacecraft. The simulation results of the controlled retraction of two-body HS-TFS are demonstrated in Figs. 8.16, 8.17, 8.18, and 8.19. Figures 8.16 and 8.17 show the optimal solutions for the controlled retraction with a relative small retracing velocity of $\dot{L} = -0.1\,\mathrm{m/s}$. The uniform retrieval operation lasts for 40 s, and the value of L decreases from 10 m to 6 m. The tension provided by sub-spacecraft is working on during all the retrieval and stops when the retrieval ends. Figure 8.17 shows that the initial value of ω is $\pi/3\,\mathrm{rad}$, it vibrates

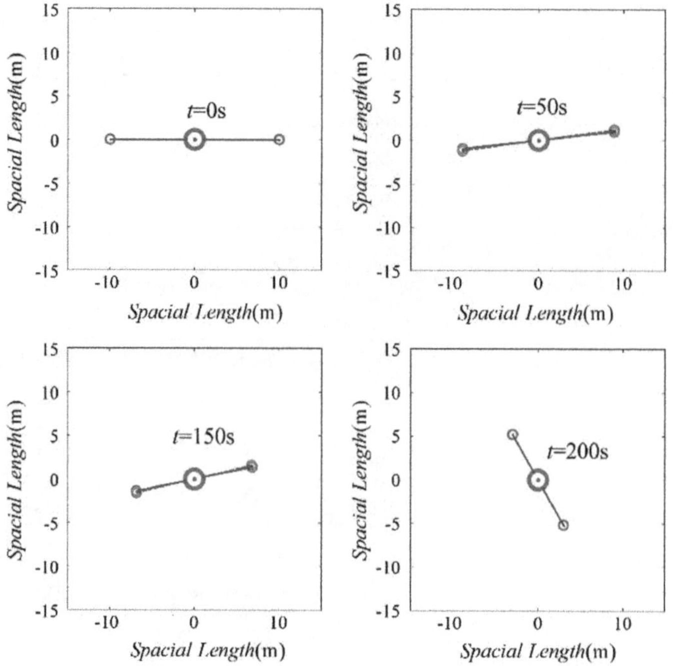

Fig. 8.11 Configuration variations for controlled retraction ($\dot{L} = -0.02$ m/s)

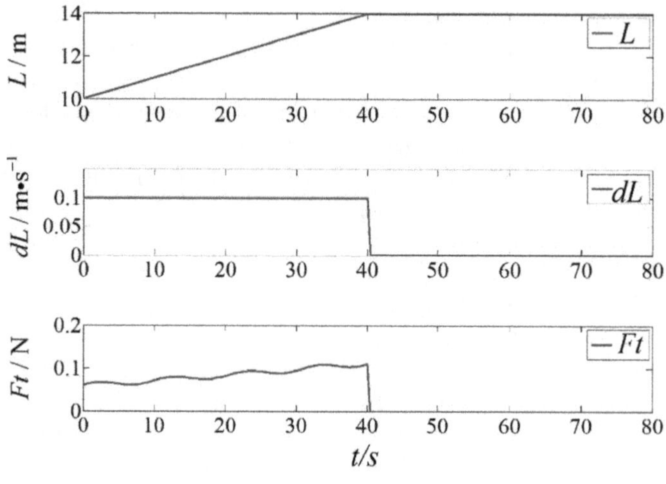

Fig. 8.12 Variations of tether length and tension of the sub-spacecraft for controlled deployment ($\dot{L} = 0.1$ m/s)

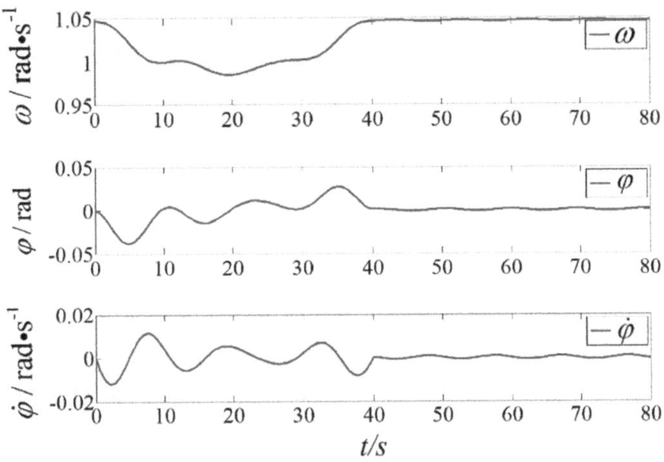

Fig. 8.13 Variations of the angular velocity of master-spacecraft, angle and angular velocity between the straight tether's direction and radial direction for controlled deployment ($\dot{L} = 0.1$ m/s)

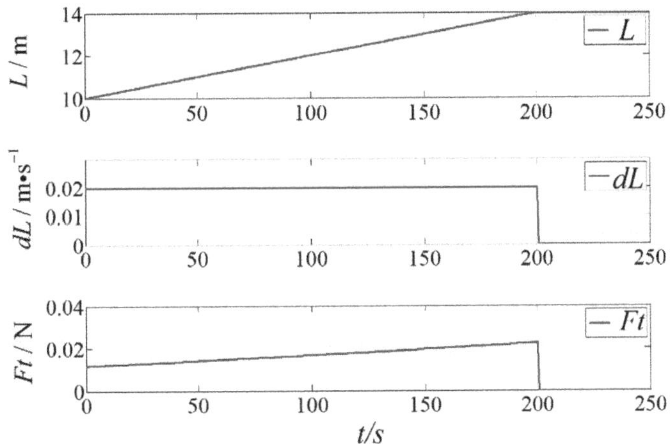

Fig. 8.14 Variations of tether length and tension of the sub-spacecraft for controlled deployment ($\dot{L} = 0.02$ m/s)

along the initial value during retrieval, and finally maintains at $\pi/3$ rad. φ, $\dot{\varphi}$ vibrate during deployment and maintain at zero. After 40 s, ω, φ, $\dot{\varphi}$ vibrate little along zero in uncontrolled situation.

Figures 8.18 and 8.19 show the results of a relatively small retrieval velocity of 0.02 m/s with tension of the sub-spacecraft. The uniform retrieval operation lasts for 200 s, and the value of L decreases from 10 to 6 m. The tension provided by sub-spacecraft is working on during all the retrieval and stops when the retrieval ends. Figure 8.19 demonstrates that ω, φ, $\dot{\varphi}$ vibrate all the time during retrieval.

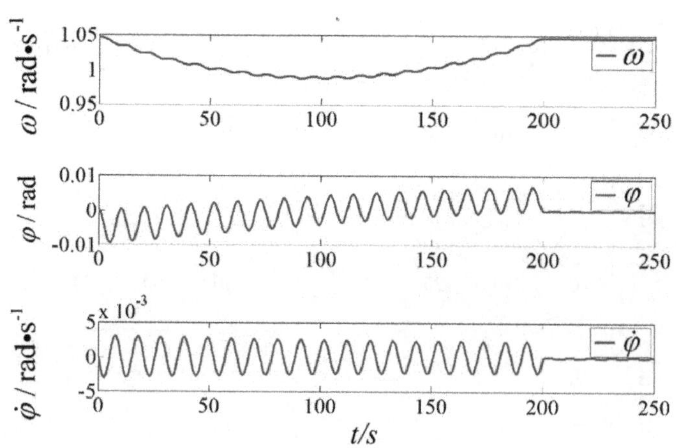

Fig. 8.15 Variations of angular velocity of master-spacecraft, angle and angular velocity between the straight tether's direction and radial direction for controlled deployment ($\dot{L} = 0.02$ m/s)

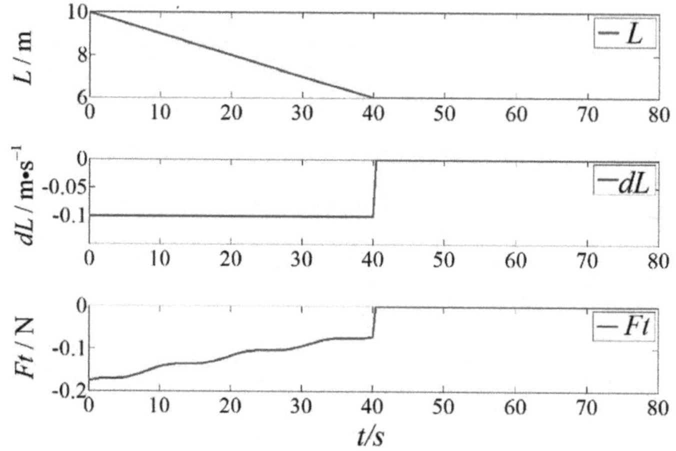

Fig. 8.16 Variations of tether length and tension of the sub-spacecraft for controlled retrieval ($\dot{L} = -0.1$ m/s)

When the retrieval phase ends, ω maintains at $\pi/3$ rad, φ, $\dot{\varphi}$ maintain at zero. Compared with the results of $\dot{L} = -0.1$ m/s, the tension of sub-spacecraft is much smaller when the retrieval velocity is $\dot{L} = -0.02$ m/s, and the amplitudes of the vibration of ω, φ, $\dot{\varphi}$ are also smaller. However, the frequencies of the vibration of ω, φ, $\dot{\varphi}$ are increased.

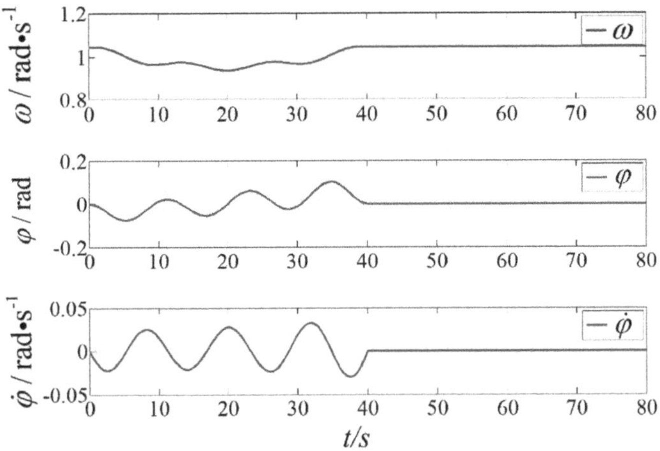

Fig. 8.17 Variations of angular velocity of master-spacecraft, angle and angular velocity between the straight tether's direction and radial direction for controlled retrieval ($\dot{L} = -0.1$ m/s)

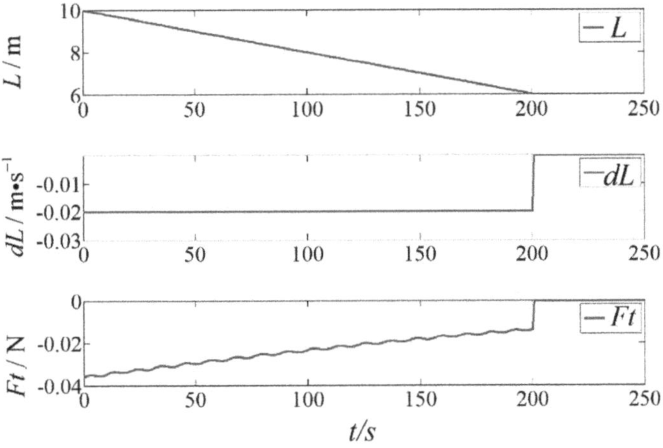

Fig. 8.18 Variations of tether length and tension of the sub-spacecraft for controlled retrieval ($\dot{L} = -0.02$ m/s)

Generally, compared with the simulation results in Sect. 8.3.1, the amplitudes of the vibration of ω, φ, $\dot{\varphi}$ are much smaller and the control consumption is also much smaller in both $\dot{L} = -0.1$ m/s and $\dot{L} = -0.02$ m/s cases. It can be concluded that the tension of sub-spacecraft is more appropriate for controlling the retrieval of HS-TFS than the torque of master-spacecraft.

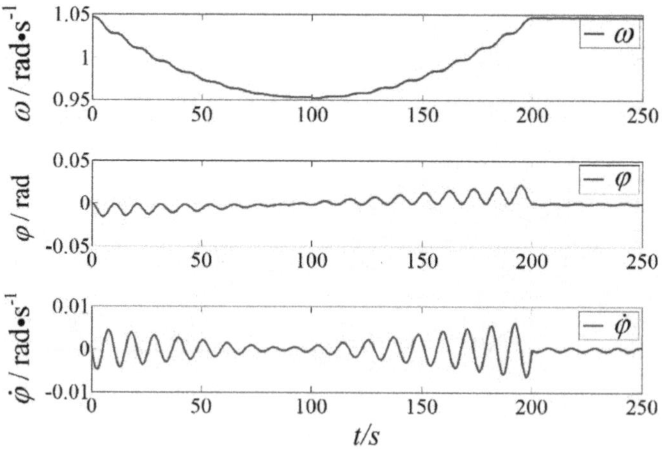

Fig. 8.19 Variations of angular velocity of master-spacecraft, angle, and angular velocity between the straight tether's direction and radial direction for controlled retrieval ($\dot{L} = -0.02$ m/s)

References

1. Enright PJ, Conway BA (1992) Discrete approximations to optimal trajectories using direct transcription and nonlinear programming. J Guid Control Dyn 15(4):994–1002
2. Gong Q, Kang W, Michael Ross I (2006) A pseudospectral method for the optimal control of constrained feedback linearizable systems. IEEE Trans Autom Control 51(7):1115–1129
3. Bonnans JF, Laurent-Varin J (2006) Computation of order conditions for symplectic partitioned Runge-Kutta schemes with application to optimal control. Numer Math 103 (1):1–10
4. Elnagar GN, Kazemi MA (1998) Pseudospectral Legendre-based optimal computation of nonlinear constrained variational problems. J Comput Appl Math 88(2):363–375
5. Rea J (2003) Launch vehicle trajectory optimization using a Legendre pseudospectral method. In: Proceedings of the AIAA guidance, navigation and control conference
6. Garg D, Patterson M, Hager WW et al (2010) A unied framework for the numerical solution of optimal control problems using pseudospectral methods. Automatica 46(11):1843–1851
7. Garg D, Hager WW, Rao AV (2011) Pseudospectral methods for solving infinite-horizon optimal control problems. Automatica 47(4):829–837
8. Reddien GW (1979) Collocation at Gauss points as a discretization in optimal control. SIAM J Control Optim 17(2):298–306
9. Hou H, Hager WW, Rao AV (2012) Convergence of a Gauss pseudospectral method for optimal control. In: AIAA guidance, navigation, and control conference and exhibit. American Institute of Aeronautics and Astronautics, Minnesota
10. Rao AV, Benson DA, Darby C et al (2010) Algorithm 902: Gpops, a matlab software for solving multiple-phase optimal control problems using the gauss pseudospectral method. ACM Trans Math Softw (TOMS) 37(2):22:1–39
11. Rao AV, Benson D, Darby CL et al (2011) User's manual for GPOPS version 4.x: a MATLAB software for solving multiple-phase optimal control problems using hp-adaptive pseudospectral methods. University of Florida, Gainesville

Chapter 9
Formation-Keeping Control of the Hub–Spoke System

The Hub–Spoke System (HSS) is usually a rotating system to acquire specific advantages, such as reduction of fuel consumption, promotion of the formation stability. An ideal Hub–Spoke configuration is necessary when the system is under the rotation motion. However, some inevitable factors, such as space environment and errors of mechanism, will lead to unpredictable formation errors of the HSS. It indicates that the ideal Hub–Spoke configuration does not exist, only if an active control strategy is introduced. Therefore, it is necessary to perform active control operations for the HSS to recover the ideal Hub–Spoke configuration when the inevitable errors occur.

The formation control equations of the Hub–Spoke Tethered Formation System are considered and controllability analysis for formation-keeping control is analyzed in Sect. 9.1. Two coordinated formation control strategies are proposed and described in detail in Sects. 9.2 and 9.3. And numerical simulations are implemented to validate the proposed coordinated formation control strategies in Sect. 9.4.

9.1 Controllability Analysis for Formation-Keeping Control

There are three feasible active control forces for the HSS, including the torque of the master spacecraft, the tension forces of the tethers connected with sub-spacecraft, and the thrusts of the sub-spacecraft. The disparities of the three active control forces can be summarized as follows:

(a) The control mechanism of tension forces is usually an electrical appliance in space missions. Because the tension control does not need any fuel consumption, it is extremely attractive for most space missions. However, to keep

P. Huang and F. Zhang, *Theory and Applications of Multi-Tethers in Space*,
Springer Tracts in Mechanical Engineering,
https://doi.org/10.1007/978-981-15-0387-0_9

the rotating motion of the formation system, the magnitude of control values provided by the tether tensions should be small.

(b) The torque of the master spacecraft and the thrusts of the sub-spacecraft can provide relatively large control values, but the fuel storage is limited.

(c) Compared to the sub-spacecraft, the master spacecraft is provided with more missions, such as formation maneuver. Thus, the fuel consumption of the master spacecraft should be used carefully. In another word, it is better to achieve the formation control operation by the thrusts of the sub-spacecraft and the tension forces.

A minimum of performance index is desirable in formation control operation such as the fuel consumption and the time consumption. For this consideration, it is possible to find a coordinated control strategy which can maximize the advantages of the three control values and improve the performance index of the formation control operation. In this paper, two coordinated formation control strategies are proposed, in which the formation errors and control time are considered separately.

There are two challenges for the formation control problem of the HSS: (1) A tether mathematical model is primary, which is used to describe the dynamic of the rotating HSS and design the controller. Although the flexible tether has attractive advantages in Tethered Formation System, it is difficult to obtain mathematic equations of motion which can be used to precisely describe the dynamics of the flexible tether and controller design at the same time. A better trade-off from the point of view of controller design should be decided. Therefore, the dynamics of the Tethered Formation System is a complicated problem due to the flexible tethers. (2) The choice of the coordinated criterion of the formation control strategy is another issue. The rationality and feasibility of the coordinated criterion should be considered.

Of more relevance to the present paper, a lot of work has been carried out on the dynamics of the Tethered Formation System, whereas comparatively few studies have been presented on the dynamics of the rotating HSS. In reference [1], the dynamics of a tethered-connected three-body system is studied. The dynamics of certain multi-tethered satellite formations containing a parent (or central) body are examined in reference [2]. In this study, the satellites are regarded as point masses; the tethers in the formation system are massless and straight, and the motion of the parent body of the formation is prescribed. The Lagrangian formulation is used to derive the equations of motion of the tethered system. Some researchers have focused on chains of N-bodies to obtain a general formulation of the dynamics [3, 4]. Others have more specifically looked at three- or four-tethered bodies as a means of modeling a micro- or variable gravity laboratory attached to the space station [1, 5, 6]. The construction of a ring of satellites around the Earth might be advantageous for future space colonization and to solution of problems associated with the high volume of space traffic [7, 8]. Certain researchers have examined the attitude control of spinning tethered formations of spacecraft modeled as rigid bodies [9]. Finally, research has been aimed at the problem of two bodies connected by multiple tethers [10]. However, most researchers introduce the Lagrangian formulation to obtain the dynamics model of tethered system. In addition, the tethers are usually seemed as rigid body and the elasticity are always neglected.

Since the Hub–Spoke Tethered Formation System is one of the configurations of the Multi-tethered Satellites (MTS), some relative literatures about the control of MTS are reviewed. Misra studied the three-dimensional motion of an N-body TTS, while all the satellites in the system are treated as mass points due to the length of tether [4]. The equilibrium configurations and stability in of Three-Body Tethered Systems, and some conclusions about the control of multi-tethered system were given [11–13]. It is proved that the characteristics of the liberation of the tethers in multi-tethered satellite system are the same as the single tether of a two-body TTS. The reel rate laws using linear pitch rate and quadratic roll rate feedback are successful in controlling the motion of the three-body system and the four-body system like TECS during deployment, retrieval, or station keeping [14]. However, in the existing papers, to our knowledge, the tension control or the reel rate control of the tether is used alone for the station keeping of the system, and all the satellites are assumed to be mass points. The tether cannot give a perfect control performance, and the other control inputs should be introduced.

In this chapter, a formation control scheme of the rotating HSS is proposed. First, an analytical model based on some reasonable assumptions is introduced to describe the dynamics of the rotating HSS. After that, two coordinated formation control strategies are proposed and described in detail, and detailed numerical simulations are implemented based on the control system.

9.1.1 Control-Based Model

Some inspirations to the dynamics of the Hub–Spoke Tethered Formation System can be found in some existing studies of the deployment of large space web [13–16]. The large space web is simplified as several straight tethers during the centrifugal deployment, and a hub in the center of the space web can provide spinning torque to control the deployment process. An analytical model based on these simplifications is introduced to describe the deployment of the large space web. From the point of view of analysis, the large space web has a similar configuration as the Hub–Spoke Tethered Formation System, and a rotating motion is also needed in the deployment phase. Therefore, the dynamics of the HSS can be described by a similar analytical model. The disparity is that the length of the tethers of the HSS is controllable.

As a common practice, a suitable low-order model of a dynamic system is employed for preliminary control system design. For instance, an inextensible tether model is often employed for single-tethered systems and has been shown to generate good results with respect to more accurate models of the system. To study the formation control of the rotating Hub–Spoke Tethered Formation System, a relatively simple analytical model is required to describe the dynamics of the formation system. For simplicity, the two-body Hub–Spoke Tethered Formation System is considered and the analytical results can be enlarged to the multi-body situation.

Although the mass of tether is important due to the gravity gradient in the traditional Multi-tethered Spacecraft System where the tethers are always longer than one kilometer, its contribution on HSS is not so that important because the tethers here are of the order of ten meters. According to this characteristic and tension forces of the tethers are used for control, the tether could be assumed to be a straight massless rigid link. Besides that, the following assumptions are used for the analytical model of the two-body Hub–Spoke Tethered Formation System.

The master spacecraft is a rigid hub, which can provide rotating torque. The sub-spacecraft are regarded as mass points, which can provide thrust.

The radial tethers are assumed to be straight, and symmetrically relative to the center of the master spacecraft, and the centrifugal forces are provided by the tether tensions.

The masses of the radial tethers are ignored.

The gravity gradient and the elasticity in the radial tethers are neglected.

The rotating motion is in the orbit plane, while out-of-plane motions are ignored.

Energy dissipation caused by deformation, friction, and environmental effects are neglected.

The two-body Hub–Spoke Tethered Formation System model considered in this paper is demonstrated in Fig. 9.1. The O_1 denotes the center of the master spacecraft, and the O_2 denotes the point from which stems the tether connected with sub-spacecraft1. Two coordinate systems are introduced to represent the dynamics, which are denoted by $O_1x_1y_1$ and $O_2x_2y_2$. The O_2x_2-axis points positively outward from O_2 along the tether connected with sub-spacecraft1. The O_2y_2-axis is perpendicular to O_2x_2. The angular velocity of the master spacecraft is ω, and the $O_2x_2y_2$ is rotating around the master spacecraft with the same angular velocity of ω. The length of the tethers connected with sub-spacecraft1 and sub-spacecratft2 are denoted by L_1 and L_2. The tension force of the tethers connected with sub-spacecraft1 and sub-spacecratft2 are denoted by T_1 and T_2. The masses of sub-spacecraft1 and sub-spacecratft2 are denoted by m_1 and m_2. The r_0 denotes the rotating radius of the master spacecraft. The I denotes the moment of inertia of master spacecraft. The torque of the master spacecraft is denoted by M. The thrusts of the sub-spacecraft are denoted by Ft_1 and Ft_2. It is supposed that the directions of Ft_1 and Ft_2 are perpendicular to the directions of L_1 and L_2.

The change in angular momentum for the central master spacecraft around its axis of rotation is

$$I\ddot{\theta} = I\dot{\omega} = r_0 T_1 \sin\varphi_1 + r_0 T_2 \sin\varphi_2 + M \tag{9.1}$$

where θ is the rotation angle of the master spacecraft, φ_1, and φ_2 are the angles between the straight tether's direction of sub-spacecrafts and radial direction. Because stiffness and damping are not included, the equation of motion for the sub-spacecraft1 is simply obtained as

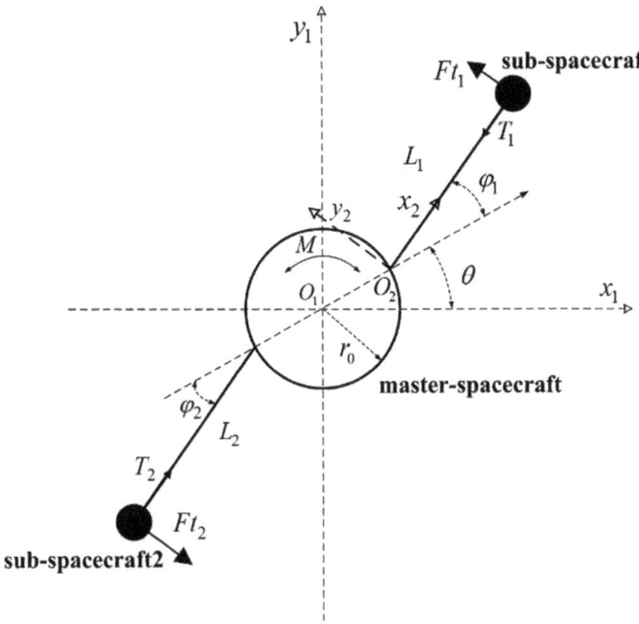

Fig. 9.1 The analytical model for the two-body HSS

$$m_1(r_0\ddot{\boldsymbol{e}}_{x1}+L_1\ddot{\boldsymbol{e}}_{x2})=T_1\boldsymbol{e}_{x2} \tag{9.2}$$

where $\boldsymbol{e}_{x1}$, $\boldsymbol{e}_{y1}$ is the unit vectors of $O_1x_1y_1$, $\boldsymbol{e}_{x2}$ and $\boldsymbol{e}_{y2}$ is the unit vectors of $O_2x_2y_2$.

In consideration of the Poisson Formula:

$$\dot{\boldsymbol{e}}_{x1}=\omega\boldsymbol{e}_{y1} \tag{9.3}$$

$$\dot{\boldsymbol{e}}_{x2}=(\omega+\dot{\varphi}_1)\boldsymbol{e}_{y2} \tag{9.4}$$

By submitting Eqs. (9.3) and (9.4) into Eq. (9.2), and projecting $\boldsymbol{e}_{x1}$ and $\boldsymbol{e}_{y1}$ in the coordinate system $O_2x_2y_2$, one can obtain as

$$\begin{cases} (r_0\dot{\omega}\sin\varphi_1-r_0\omega^2\cos\varphi_1-L_1(\omega+\dot{\varphi}_1)^2+\dot{L}_1)\mathbf{e}_{x2} \\ +(r_0\dot{\omega}\cos\varphi_1+r_0\omega^2\sin\varphi_1+L_1(\dot{\omega}+\ddot{\varphi}_1)+2\dot{L}_1(\omega+\dot{\varphi}_1))\mathbf{e}_{y2} \\ =-T_1/m_1\boldsymbol{e}_{x2}+F_{t1}/m_1\boldsymbol{e}_{y2} \end{cases} \tag{9.5}$$

From Eq. (9.5), we obtain the following equations of motion for sub-spacecraft 1:

$$\begin{cases} r_0\dot{\omega}\sin\varphi_1-r_0\omega^2\cos\varphi_1-L_1(\omega+\dot{\varphi}_1)^2+\ddot{L}_1=-T_1/m_1 \\ r_0\dot{\omega}\cos\varphi_1+r_0\omega^2\sin\varphi_1+L_1(\dot{\omega}+\ddot{\varphi}_1)+2\dot{L}_1(\omega+\dot{\varphi}_1)=F_{t1}/m_1 \end{cases} \tag{9.6}$$

Similarly, the equations of motion for sub-spacecraft 2 can be obtained as

$$\begin{cases} r_0\dot{\omega}\sin\varphi_2 - r_0\omega^2\cos\varphi_2 - L_2(\omega+\dot{\varphi}_2)^2 + \ddot{L}_2 = -T_2/m_2 \\ r_0\dot{\omega}\cos\varphi_2 + r_0\omega^2\sin\varphi_2 + L_2(\dot{\omega}+\ddot{\varphi}_2) + 2\dot{L}_2(\omega+\dot{\varphi}_2) = F_{t2}/m_2 \end{cases} \tag{9.7}$$

The following equations of motion for the two-body Hub–Spoke Tethered Formation System are obtained as

$$\begin{cases} r_0\dot{\omega}\sin\varphi_1 - r_0\omega^2\cos\varphi_1 - L_1(\omega+\dot{\varphi}_1)^2 + \ddot{L}_1 = -T_1/m_1 \\ r_0\dot{\omega}\cos\varphi_1 + r_0\omega^2\sin\varphi_1 + L_1(\dot{\omega}+\ddot{\varphi}_1) + 2\dot{L}_1(\omega+\dot{\varphi}_1) = F_{t1}/m_1 \\ r_0\dot{\omega}\sin\varphi_2 - r_0\omega^2\cos\varphi_2 - L_2(\omega+\dot{\varphi}_2)^2 + \ddot{L}_2 = -T_2/m_2 \\ r_0\dot{\omega}\cos\varphi_2 + r_0\omega^2\sin\varphi_2 + L_2(\dot{\omega}+\ddot{\varphi}_2) + 2\dot{L}_2(\omega+\dot{\varphi}_2) = F_{t2}/m_2 \\ I\dot{\omega} = r_0T_1\sin\varphi_1 + r_0T_2\sin\varphi_2 + M \end{cases} \tag{9.8}$$

A similar model has been derived and used for deployment control of rotating space webs [10–16]. Even though the Hub–Spoke Tethered Formation System and the space webs are different in practice, the mathematical formulations of the rotating motion are similar. The disparity is that the length of the tethers in the Hub–Spoke Tethered Formation System is controllable, whereas the length of the arms of the space webs is decided by the folding pattern and deployment velocity of the space webs. For simplification, the following assumptions are given as

$$\begin{cases} m_1 = m_2 = m \\ L_1 = L_2 = L \end{cases} \tag{9.9}$$

Then Eq. (9.8) can be rewritten as

$$\begin{cases} r_0\dot{\omega}\sin\varphi_1 - r_0\omega^2\cos\varphi_1 - L(\omega+\dot{\varphi}_1)^2 = -T_1/m_1 \\ r_0\dot{\omega}\cos\varphi_1 + r_0\omega^2\sin\varphi_1 + L(\dot{\omega}+\ddot{\varphi}_1) = F_{t1}/m_1 \\ r_0\dot{\omega}\sin\varphi_2 - r_0\omega^2\cos\varphi_2 - L(\omega+\dot{\varphi}_2)^2 = -T_2/m_2 \\ r_0\dot{\omega}\cos\varphi_2 + r_0\omega^2\sin\varphi_2 + L(\dot{\omega}+\ddot{\varphi}_2) = F_{t2}/m_2 \\ I\dot{\omega} = r_0T_1\sin\varphi_1 + r_0T_2\sin\varphi_2 + M \end{cases} \tag{9.10}$$

A uniform rotating motion is needed in most space applications of the HSS, and the following conditions can be obtained in the uniform rotating motion:

$$\begin{cases} \varphi_1 = \bar{\varphi}_1 = 0 \\ \varphi_2 = \bar{\varphi}_2 = 0 \\ \omega = \bar{\omega} \\ \dot{\omega} = \dot{\bar{\omega}} = 0 \\ M = \bar{M} = 0 \\ F_{t1} = \bar{F}_{t1} = 0 \\ F_{t2} = \bar{F}_{t2} = 0 \\ T_1 = \bar{T}_1 \\ T_2 = \bar{T}_2 \end{cases} \tag{9.11}$$

where $\bar{\varphi}_1$, $\bar{\varphi}_2$, $\bar{\omega}$, $\dot{\bar{\omega}}$, $\bar{M}$, $\bar{F}_{t1}$, $\bar{F}_{t2}$, $\bar{T}_1$, and $\bar{T}_2$ denote the values of parameters of the HSS in the uniform rotating motion. By submitting Eq. (9.11) into Eq. (9.10), the ideal uniform rotating motion of the HSS can be described as

$$\begin{cases} \bar{\omega}^2(L+r_0) = \bar{T}_1/m \\ \bar{\omega}^2(L+r_0) = \bar{T}_2/m \end{cases} \tag{9.12}$$

Actually $\bar{T}_1 = \bar{T}_2$ is necessary for the uniform rotating motion.

It is expected that the uniform rotating motion of the HSS is a durative motion in applications. But this is almost impossible. Errors of φ_1, φ_2, and ω will occurs because of effecting of space environment, and the ideal rotating motion of the HSS will be destroyed. The small errors of φ_1, φ_2, and ω are introduced:

$$\begin{cases} \varphi_1 = \bar{\varphi}_1 + \Delta\varphi_1 = \Delta\varphi_1 \\ \varphi_2 = \bar{\varphi}_2 + \Delta\varphi_2 = \Delta\varphi_2 \\ \omega = \bar{\omega} + \Delta\omega \end{cases} \tag{9.13}$$

where $\Delta\varphi_1$, $\Delta\varphi_2$, and $\Delta\omega$ denote the small errors of φ_1, φ_2, and ω. Due to these small errors, the values of T_1 and T_2 will be changed. It is supposed that the T_1, T_2, M, F_{t1}, and F_{t2} can provide control values, though the control values provided by T_1 and T_2 are inevitable small. These control values can be described by

$$\begin{cases} T_1 = \bar{T}_1 + \Delta T_1 \\ T_2 = \bar{T}_2 + \Delta T_2 \\ M = \bar{M} + \Delta M = \Delta M \\ F_{t1} = \bar{F}_{t1} + \Delta F_{t1} = \Delta F_{t1} \\ F_{t2} = \bar{F}_{t2} + \Delta F_{t2} = \Delta F_{t2} \end{cases} \tag{9.14}$$

where ΔT_1, ΔT_2, ΔM, ΔF_{t1}, and ΔF_{t2} denote the control values provided by T_1, T_2, M, F_{t1}, and F_{t2}. By submitting Eqs. (9.13) and (9.14) into Eq. (9.10), and considering Eqs. (9.12) and (9.13), the following linear system equations are obtained (the higher order terms of $\Delta\varphi_1$, $\Delta\varphi_2$, and $\Delta\omega$ are neglected):

$$\begin{cases} \Delta\dot{\varphi}_1 = c\Delta\omega + d\Delta T_1 \\ \Delta\dot{\varphi}_2 = c\Delta\omega + d\Delta T_2 \\ \Delta\ddot{\varphi}_1 = (a+b)\Delta\varphi_1 + a\Delta\varphi_2 + e\Delta F_{t1} + c/I\Delta M \\ \Delta\ddot{\varphi}_2 = a\Delta\varphi_1 + (a+b)\Delta\varphi_2 + e\Delta F_{t2} + c/I\Delta M \\ \Delta\dot{\omega} = (a/c)\Delta\varphi_1 + (a/c)\Delta\varphi_2 + (1/I)\Delta M \end{cases} \tag{9.15}$$

where

$$\begin{cases} a = -m\bar{\omega}^2 r_0 (r_0 + L)^2 \big/ (IL) \\ b = -\bar{\omega}^2 r_0 / L \\ c = -(r_0 + L)/L \\ d = 1/(2mL\bar{\omega}) \\ e = 1/(mL) \end{cases} \tag{9.16}$$

Equation (9.15) can be used to describe the formation errors of the HSS. But the controllability of Eq. (9.15) should be analyzed prior to discussions of the formation control problems. The vector of state variables is

$$\boldsymbol{x} = [\Delta\varphi_1 \ \Delta\varphi_2 \ \Delta\dot{\varphi}_1 \ \Delta\dot{\varphi}_2 \ \Delta\omega]^{\mathrm{T}} \tag{9.17}$$

The vector of control variables can take seven forms

$$\boldsymbol{u} = [\Delta T_1 \ \Delta T_2]^{\mathrm{T}} \tag{9.18}$$

$$\boldsymbol{u} = [\Delta F_{t1} \ \Delta F_{t2}]^{\mathrm{T}} \tag{9.19}$$

$$\boldsymbol{u} = [\Delta M]^{\mathrm{T}} \tag{9.20}$$

$$\boldsymbol{u} = [\Delta F_{t1} \ \Delta F_{t2} \ \Delta M]^{\mathrm{T}} \tag{9.21}$$

$$\boldsymbol{u} = [\Delta T_1 \ \Delta T_2 \ \Delta F_{t1} \ \Delta F_{t2}]^{\mathrm{T}} \tag{9.22}$$

$$\boldsymbol{u} = [\Delta T_1 \ \Delta T_2 \ \Delta M]^{\mathrm{T}} \tag{9.23}$$

$$\boldsymbol{u} = [\Delta T_1 \ \Delta T_2 \ \Delta F_{t1} \ \Delta F_{t2} \ \Delta M]^{\mathrm{T}} \tag{9.24}$$

The controllability of Eq. (9.15) in different vector of control variables (Eqs. (9.18)–(9.24)) will be analyzed. Considering Eqs. (9.17) and (9.18), the linear system in (9.15) can be transformed into the form $\dot{\boldsymbol{x}} = \boldsymbol{A}\boldsymbol{x} + \boldsymbol{B}\boldsymbol{u}$. where

$$\boldsymbol{A} = \begin{bmatrix} 0 & 0 & 0 & 0 & c \\ 0 & 0 & 0 & 0 & c \\ a+b & a & 0 & 0 & 0 \\ a & a+b & 0 & 0 & 0 \\ a/c & a/c & 0 & 0 & 0 \end{bmatrix} \tag{9.25}$$

$$B = \begin{bmatrix} d & 0 \\ 0 & d \\ 0 & 0 \\ 0 & 0 \\ 0 & 0 \end{bmatrix}$$

The controllability matrix turns out to be

$$C = \begin{bmatrix} B & AB & A^2B & A^3B & A^4B \end{bmatrix}$$
$$= \begin{bmatrix} d & 0 & 0 & 0 & ad & ad & 0 & 0 & 2a^2d & 2a^2d \\ 0 & d & 0 & 0 & ad & ad & 0 & 0 & 2a^2d & 2a^2d \\ 0 & 0 & (a+b)d & ad & 0 & 0 & (2a+b)acd & (2a+b)acd & 0 & 0 \\ 0 & 0 & ad & (a+b)d & 0 & 0 & (2a+b)acd & (2a+b)acd & 0 & 0 \\ 0 & 0 & ad/c & ad/c & 0 & 0 & 2a^2d/c & 2a^2d/c & 0 & 0 \end{bmatrix} \tag{9.26}$$

Then rank(C) = 4 is obtained by considering the values of a, b c and d. Thus the linear system in Eq. (9.15) is uncontrollable when $u = [\Delta T_1 \ \Delta T_2]^{\mathrm{T}}$.

Considering Eqs. (9.17) and (9.19), the linear system in (9.15) can be transformed into the form $\dot{x}=Ax+Bu$. Where A is the same as Eq. (9.25) and

$$B = \begin{bmatrix} 0 & 0 \\ 0 & 0 \\ e & 0 \\ 0 & e \\ 0 & 0 \end{bmatrix} \tag{9.27}$$

The controllability matrix turns out to be

$$C = \begin{bmatrix} B & AB & A^2B & A^3B & A^4B \end{bmatrix}$$
$$= \begin{bmatrix} 0 & 0 & 0 & 0 & 0 & 0 & 0 & 0 & 0 & 0 \\ 0 & 0 & 0 & 0 & 0 & 0 & 0 & 0 & 0 & 0 \\ e & 0 & 0 & 0 & 0 & 0 & 0 & 0 & 0 & 0 \\ 0 & e & 0 & 0 & 0 & 0 & 0 & 0 & 0 & 0 \\ 0 & 0 & 0 & 0 & 0 & 0 & 0 & 0 & 0 & 0 \end{bmatrix} \tag{9.28}$$

Then rank(C) = 2 is obtained by considering the values of e. Thus, the linear system in Eq. (9.15) is uncontrollable when $u = [\Delta F_{t1} \ \Delta F_{t2}]^{\mathrm{T}}$.

Considering Eqs. (9.17) and (9.20), the linear system in (9.15) can be transformed into the form $\dot{x}=Ax+Bu$. Where A is the same as Eq. (9.25) and

$$\boldsymbol{B} = \begin{bmatrix} 0 \\ 0 \\ c/I \\ c/I \\ 1/I \end{bmatrix} \tag{9.29}$$

The controllability matrix turns out to be

$$\boldsymbol{C} = \begin{bmatrix} \boldsymbol{B} & \boldsymbol{AB} & \boldsymbol{A^2B} & \boldsymbol{A^3B} & \boldsymbol{A^4B} \end{bmatrix} = \begin{bmatrix} 0 & \frac{c}{I} & 0 & \frac{2ac}{I} & 0 \\ 0 & \frac{c}{I} & 0 & \frac{2ac}{I} & 0 \\ \frac{c}{I} & 0 & \frac{(2a+b)c}{I} & 0 & \frac{2(2a+b)ac^2}{I} \\ \frac{c}{I} & 0 & \frac{(2a+b)c}{I} & 0 & \frac{2(2a+b)ac^2}{I} \\ \frac{1}{I} & 0 & \frac{2a}{I} & 0 & \frac{4a^2}{I} \end{bmatrix} \tag{9.30}$$

Then rank$(\mathbf{C}) = 3$ is obtained by considering the values of a, b, and c. Thus, the linear system in Eq. (9.15) is uncontrollable when $\boldsymbol{u} = [\Delta M]^{\mathrm{T}}$.

Considering Eqs. (9.17) and (9.21), the linear system in (9.15) can be transformed into the form $\dot{\boldsymbol{x}}=\boldsymbol{Ax}+\boldsymbol{Bu}$. Where $\boldsymbol{A}$ is the same as Eq. (9.25) and

$$\boldsymbol{B} = \begin{bmatrix} 0 & 0 & 0 \\ 0 & 0 & 0 \\ e & 0 & c/I \\ 0 & e & c/I \\ 0 & 0 & 1/I \end{bmatrix} \tag{9.31}$$

The controllability matrix turns out to be

$$\boldsymbol{C} = \begin{bmatrix} \boldsymbol{B} & \boldsymbol{AB} & \boldsymbol{A^2B} & \boldsymbol{A^3B} & \boldsymbol{A^4B} \end{bmatrix}$$
$$= \begin{bmatrix} 0 & 0 & 0 & 0 & 0 & \frac{c}{I} & 0 & 0 & 0 & 0 & 0 & \frac{2ac}{I} & 0 & 0 & 0 \\ 0 & 0 & 0 & 0 & 0 & \frac{c}{I} & 0 & 0 & 0 & 0 & 0 & \frac{2ac}{I} & 0 & 0 & 0 \\ e & 0 & \frac{c}{I} & 0 & 0 & 0 & 0 & 0 & \frac{(2a+b)c}{I} & 0 & 0 & 0 & 0 & 0 & \frac{2(2a+b)ac^2}{I} \\ 0 & e & \frac{c}{I} & 0 & 0 & 0 & 0 & 0 & \frac{(2a+b)c}{I} & 0 & 0 & 0 & 0 & 0 & \frac{2(2a+b)ac^2}{I} \\ 0 & 0 & \frac{1}{I} & 0 & 0 & 0 & 0 & 0 & \frac{2a}{I} & 0 & 0 & 0 & 0 & 0 & \frac{4a^2}{I} \end{bmatrix} \tag{9.32}$$

Then rank$(\boldsymbol{C}) = 4$ is obtained by considering the values of a, b, and c. Thus, the linear system in Eq. (9.15) is uncontrollable when $\boldsymbol{u} = [\Delta F_{t1} \quad \Delta F_{t2} \quad \Delta M]^{\mathrm{T}}$.

Considering Eqs. (9.17) and (9.22), the linear system in (9.15) can be transformed into the form $\dot{\boldsymbol{x}}=\boldsymbol{A}\boldsymbol{x}+\boldsymbol{B}\boldsymbol{u}$. Where $\boldsymbol{A}$ is the same as Eq. (9.25) and:

$$\boldsymbol{B}=\begin{bmatrix} d & 0 & 0 & 0 \\ 0 & d & 0 & 0 \\ 0 & 0 & e & 0 \\ 0 & 0 & 0 & e \\ 0 & 0 & 0 & 0 \end{bmatrix} \tag{9.33}$$

The controllability matrix turns out to be

$$\begin{aligned} \boldsymbol{C} &= \begin{bmatrix} \boldsymbol{B} & \boldsymbol{AB} & \boldsymbol{A}^2\boldsymbol{B} & \boldsymbol{A}^3\boldsymbol{B} & \boldsymbol{A}^4\boldsymbol{B} \end{bmatrix} \\ &= \begin{bmatrix} d & 0 & 0 & 0 & 0 & 0 & 0 & 0 & f_5 & f_5 & 0 & 0 & 0 & 0 & 0 & 0 & f_6 & f_6 & 0 & 0 \\ 0 & d & 0 & 0 & 0 & 0 & 0 & 0 & f_5 & f_5 & 0 & 0 & 0 & 0 & 0 & 0 & f_6 & f_6 & 0 & 0 \\ 0 & 0 & e & 0 & f_1 & f_5 & 0 & 0 & 0 & 0 & 0 & 0 & f_2 & f_2 & 0 & 0 & 0 & 0 & 0 & 0 \\ 0 & 0 & 0 & e & f_5 & f_1 & 0 & 0 & 0 & 0 & 0 & 0 & f_2 & f_2 & 0 & 0 & 0 & 0 & 0 & 0 \\ 0 & 0 & 0 & 0 & f_3 & f_3 & 0 & 0 & 0 & 0 & 0 & 0 & f_4 & f_4 & 0 & 0 & 0 & 0 & 0 & 0 \end{bmatrix} \end{aligned} \tag{9.34}$$

where

$$\begin{cases} f_1 = (a+b)d \\ f_2 = (2a+b)acd \\ f_3 = ad/c \\ f_4 = 2a^2d/c \\ f_5 = ad \\ f_6 = -2a^2d \end{cases} \tag{9.35}$$

Then $\mathrm{rank}(\boldsymbol{C}) = 5$ is obtained by considering the values of a, b, and c. Thus, the linear system in Eq. (9.15) is controllable when $\boldsymbol{u} = [\Delta T_1 \quad \Delta T_2 \quad \Delta F_{t1} \quad \Delta F_{t2}]^{\mathrm{T}}$.

Considering Eqs. (9.17) and (9.23), the linear system in (9.15) can be transformed into the form $\dot{\boldsymbol{x}}=\boldsymbol{A}\boldsymbol{x}+\boldsymbol{B}\boldsymbol{u}$. Where $\boldsymbol{A}$ is the same as Eq. (9.25) and

$$\boldsymbol{B}=\begin{bmatrix} d & 0 & 0 \\ 0 & d & 0 \\ 0 & 0 & c/I \\ 0 & 0 & c/I \\ 0 & 0 & 1/I \end{bmatrix} \tag{9.36}$$

The controllability matrix turns out to be

$$C = \begin{bmatrix} B & AB & A^2B & A^3B & A^4B \end{bmatrix} = \begin{bmatrix} d & 0 & 0 & 0 & 0 & \frac{c}{I} & ad & ad & 0 & 0 & 0 & g_6 & g_5 & g_5 & 0 \\ 0 & d & 0 & 0 & 0 & \frac{c}{I} & ad & ad & 0 & 0 & 0 & g_6 & g_5 & g_5 & 0 \\ 0 & 0 & \frac{c}{I} & g_1 & ad & 0 & 0 & 0 & g_2 & g_3 & g_3 & 0 & 0 & 0 & g_4 \\ 0 & 0 & \frac{c}{I} & ad & g_1 & 0 & 0 & 0 & g_2 & g_3 & g_3 & 0 & 0 & 0 & g_4 \\ 0 & 0 & \frac{1}{I} & \frac{ad}{c} & \frac{ad}{c} & 0 & 0 & 0 & \frac{2a}{I} & \frac{g_5}{c} & \frac{g_5}{c} & 0 & 0 & 0 & \frac{4a^2}{I} \end{bmatrix} \tag{9.37}$$

where

$$\begin{cases} g_1 = (a+b)d \\ g_2 = (2a+b)c/I \\ g_3 = (2a+b)acd \\ g_4 = 2(2a+b)ac^2/I \\ g_5 = 2a^2d \\ g_6 = 2ac/I \end{cases} \tag{9.38}$$

Then $\text{rank}(C) = 5$ is obtained by considering the values of a, b, and c. Thus, the linear system in Eq. (9.15) is controllable when $\boldsymbol{u} = [\Delta T_1 \quad \Delta T_2 \quad \Delta M]^{\mathrm{T}}$. In addition, it is apparent that the linear system in Eq. (9.15) is controllable when $\boldsymbol{u} = [\Delta T_1 \quad \Delta T_2 \quad \Delta F_{t1} \quad \Delta F_{t2} \quad \Delta M]^{\mathrm{T}}$. Therefore, the following conclusions are made:

(a) The torque of the master spacecraft (ΔM), the tension forces of the tethers connected with sub-spacecraft (ΔT_1 and ΔT_2), and the thrusts of the sub-spacecraft (ΔF_{t1} and ΔF_{t2}) can't achieve the formation control operation separately.
(b) The ΔM, ΔF_{t1}, and ΔF_{t2} can't achieve the formation control operation corporately.
(c) The formation control operation can be achieved by two combinations: (1) ΔT_1, ΔT_2, and ΔM. (2) ΔF_{t1}, ΔF_{t2}, ΔT_1, and ΔT_2.

9.1.2 Numerical Simulation

Because the main purpose of the simulation is to verify that the control model of formation maintenance is controllable in the two control modes, the amplitude and continuity of various control quantities will not be considered in the simulation process. In addition, the state feedback controller is used to control the initial error in both control modes, that is, the design controller is

Table 9.1 Simulation parameters

Parameter	Value
L_1	10.0 m
L_2	10.0 m
m_1	5.0 kg
m_2	5.0 kg
r_0	1.0 m
I	500.0 kg m^2

$$\boldsymbol{u} = -\boldsymbol{Kx} \tag{9.39}$$

where $\boldsymbol{K}$ is the state feedback matrix. The closed-loop system can be obtained as

$$\dot{\boldsymbol{x}} = (\boldsymbol{A} - \boldsymbol{BK})\boldsymbol{x} \tag{9.40}$$

The parameters of the state feedback matrix will be adjusted according to the actual control effect in the simulation. Other main simulation parameters are shown in Table 9.1.

(a) ΔT_1, ΔT_2, ΔF_{t1}, and ΔF_{t2} as control quantity

First, the tether force of the spiral arm and the autonomous machine power of the sub-TSR are both as the control quantity are simulated. Suppose the initial value of the angle error and spin speed error of the spiral arm is

$$\begin{gathered}[\Delta\varphi_{10} \quad \Delta\varphi_{20} \quad \Delta\dot{\varphi}_1 \quad \Delta\dot{\varphi}_2 \quad \Delta\omega_0]^{\mathrm{T}} = [0.1\,\text{rad} \quad -0.1\,\text{rad} \quad 0.1\,\text{rad s}^{-1} \\ -0.1\,\text{rad s}^{-1} \quad 0.1\,\text{rad s}^{-1}]^{\mathrm{T}}\end{gathered} \tag{9.41}$$

The tether tensile force of the spiral arm and the autonomous machine power of sub-TSR are used as the control quantities, and the simulation result is as shown in Figs. 9.2, 9.3, 9.4, 9.5, and 9.6. The state variables change with time when the state feedback controller is used. The simulation curves show that $\Delta\varphi_1$, $\Delta\varphi_2$, $\Delta\dot{\varphi}_1$, $\Delta\dot{\varphi}_2$, and $\Delta\omega$ are in the end with time approaching 0, which is in accordance with the conclusion of controllability analysis. The control inputs are shown in Figs. 9.7, 9.8, 9.9, and 9.10.

(b) ΔT_1, ΔT_2, and ΔM as control quantity

The tension of the tether system and the spin torque of the master-TSR are simulated as the control quantity. Suppose the initial value of the angle error and spin speed error of the spiral arm is

$$\begin{gathered}[\Delta\varphi_{10} \quad \Delta\varphi_{20} \quad \Delta\dot{\varphi}_1 \quad \Delta\dot{\varphi}_2 \quad \Delta\omega_0]^{\mathrm{T}} = [0.1\,\text{rad} \quad -0.1\,\text{m} \quad 0.1\,\text{m s}^{-1} \\ -0.1\,\text{m s}^{-1} \quad 0.1\,\text{m s}^{-1}]^{\mathrm{T}}\end{gathered}$$

Fig. 9.2 The value of $\triangle\varphi_1$

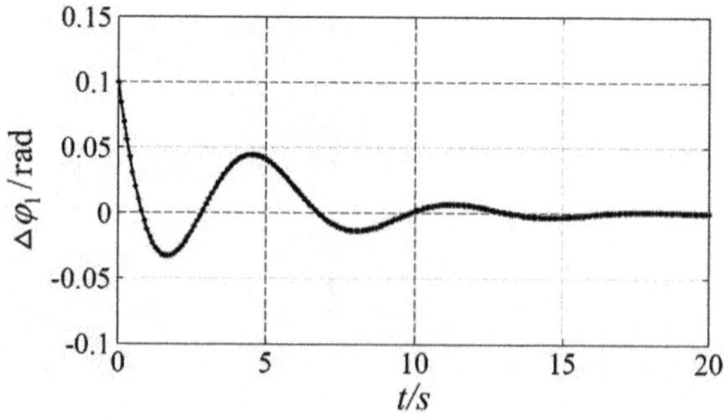

Fig. 9.3 The value of $\triangle\varphi_2$

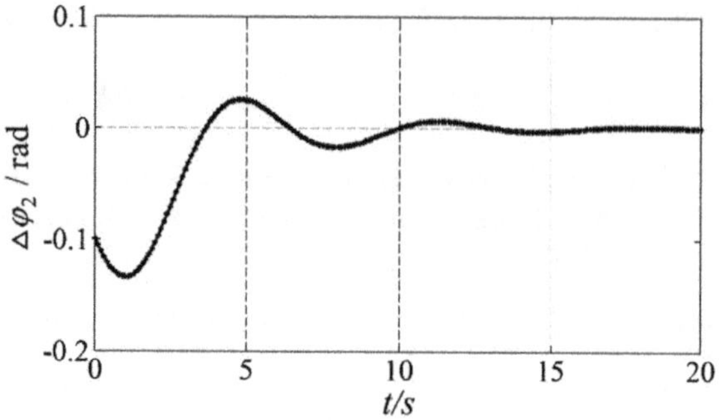

Fig. 9.4 The value of $\triangle\dot{\varphi}_1$

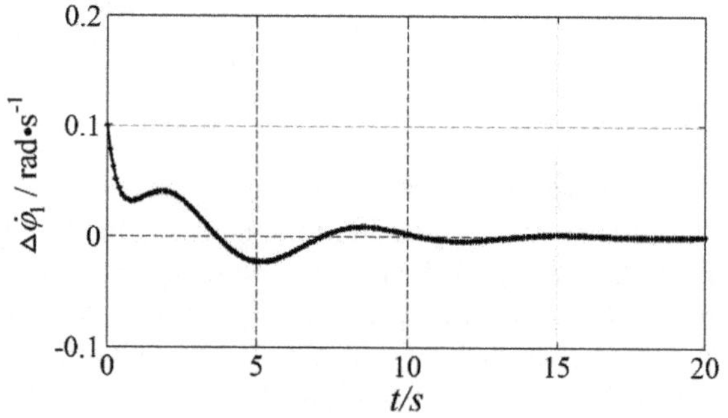

Fig. 9.5 The value of $\triangle\dot{\varphi}_2$

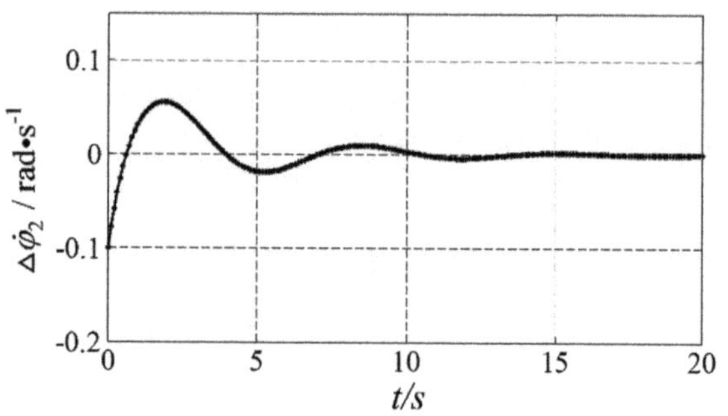

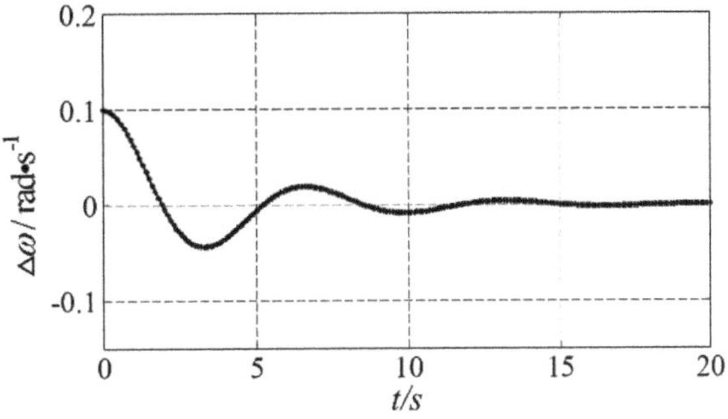

Fig. 9.6 The value of $\Delta\omega$

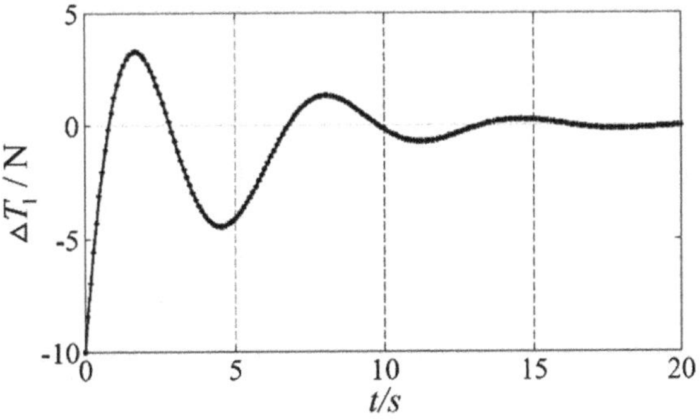

Fig. 9.7 The value of ΔT_1

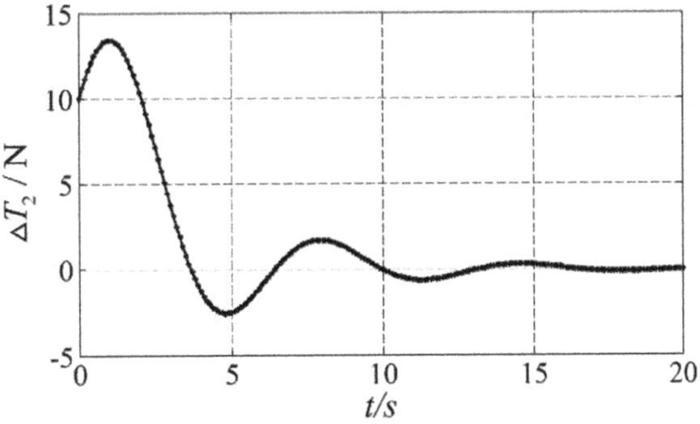

Fig. 9.8 The value of ΔT_2

The tether pull of the spiral arm and the spin torque of the MASTER-TSR are combined as control quantities, and the simulation result is shown in Figs. 9.11, 9.12, 9.13, 9.14, and 9.15, the state variables change with time when the state feedback controller is used. The simulation curves show that $\Delta\varphi_1$, $\Delta\varphi_2$, $\Delta\dot{\varphi}_1$, $\Delta\dot{\varphi}_2$, and $\Delta\omega$ are in the end with time approaching 0, which is in accordance with the conclusion of controllability analysis. The control inputs are shown in Figs. 9.16, 9.17, and 9.18.

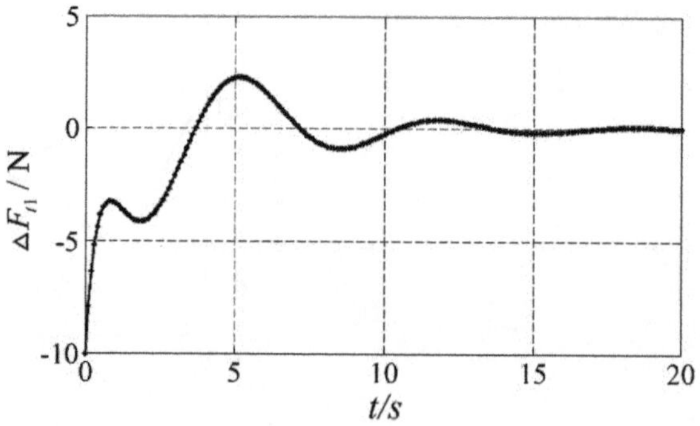

Fig. 9.9 The value of $\triangle F_{t1}$

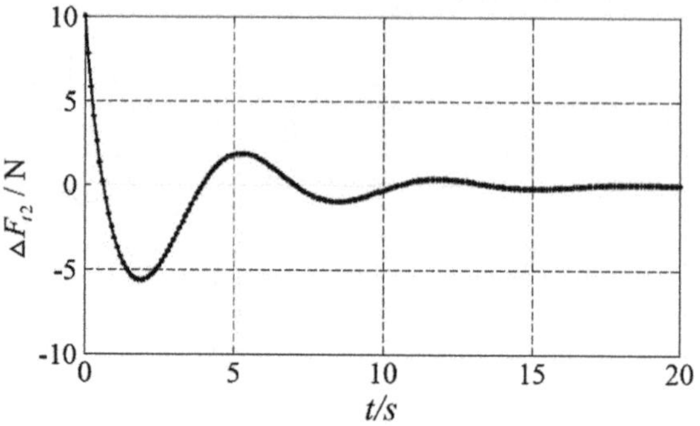

Fig. 9.10 The value of $\triangle F_{t2}$

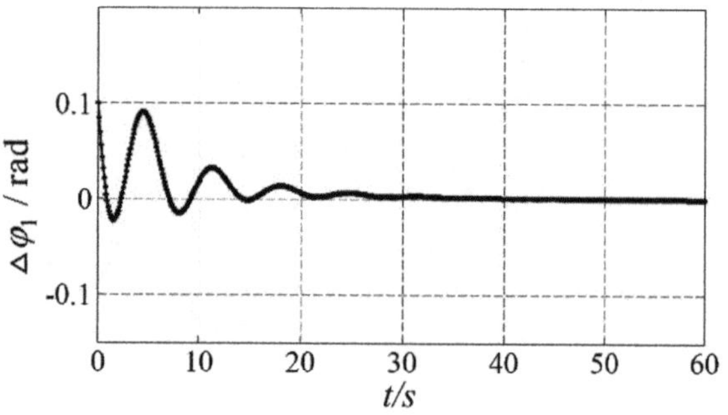

Fig. 9.11 The value of $\triangle \varphi_1$

(c) $\triangle M$ as the control quantity, and $\triangle \varphi_1 = \triangle \varphi_2$

Figures 9.19, 9.20, and 9.21 are the angular error of the rotor in this special spin torque control mode, and the change of spin speeds error under the action of State feedback controller over time. It is assumed that the initial value of the angle error and spin speed error in the simulation process is

Fig. 9.12 The value of $\triangle\varphi_2$

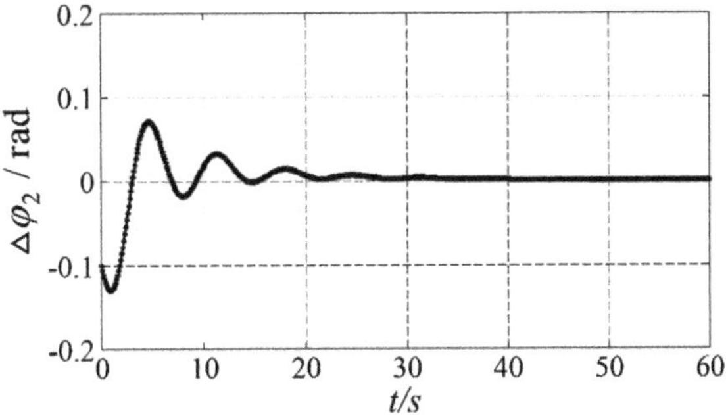

Fig. 9.13 The value of $\triangle\dot{\varphi}_1$

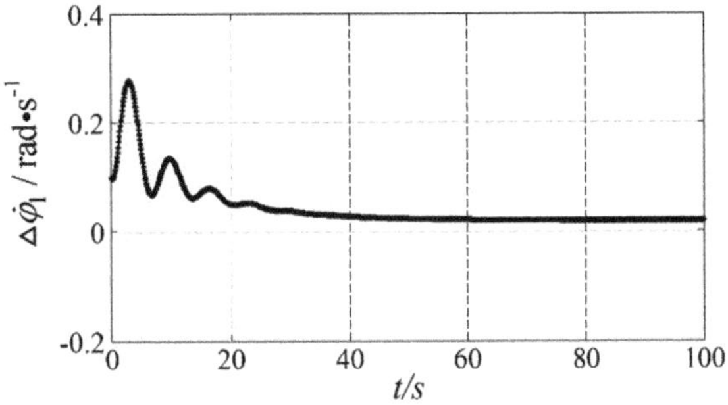

Fig. 9.14 The value of $\triangle\dot{\varphi}_2$

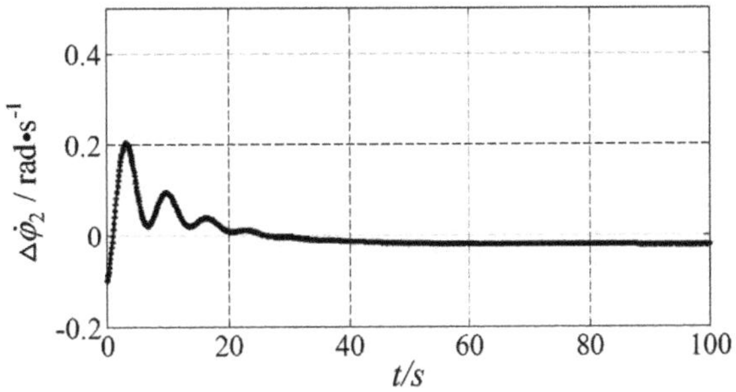

Fig. 9.15 The value of $\triangle\omega$

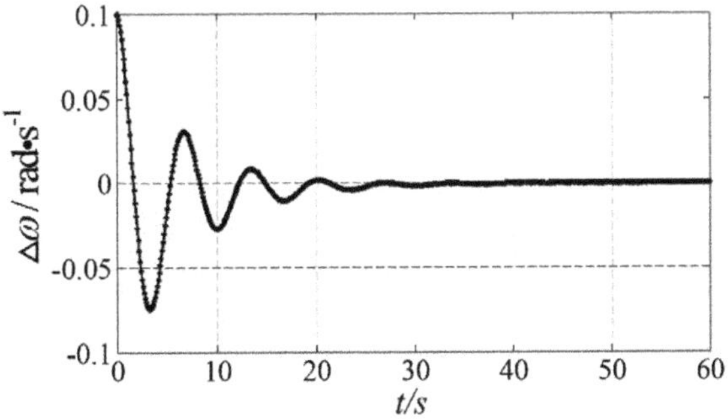

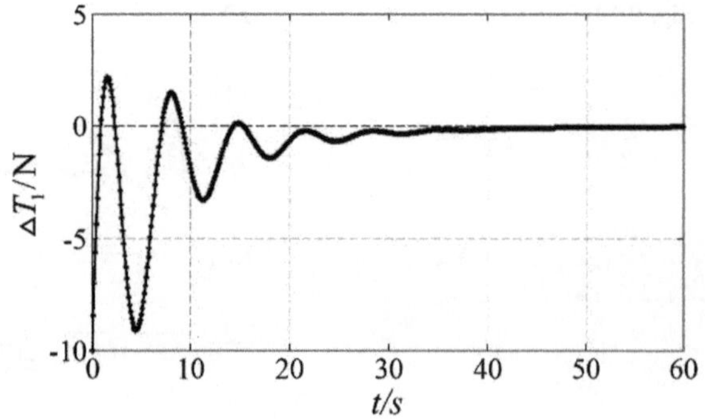

Fig. 9.16 The value of $\triangle T_1$

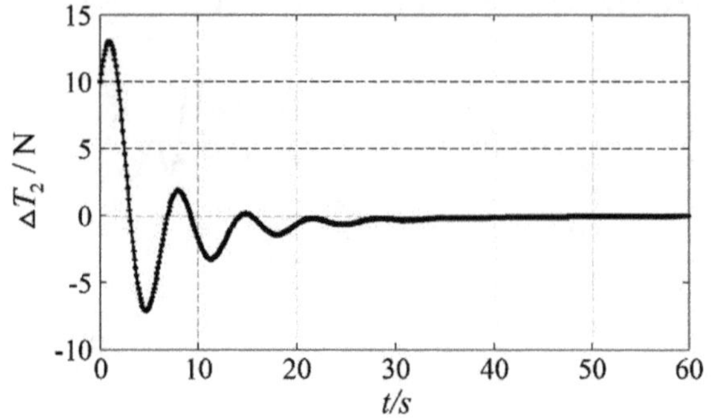

Fig. 9.17 The value of $\triangle T_2$

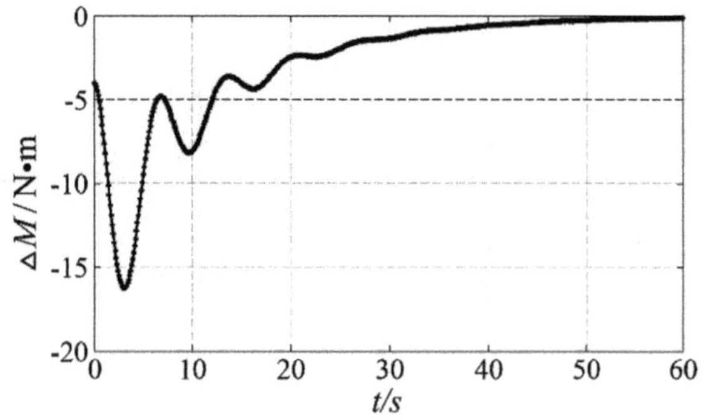

Fig. 9.18 The value of $\triangle M$

$$[\triangle\varphi_0 \quad \triangle\dot{\varphi}_0 \quad \triangle\omega_0]^{\mathrm{T}} = [0.1\,\mathrm{rad} \quad -0.1\,\mathrm{rad\,s^{-1}} \quad 0.1\,\mathrm{rad\,s^{-1}}]^{\mathrm{T}} \tag{9.42}$$

In the special case $\triangle\varphi_1 = \triangle\varphi_2$, the master-TSR spin torque can be used as the control quantity alone, and it can realize the angle error and the spin speed error control. The control inputs are shown in Fig. 9.22.

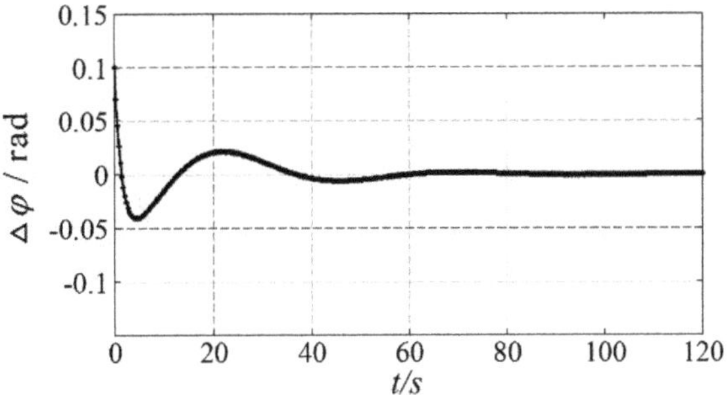

Fig. 9.19 The value of $\triangle\varphi$

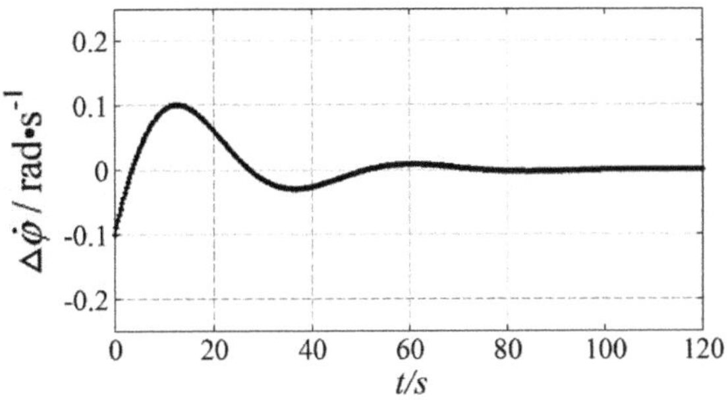

Fig. 9.20 The value of $\triangle\dot{\varphi}$

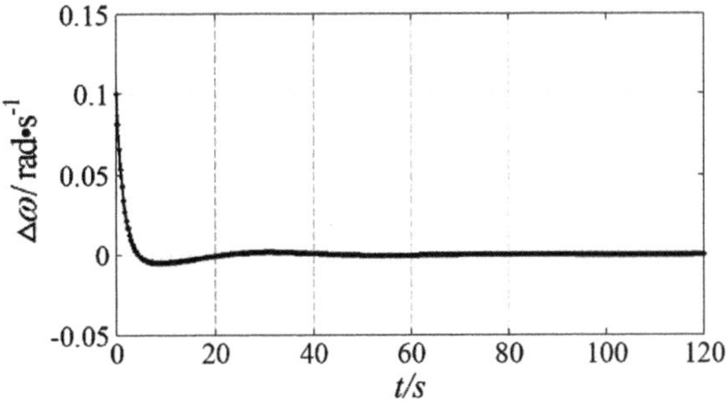

Fig. 9.21 The value of $\triangle\omega$

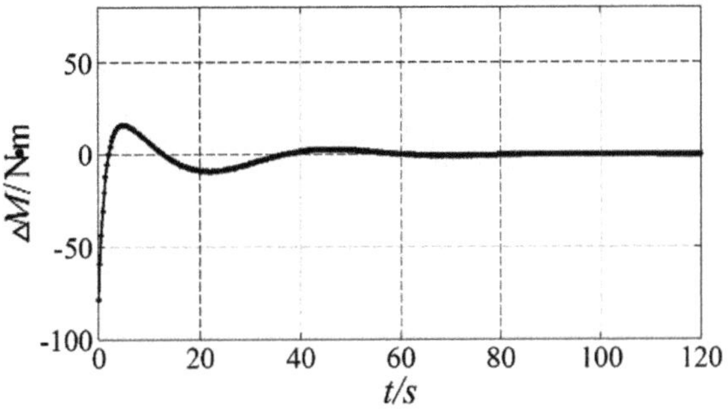

Fig. 9.22 The value of $\triangle M$

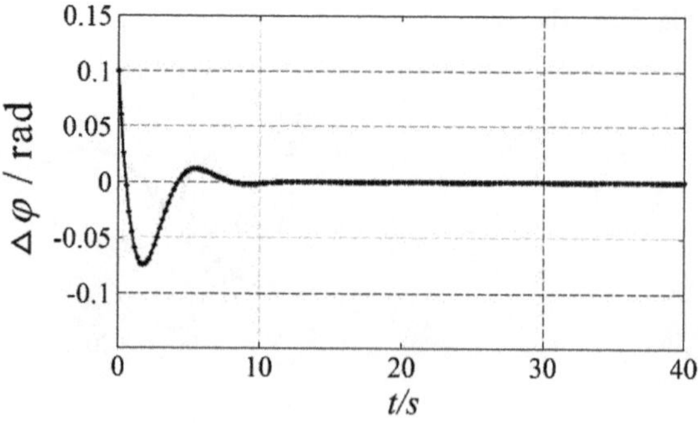

Fig. 9.23 The value of $\Delta\varphi$

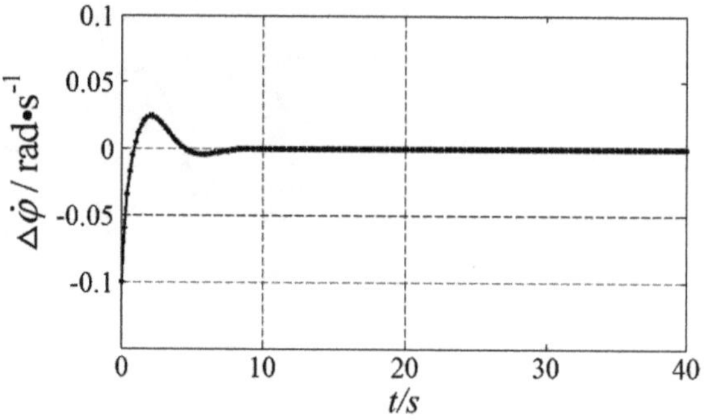

Fig. 9.24 The value of $\Delta\dot{\varphi}$

(d) ΔF_t and ΔT as the control quantity, and $\Delta\varphi_1 = \Delta\varphi_2$

In a special case $\Delta\varphi_1 = \Delta\varphi_2$, the sub-TSR's autonomous machine power and the tether force of the spiral arm together as the control quantity, the rotation arm angle error, as well as spin speed error under the action of State feedback controller over time, as shown in Figs. 9.23, 9.24, and 9.25. It is assumed that the initial value of the angle error and spin speed error in the simulation process is

$$[\Delta\varphi_0 \quad \Delta\dot{\varphi}_0 \quad \Delta\omega_0]^{\mathrm{T}} = [0.1\,\mathrm{rad} \quad -0.1\,\mathrm{rad\,s^{-1}} \quad 0.1\,\mathrm{rad\,s^{-1}}]^{\mathrm{T}} \tag{9.43}$$

From the simulation curve, we can see that, under special circumstances $\Delta\varphi_1 = \Delta\varphi_2$, the angle error of the rotor $\Delta\varphi_1$ and the spin speed error $\Delta\dot{\varphi}$ can be approximated to the 0 value gradually under the action of the controller. The simulation results show that the control of the rotor angle error and the spin speed error can be realized when the sub-TSR power and the tether force of the spiral arm are combined as control quantities. The simulation results verify the correctness of the conclusion of controllability analysis. The control inputs are shown in Figs. 9.26 and 9.27.

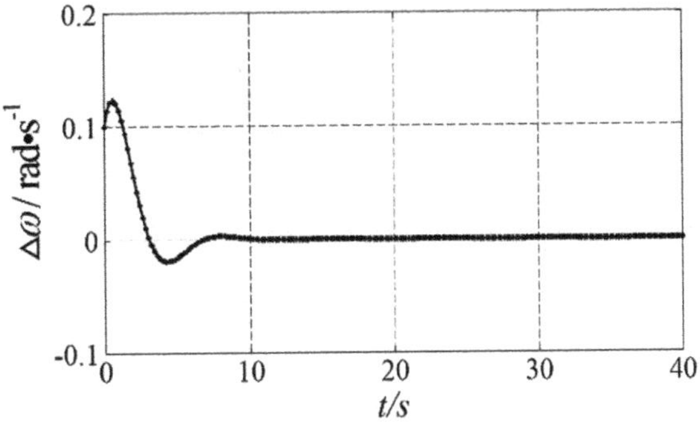

Fig. 9.25 The value of $\triangle\omega$

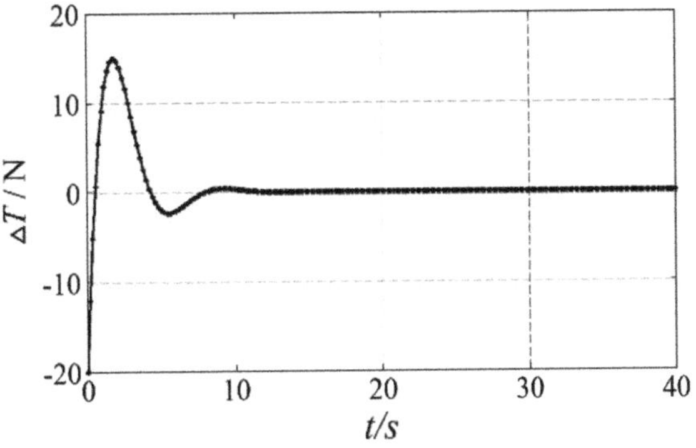

Fig. 9.26 The value of $\triangle T$

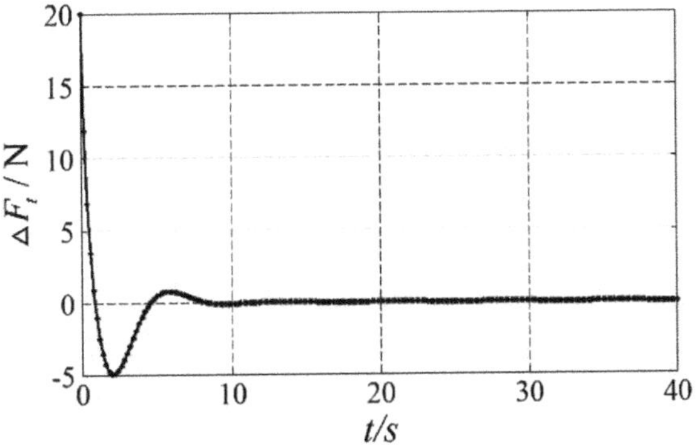

Fig. 9.27 The value of $\triangle F_t$

9.2 Error-Based Coordinated Control for Formation-Keeping

9.2.1 Problem Statement

The objective of formation control of HSS is to recover the ideal Hub–Spoke configuration when $\Delta\varphi_1$, $\Delta\varphi_2$, and $\Delta\omega$ occurs. The magnitudes of $\Delta\varphi_1$, $\Delta\varphi_2$, and $\Delta\omega$ can be regarded as a coordinated criterion for the formation control operation. That is the error-based coordinated control strategy, which involves two codes:

(a) The ΔF_{t1}, ΔF_{t2}, ΔT_1, and ΔT_2 are performed as control quantity when $\Delta\varphi_1$, $\Delta\varphi_2$, $\Delta\dot{\varphi}_1$, $\Delta\dot{\varphi}_2$, and $\Delta\omega$ are relatively small values. The ΔM is useless.
(b) The ΔF_{t1}, ΔF_{t2}, ΔT_1, ΔT_2, and ΔM are all used as control quantity when $\Delta\varphi_1$, $\Delta\varphi_2$, $\Delta\dot{\varphi}_1$, $\Delta\dot{\varphi}_2$, and $\Delta\omega$ exceed a threshold value, which is a regulable and limited value.

Under the error-based coordinated control strategy, the relative large control value provided by torque of the master spacecraft is well used. The fuel consumption of the master spacecraft will also be economized by adjusting the threshold value during the formation control operation of HSS. To describe the error-based coordinated control strategy mathematically, the vector of state variables can be written as

$$\boldsymbol{x} = [\Delta\varphi_1 \quad \Delta\varphi_2 \quad \Delta\dot{\varphi}_1 \quad \Delta\dot{\varphi}_2 \quad \Delta\omega]^{\mathrm{T}} \tag{9.44}$$

The vector of control variables is

$$\boldsymbol{u} = [\Delta T_1 \quad \Delta T_2 \quad \Delta F_{t1} \quad \Delta F_{t2} \quad \alpha\Delta M]^{\mathrm{T}} \tag{9.45}$$

where

$$\alpha = \begin{cases} 0 & \|\mathbf{x}\| \le E \\ 1 & \|\mathbf{x}\| > E \end{cases} \tag{9.46}$$

where $\|\mathbf{x}\|$ is the 2-norm of the vector of control variables, and E is defined as error valve coefficient. It can be realized that only ΔF_{t1}, ΔF_{t2}, ΔT_1, and ΔT_2 are used when $\|\mathrm{x}\| \le E$. The ΔM will come into use as a redundant control value when $\|\mathrm{x}\| > E$. Then the linear system in (9.15) can be transformed into the form $\dot{\boldsymbol{x}}=\boldsymbol{A}\boldsymbol{x}+\boldsymbol{B}\boldsymbol{u}$, where

$$\boldsymbol{A} = \begin{bmatrix} 0 & 0 & 0 & 0 & c \\ 0 & 0 & 0 & 0 & c \\ a+b & a & 0 & 0 & 0 \\ a & a+b & 0 & 0 & 0 \\ a/c & a/c & 0 & 0 & 0 \end{bmatrix} \tag{9.47}$$

$$B = \begin{bmatrix} d & 0 & 0 & 0 & 0 \\ 0 & d & 0 & 0 & 0 \\ 0 & 0 & e & 0 & c/I \\ 0 & 0 & 0 & e & c/I \\ 0 & 0 & 0 & 0 & 1/I \end{bmatrix} \tag{9.48}$$

where a, b, c, d, and e are demonstrated in Eq. (9.16).

9.2.2 Controller Design

According to the coordinated control strategy based on error, take the state variable as

$$\boldsymbol{x} = \begin{bmatrix} \Delta\varphi_1 & \Delta\varphi_2 & \Delta\dot{\varphi}_1 & \Delta\dot{\varphi}_2 & \Delta\omega \end{bmatrix}^{\mathrm{T}} \tag{9.49}$$

Take the control quantity as follows:

$$\boldsymbol{u} = \begin{bmatrix} \Delta T_1 & \Delta T_2 & \Delta F_{t1} & \Delta F_{t2} & \alpha\Delta M \end{bmatrix}^{\mathrm{T}} \tag{9.50}$$

The following model of formation maintenance control based on error can be obtained as

$$\dot{\boldsymbol{x}} = \boldsymbol{A}\boldsymbol{x} + \boldsymbol{B}\boldsymbol{u} \tag{9.51}$$

where

$$\boldsymbol{A} = \begin{bmatrix} 0 & 0 & 0 & 0 & c \\ 0 & 0 & 0 & 0 & c \\ a+b & a & 0 & 0 & 0 \\ a & a+b & 0 & 0 & 0 \\ a/c & a/c & 0 & 0 & 0 \end{bmatrix} \tag{9.52}$$

$$\boldsymbol{B} = \begin{bmatrix} d & 0 & 0 & 0 & 0 \\ 0 & d & 0 & 0 & 0 \\ 0 & 0 & e & 0 & \frac{c}{I} \\ 0 & 0 & 0 & e & \frac{c}{I} \\ 0 & 0 & 0 & 0 & \frac{1}{I} \end{bmatrix} \tag{9.53}$$

$$\alpha = \begin{cases} 0 & \|\mathbf{x}\| \leq E \\ 1 & \|\mathbf{x}\| > E \end{cases} \tag{9.54}$$

where $\|\boldsymbol{x}\|$ is the 2-norm of state vector $\boldsymbol{x}$.

When the 2-norm of the state vector $\boldsymbol{x}$ is less than E, the error is controlled by the tether tensile force of the spiral arm and the sub-TSR power of the master-TSR, and the spin torque of the rotor is not working; when the norm of state vector $\boldsymbol{x}$ is greater than E, the tether of the spiral arm is used The master-TSR torque and sub-TSR power are combined as control quantities to control the formation error. The coordinated control strategy of the line tension of the spiral arm, the spin torque of the master-TSR and the sub-TSR of the autonomous machine power determined by the error. In addition, it is necessary to design controller to realize error control. A full state feedback controller is designed to complete the control loop under the action of the coordinated control strategy based on the error. The state feedback controller is designed as

$$\boldsymbol{u} = -\boldsymbol{K}\boldsymbol{x} \tag{9.55}$$

The state feedback matrix $\boldsymbol{K}$ is as follows:

$$\boldsymbol{K} = \mathrm{diag}(k_1, k_2, k_3, k_4, k_5) \tag{9.56}$$

The parameters of the state feedback matrix $\boldsymbol{K}$ are adjusted according to the actual control effect, and finally, the closed-loop system is obtained as

$$\dot{\boldsymbol{x}} = (\boldsymbol{A} - \boldsymbol{B}\boldsymbol{K})\boldsymbol{x} \tag{9.57}$$

Based on the above content, the scheme block diagram of coordinated control strategy is shown in Fig. 9.28.

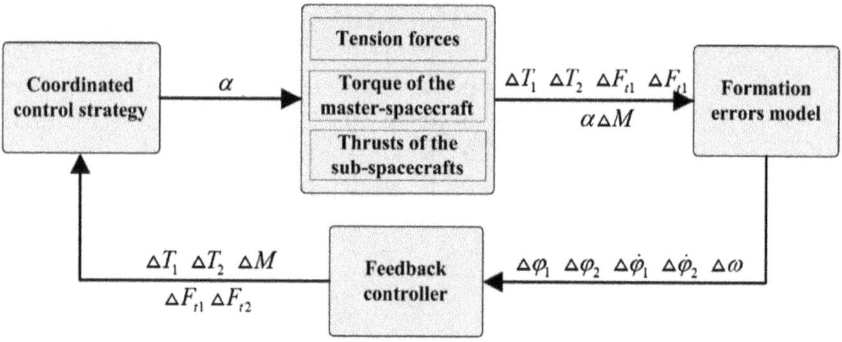

Fig. 9.28 A block diagram of coordinated control strategy based on error

9.3 Time-Based Coordinated Control for Formation-Keeping

9.3.1 Problem Statement

Adjusting time is one of the important indexes to measure the general control system, for the formation configuration keeping control of the radial open-loop space tether robot formation, when the formation error (the angular velocity of the rotor angle is extremely angular and the rotational speed error) is expected, it is hoped that it can be controlled quickly to meet the requirement of formation configuration maintenance. Similar to the coordinated control strategy based on error, the sub-TSR of the spiral arm can be used as the control quantity in general condition, and the spin torque of the master-TSR is used as the redundancy control quantity. This section is designed to maintain a coordinated control strategy based on time, which can be described as follows:

When $\Delta\varphi_1$, $\Delta\varphi_2$, $\Delta\dot{\varphi}_1$, $\Delta\dot{\varphi}_2$, and $\Delta\omega$ appear, the tether force of the spiral arm, the spin torque of the master-TSR and the sub-TSR's autonomic power work together as control quantity, after a period of time, the master-TSR spin torque stops working to save the master-TSR fuel consumption. The control of formation error is continued by using the tether tensile force of the spiral arm and the sub-TSR power of the robot.

9.3.2 Controller Design

According to the coordinated control strategy based on time, take the state variable as

$$\boldsymbol{x} = [\Delta\varphi_1\ \Delta\varphi_2\ \Delta\dot{\varphi}_1\ \Delta\dot{\varphi}_2\ \Delta\omega]^{\mathrm{T}} \tag{9.58}$$

Take the control quantity as follows:

$$\boldsymbol{u} = [\Delta T_1\ \Delta T_2\ \Delta F_{t1}\ \Delta F_{t2}\ \beta\Delta M]^{\mathrm{T}} \tag{9.59}$$

The control model for formation configurations based on time are as follows:

$$\dot{\boldsymbol{x}} = \boldsymbol{A}\boldsymbol{x} + \boldsymbol{B}\boldsymbol{u} \tag{9.60}$$

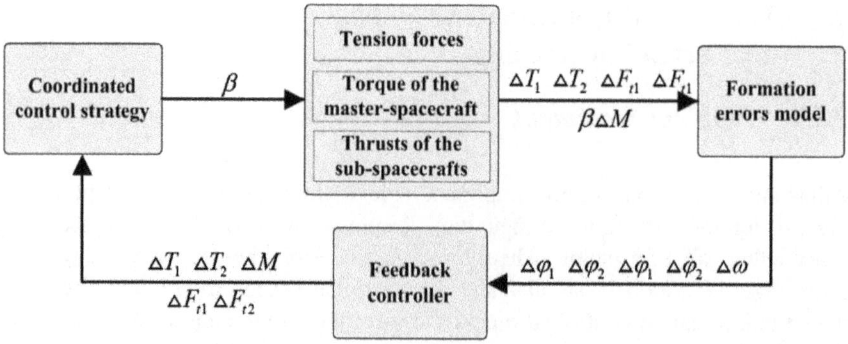

Fig. 9.29 A block diagram of coordinated control strategy based on time

where

$$\boldsymbol{A} = \begin{bmatrix} 0 & 0 & 0 & 0 & c \\ 0 & 0 & 0 & 0 & c \\ a+b & a & 0 & 0 & 0 \\ a & a+b & 0 & 0 & 0 \\ a/c & a/c & 0 & 0 & 0 \end{bmatrix} \tag{9.61}$$

$$\beta = \begin{cases} 0 \ t \geq Ts \\ 1 \ t < Ts \end{cases} \tag{9.62}$$

Define Ts as the time valve value factor. When the control time is less than Ts, the tether tensile force of the spiral arm, the spin torque of the master-TSR and the sub-TSR power are combined as the control quantity; When the control time reaches Ts, the master-TSR's spin torque stops working, The control task of the formation configuration is continued by the tether strain of the spiral arm and the sub-TSR power of the robot.

Combined with the above, the time-based coordinated control strategy block diagram is shown in Fig. 9.29. In addition, the controller used in the control loop design of coordinated control strategy based on time is the same as that of the coordinated control strategy based on error, that is, the full state feedback controller.

9.4 Numerical Simulation

9.4.1 Tension-and-Thruster-Based Simulation

The first simulation involved the ΔF_{t1}, ΔF_{t2}, ΔT_1, and ΔT_2 control, in which the ΔM was unused. This simulation was implemented based on the closed-loop coordinated formation control system, but the $\alpha = \beta = 0$ was satisfied during the simulation. The feedback matrix $\boldsymbol{K}$ is as follows:

$$\boldsymbol{K} = \mathrm{diag}(k_1, k_2, k_3, k_4) = \mathrm{diag}(-250, -250, -250, -250) \tag{9.63}$$

The vector of the initial formation errors is as follows:

$$[\Delta\varphi_1 \quad \Delta\varphi_2 \quad \Delta\dot{\varphi}_1 \quad \Delta\dot{\varphi}_2 \quad \Delta\omega]^{\mathrm{T}} = [0.1 \quad -0.1 \quad 0.1 \quad -0.1 \quad 0.1]^{\mathrm{T}} \tag{9.64}$$

The simulation results are demonstrated in Figs. 9.30, 9.31, and 9.32. The variations with time of the formation errors are presented in Fig. 9.4. The $\Delta\varphi_1$, $\Delta\varphi_2$, $\Delta\dot{\varphi}_1$, $\Delta\dot{\varphi}_2$, and $\Delta\omega$ are gradually controlled, and the convergent time is about 45 s. It is concluded that formation errors can be controlled just by the ΔF_{t1}, ΔF_{t2}, ΔT_1, and ΔT_2, though the convergent time and oscillatory behaviors of the simulation curves are not good enough. Figures 9.31 and 9.32 demonstrate the

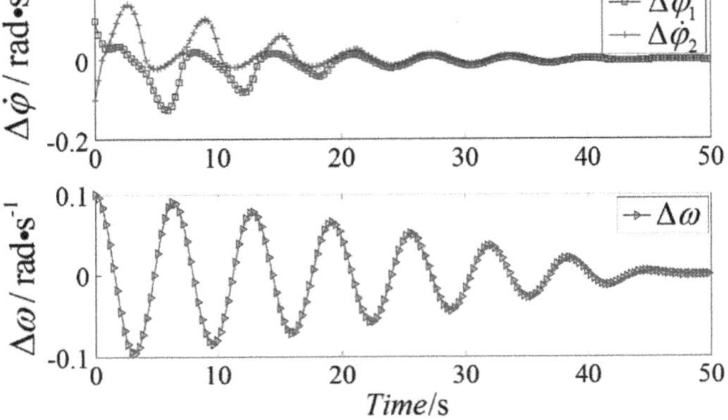

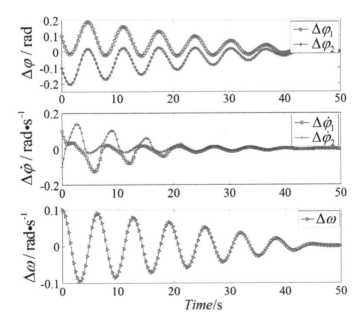

Fig. 9.30 Error variations with time

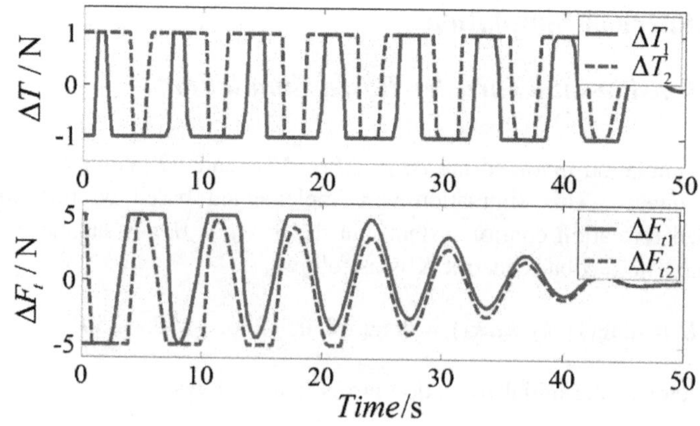

Fig. 9.31 ΔT_1, ΔT_2, ΔF_{t2}, and ΔF_{t1} variations with time

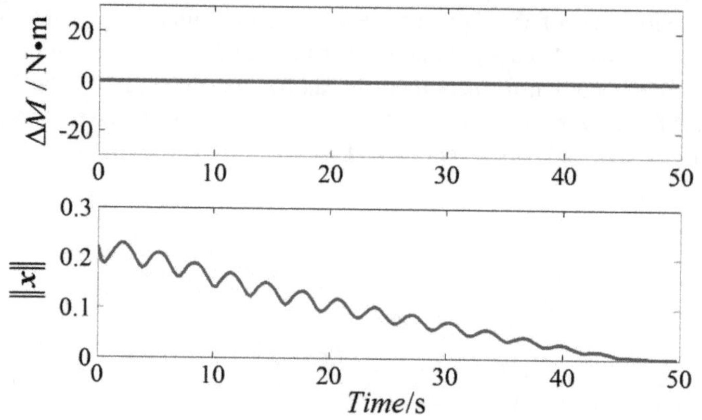

Fig. 9.32 ΔM and $\|\boldsymbol{x}\|$ variations with time

variations with time of the control variables and $\|\boldsymbol{x}\|$. The ΔM is unused and the $\|\boldsymbol{x}\|$ is decreasing with time.

In addition, it can be seen from Fig. 9.32 that the 2-norm of formation error decreases with time and eventually approaches 0. In a word, it is possible to maintain the control task of formation configuration by using the tether tensile force of the spiral arm and the autonomous machine power of the sub-TSR, but the convergence time of the initial error is longer, and the convergence curve has the oscillation phenomenon.

9.4.2 Error-Based Simulation

For the error-based coordinated control, the torque of the master-TSR ΔM is applied during the formation control operation. The value of α is decided by the 2-norm of the vector of control variables $\|\boldsymbol{x}\|$. The feedback matrix $\boldsymbol{K}$ and initial formation errors are illustrated in Eqs. (9.63) and (9.64). The error valve coefficient $E = 0.05$.

Figure 9.33 demonstrates the variations with time of the formation errors in case of error-based coordinated control strategy. Compared to the ΔF_{t1}, ΔF_{t2}, ΔT_1, and ΔT_2 control, the curves of $\Delta\varphi_1$, $\Delta\varphi_2$, $\Delta\dot{\varphi}_1$, $\Delta\dot{\varphi}_2$, and $\Delta\omega$ are greatly improved based on the same matrix $\boldsymbol{K}$. The convergent time decreased to about 20 s, and the oscillations of the curves are reduced. Figures 9.34 and 9.35 demonstrate the variations with time of the control variables and $\|\boldsymbol{x}\|$. It is noted that the ΔM operated when $\|\boldsymbol{x}\| \geq 0.05$, and stopped when $\|\boldsymbol{x}\| < 0.05$. Thus, the error-based coordinated control strategy is effective for the formation control operation of HSS.

Also included in Fig. 9.36 are curves demonstrating the variations with time of the formation errors in case of $E = 0.05$ and $E = 0.15$. It is noted that the vibration amplitude, vibration numbers, and convergence time of the error curves are decreasing with the decrease of the error coefficient. It is concluded that the efficiency of the error-based coordinated control strategy will get better when E decreases. However, smaller E will make the torque of master-TSR provide more control effect, and fuel consumption of the master-TSR does inevitably increase.

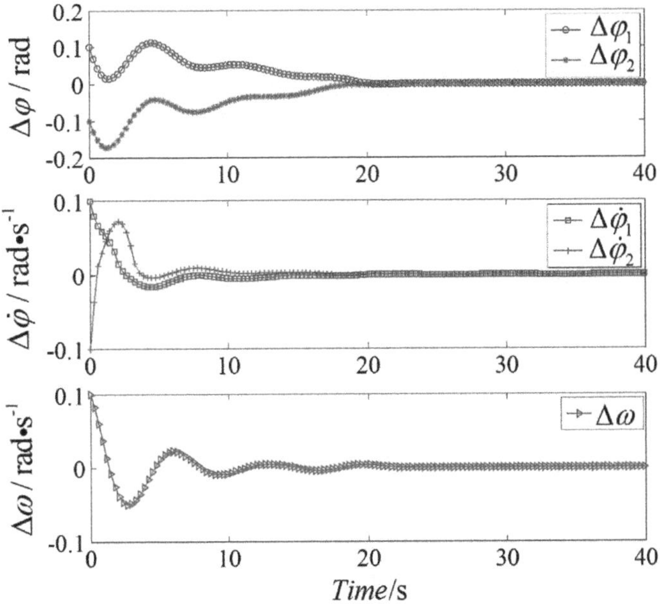

Fig. 9.33 Error variations with time

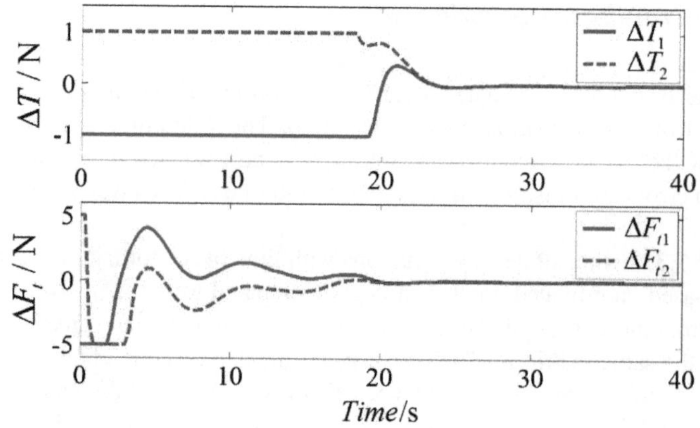

Fig. 9.34 ΔT_1, ΔT_2, ΔF_{t2}, and ΔF_{t1} variations with time

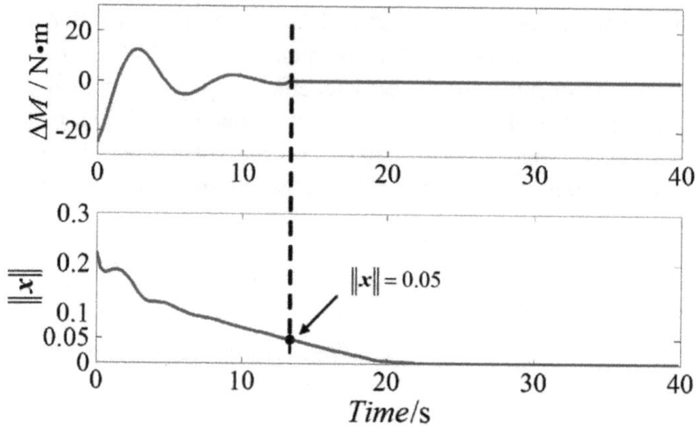

Fig. 9.35 ΔM and $\|x\|$ variations with time

9.4.3 Time-Based Simulation

For the time-based coordinated control, the torque of the master-TSR ΔM is also applied during the formation control operation. But the coordinated criterion for the formation control operation is based on control time. The value of β is decided by the control time t. The feedback matrix $\boldsymbol{K}$ and initial formation errors are demonstrated in Eqs. (9.63) and (9.64). The time valve coefficient $Ts = 5\text{s}$.

Figure 9.37 demonstrates the variations with time of the formation errors in case of time-based coordinated control strategy. The convergent time of the curves is about 28 s, and the oscillations of the curves are also reduced. Figure 9.38 demonstrates the variations with time of the control variables. It is noted that the ΔM operated when $t < 5$ s and stopped when $t \geq 5$ s.

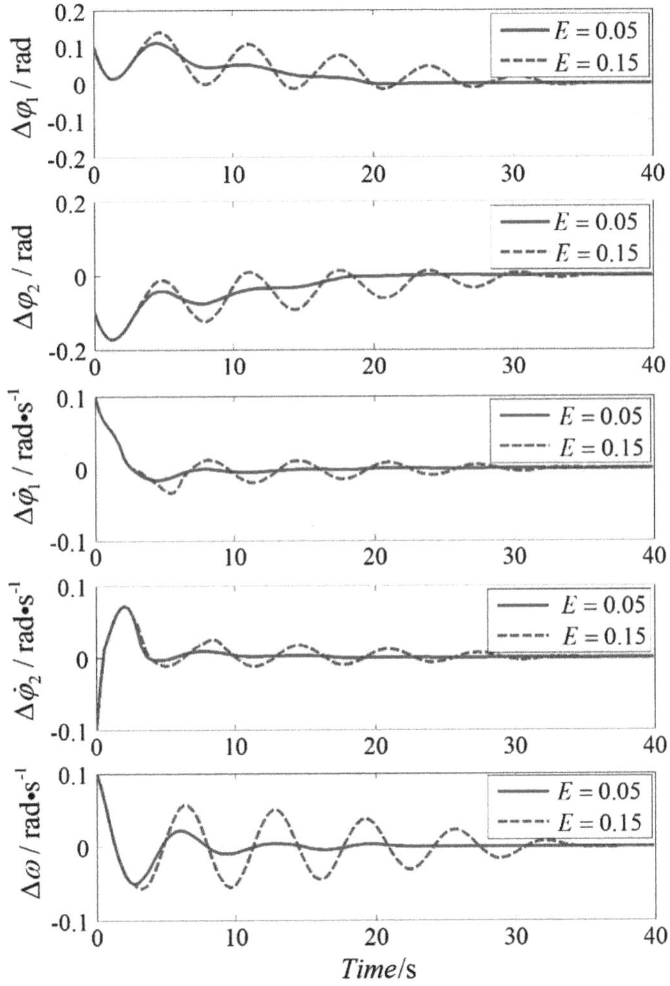

Fig. 9.36 Error variations with different E

The simulations of time-based coordinated control strategy with different time valve coefficient Ts are also performed (Fig. 9.39). It is concluded that the efficiency of the time-based coordinated control strategy will get better when Ts increases. This is because larger Ts can increase the operating time of ΔM. However, fuel consumption of the master-TSR does inevitably increase.

All simulation results suggest that the proposed error-based coordinated control strategy and time-based coordinated control strategy can effectively improve the formation control operation of the HSS. The operating time of ΔM can be adjusted by changing the error valve coefficient or the time valve coefficient. Thus, it is possible to obtain desired formation control efficiency with reasonable fuel

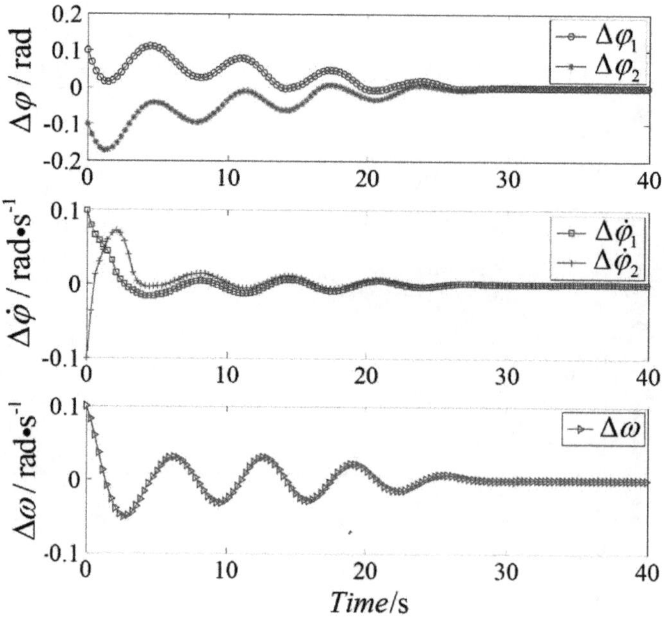

Fig. 9.37 Error variations with time

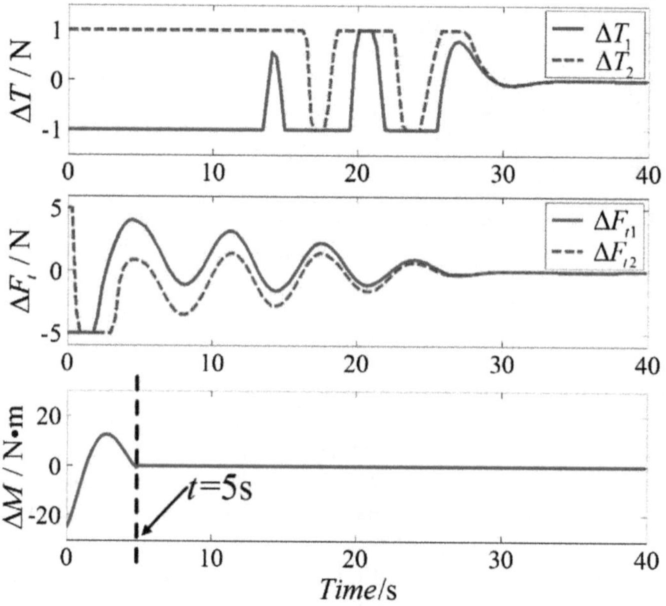

Fig. 9.38 Variations of control values with time

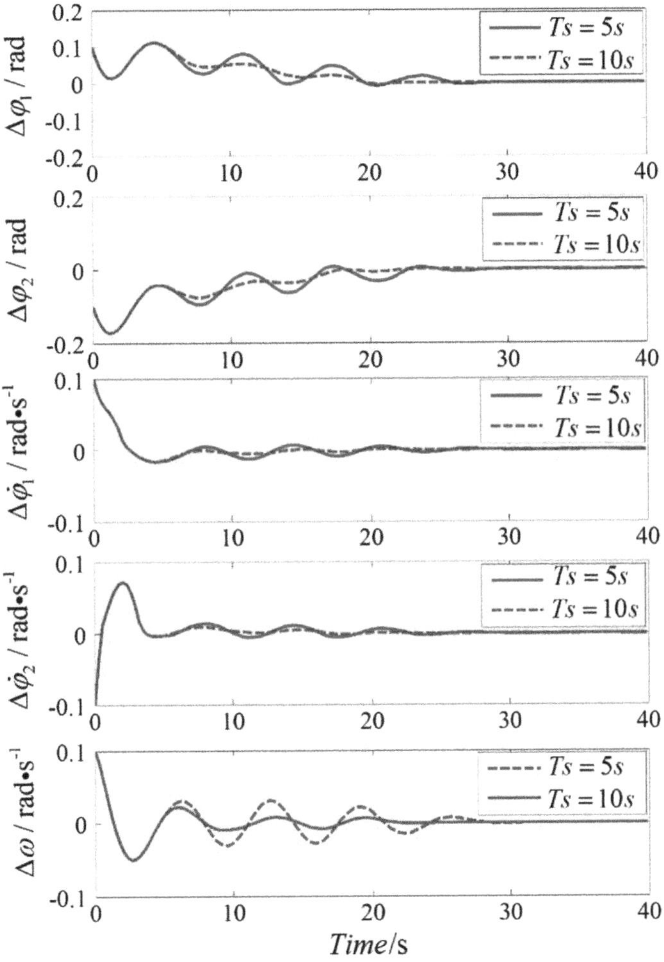

Fig. 9.39 Error variations with different *Ts*

consumption of the master spacecraft. Some useful conclusions are also obtained: The smaller the error valve coefficient, the more effective the formation control operation. Contrarily, the larger the time valve coefficient, the more effective the formation control operation. The fuel consumption of the master spacecraft does inevitably increase if the two proposed coordinated control strategies are applied during the formation control operation of the HSS.

References

1. Misra AK, Amier Z, Modi VJ (1988) Attitude dynamics of three-body tethered systems. Acta Astronaut 17(10):1059–1068
2. Pizarro-Chong A, Misra AK (2008) Dynamics of multi-tethered satellite formations containing a parent body. Acta Astronaut 63(11):1188–1202
3. Keshmiri M, Misra AK, Modi VJ (1996) General formulation for N-body tethered satellite system dynamics. J Guid Control Dyn 19(1):75–83
4. Misra AK, Modi VJ (1992) Three-dimensional dynamics and control of tether-connected N-body systems. Acta Astronaut 26(2):77–84
5. Lorenzini EC (1987) A three-mass tethered system for micro-g/variable-g applications. J Guid Control Dyn 10(3):242–249
6. Lorenzini EC, Cosmo M, Vetrella S et al (1988) Acceleration levels on board the space station and a tethered elevator for micro and variable-gravity applications. In: Space tethers for science in the space station era, vol 1, pp 513–522
7. Breakwell JV (1981) Stability of an orbiting ring. J Guid Control Dyn 4(2):197–200
8. Beletsky VV, Levin EM (1985) Stability of a ring of connected satellites. Acta Astronaut 12 (10):765–769
9. Menon C, Bombardelli C, Bianchini G (2005) Spinning tethered formation with self-stabilising attitude control. In: International Astronautical Congress, Fukuoka, Japan
10. Pengelley CD (1966) Preliminary survey of dynamic stability of a cable-connected spinning space station. J Spacecr Rocket 3(10):1456–1462
11. Amour AS, Misra AK, Modi VJ (2001) Equilibrium configurations and their stability in three-dimensional motion of three-body tethered systems. In: IAF, International Astronautical Congress, 52 nd, Toulouse, France
12. Lavagna M, Ercoli Finzi A (2003) Equilibrium analysis of a large multi-hinged space system. Acta Astronaut 53(1):9–20
13. Liu G, Huang J, Ma G et al (2013) Nonlinear dynamics and station-keeping control of a rotating tethered satellite system in halo orbits. Chin J Aeronaut 26(5):1227–1237
14. Kumar KD (2006) Review on dynamics and control of nonelectrodynamic tethered satellite systems. J Spacecr Rocket 43(4):705–720
15. Gärdsback M, Tibert G (2009) Deployment control of spinning space webs. J Guid Control Dyn 32(1):40–50
16. Gärdsback M, Tibert G (2009) Optimal deployment control of spinning space webs and membranes. J Guid Control Dyn 32(5):1519–1530

Appendices

Appendix A: Equations of Motion via Lagrangian Dynamics

The generalized coordinates are selected as $(\alpha_i \quad \beta_i)$ $i = 1, 2, 3, 4$, which denotes four in-plane angles, out-of-plane angles of each mass points, namely the four MUs in this simplified case. Based on the same assumptions and defined coordinates in dynamics modeling, the Lagrangian dynamics are derived as:

$$
\begin{aligned}
L &= T_{rans} + T_{rot} - V \\
&= \frac{1}{2}\sum_{i=1}^{n=4} m_i \dot{\boldsymbol{R}}_i \dot{\boldsymbol{R}}_i + 0 - \mu \sum_{i=1}^{n=4} m_i \frac{1}{|\boldsymbol{R}_i|} \\
&= \frac{1}{2} M R_c^2 \dot{\gamma}^2 + \frac{1}{2}\frac{l^2}{M}\sum_{i=1}^{4}\left\{ m_i (M - m_i)\left[(\dot{\gamma} + \dot{\alpha}_i)^2 \cos^2\beta_i + \dot{\beta}_i^2 \right]\right\} \\
&= -\frac{l^2}{M}\sum_{i=1}^{3}\left\{ m_i(\dot{\gamma} + \dot{\alpha}_i)\cos\beta_i \sum_{j=i+1}^{4} m_j(\dot{\gamma} + \dot{\alpha}_j)\cos\beta_j \right\} - \frac{l^2}{M}\sum_{i=1}^{3}\left[m_i\dot{\beta}_i \sum_{j=i+1}^{4} m_j\dot{\beta}_j \right] \\
&= -\mu M R_c^{-1} - \mu R_c^{-3}\frac{l^2}{M}\left\{
\begin{array}{l}
\sum\limits_{i=1}^{4}[m_i(M - m_i)] \\
-2\sum\limits_{i=1}^{3}\left[m_i \cos\alpha_i \sum\limits_{j=i+1}^{4} m_j \cos\alpha_j \cos(\beta_i - \beta_j) \right] \\
-2\sum\limits_{i=1}^{3}\left[m_i \sin\alpha_i \sum\limits_{j=i+1}^{4} m_j \sin\alpha_j \right]
\end{array}
\right\}
\end{aligned}
\tag{A.1}
$$

P. Huang and F. Zhang, *Theory and Applications of Multi-Tethers in Space*, Springer Tracts in Mechanical Engineering,
https://doi.org/10.1007/978-981-15-0387-0

All the variables used in Eq. (A.1) are the same with the ones in dynamics modeling. According to the Lagrangian scheme $\frac{\mathrm{d}}{\mathrm{d}t}\frac{\partial L}{\partial \dot{\theta}_i} - \frac{\partial L}{\partial \theta_i} = Q_i$, where θ_i denotes each generalized coordinates and Q_i denotes the generalized force, we can acquired the dimensional dynamics equations governing each generalized coordinate as:

$$
\begin{aligned}
&\frac{m_1}{M}(m_2+m_3+m_4)l^2\ddot{\alpha}_1\cos^2\beta_1 - \frac{m_1m_2}{M}l^2\ddot{\alpha}_2\cos\beta_1\cos\beta_2 \\
&\quad - \frac{m_1m_3}{M}l^2\ddot{\alpha}_3\cos\beta_1\cos\beta_3 - \frac{m_1m_4}{M}l^2\ddot{\alpha}_4\cos\beta_1\cos\beta_4 \\
&\quad - 2\frac{m_1}{M}(m_2+m_3+m_4)l^2(\dot{\gamma}+\dot{\alpha}_1)\dot{\beta}_1\sin\beta_1\cos\beta_1 \\
&\quad + \frac{m_1m_2}{M}l^2(\dot{\gamma}+\dot{\alpha}_2)\dot{\beta}_1\sin\beta_1\cos\beta_2 + \frac{m_1m_2}{M}l^2(\dot{\gamma}+\dot{\alpha}_2)\dot{\beta}_2\cos\beta_1\sin\beta_2 \\
&\quad + \frac{m_1m_3}{M}l^2(\dot{\gamma}+\dot{\alpha}_3)\dot{\beta}_1\sin\beta_1\cos\beta_3 + \frac{m_1m_3}{M}l^2(\dot{\gamma}+\dot{\alpha}_3)\dot{\beta}_3\cos\beta_1\sin\beta_3 \\
&\quad + \frac{m_1m_4}{M}l^2(\dot{\gamma}+\dot{\alpha}_4)\dot{\beta}_1\sin\beta_1\cos\beta_4 + \frac{m_1m_4}{M}l^2(\dot{\gamma}+\dot{\alpha}_4)\dot{\beta}_4\cos\beta_1\sin\beta_4 \\
&\quad - \mu R_c^{-3}\frac{1}{M}l^2\left\{\begin{array}{l} 2m_1m_2[\sin\alpha_1\cos\alpha_2\cos(\beta_1-\beta_2)-\cos\alpha_1\sin\alpha_2] \\ 2m_1m_3[\sin\alpha_1\cos\alpha_3\cos(\beta_1-\beta_3)-\cos\alpha_1\sin\alpha_3] \\ 2m_1m_4[\sin\alpha_1\cos\alpha_4\cos(\beta_1-\beta_4)-\cos\alpha_1\sin\alpha_4] \end{array}\right\} = Q_{\alpha_1}
\end{aligned}
\tag{A.2}
$$

$$
\begin{aligned}
&\frac{m_1}{M}(m_2+m_3+m_4)l^2\ddot{\beta}_1 - \frac{m_1m_2}{M}l^2\ddot{\beta}_2 - \frac{m_1m_3}{M}l^2\ddot{\beta}_3 - \frac{m_1m_4}{M}l^2\ddot{\beta}_4 \\
&\quad + \frac{m_1}{M}(m_2+m_3+m_4)l^2(\dot{\gamma}+\dot{\alpha}_1)^2\sin\beta_1\cos\beta_1 - \frac{m_1m_2}{M}l^2(\dot{\gamma}+\dot{\alpha}_1)(\dot{\gamma}+\dot{\alpha}_2)\sin\beta_1\cos\beta_2 \\
&\quad - \frac{m_1m_3}{M}l^2(\dot{\gamma}+\dot{\alpha}_1)(\dot{\gamma}+\dot{\alpha}_3)\sin\beta_1\cos\beta_3 - \frac{m_1m_4}{M}l^2(\dot{\gamma}+\dot{\alpha}_1)(\dot{\gamma}+\dot{\alpha}_4)\sin\beta_1\cos\beta_4 \\
&\quad - \mu R_c^{-3}\frac{1}{M}l^2\begin{bmatrix} +2m_1m_2[\cos\alpha_1\cos\alpha_2\sin(\beta_1-\beta_2)] \\ +2m_1m_3[\cos\alpha_1\cos\alpha_3\sin(\beta_1-\beta_3)] \\ +2m_1m_4[\cos\alpha_1\cos\alpha_4\sin(\beta_1-\beta_4)] \end{bmatrix} = Q_{\beta_1}
\end{aligned}
\tag{A.3}
$$

$$
\begin{aligned}
&\frac{m_2}{M}(m_1+m_3+m_4)l^2\ddot{\alpha}_2\cos^2\beta_2 - \frac{m_1m_2}{M}l^2\ddot{\alpha}_1\cos\beta_1\cos\beta_2 \\
&\quad - \frac{m_2m_3}{M}l^2\ddot{\alpha}_3\cos\beta_2\cos\beta_3 - \frac{m_2m_4}{M}l^2\ddot{\alpha}_4\cos\beta_2\cos\beta_4 \\
&\quad - 2\frac{m_2}{M}(m_1+m_3+m_4)l^2(\dot{\gamma}+\dot{\alpha}_2)\dot{\beta}_2\sin\beta_2\cos\beta_2 \\
&\quad + \frac{m_1m_2}{M}l^2(\dot{\gamma}+\dot{\alpha}_1)\dot{\beta}_1\sin\beta_1\cos\beta_2 + \frac{m_1m_2}{M}l^2(\dot{\gamma}+\dot{\alpha}_1)\dot{\beta}_2\cos\beta_1\sin\beta_2 \\
&\quad + \frac{m_2m_3}{M}l^2(\dot{\gamma}+\dot{\alpha}_3)\dot{\beta}_2\sin\beta_2\cos\beta_3 + \frac{m_2m_3}{M}l^2(\dot{\gamma}+\dot{\alpha}_3)\dot{\beta}_3\cos\beta_2\sin\beta_3 \\
&\quad + \frac{m_2m_4}{M}l^2(\dot{\gamma}+\dot{\alpha}_4)\dot{\beta}_2\sin\beta_2\cos\beta_4 + \frac{m_2m_4}{M}l^2(\dot{\gamma}+\dot{\alpha}_4)\dot{\beta}_4\cos\beta_2\sin\beta_4 \\
&\quad - \mu R_c^{-3}\frac{1}{M}l^2\left\{\begin{array}{l} 2m_1m_2[\cos\alpha_1\sin\alpha_2\cos(\beta_1-\beta_2)-\sin\alpha_1\cos\alpha_2] \\ 2m_2m_3[\sin\alpha_2\cos\alpha_3\cos(\beta_2-\beta_3)-\cos\alpha_2\sin\alpha_3] \\ 2m_2m_4[\sin\alpha_2\cos\alpha_4\cos(\beta_2-\beta_4)-\cos\alpha_2\sin\alpha_4] \end{array}\right\} = Q_{\alpha_2}
\end{aligned}
\tag{A.4}
$$

$$\begin{aligned}
&\frac{m_2}{M}(m_1+m_3+m_4)l^2\ddot{\beta}_2 - \frac{m_1m_2}{M}l^2\ddot{\beta}_1 - \frac{m_2m_3}{M}l^2\ddot{\beta}_3 - \frac{m_2m_4}{M}l^2\ddot{\beta}_4 \\
&\quad + \frac{m_2}{M}(m_1+m_3+m_4)l^2(\dot{\gamma}+\dot{\alpha}_2)^2\sin\beta_2\cos\beta_2 - \frac{m_1m_2}{M}l^2(\dot{\gamma}+\dot{\alpha}_1)(\dot{\gamma}+\dot{\alpha}_2)\cos\beta_1\sin\beta_2 \\
&\quad - \frac{m_2m_3}{M}l^2(\dot{\gamma}+\dot{\alpha}_2)(\dot{\gamma}+\dot{\alpha}_3)\sin\beta_2\cos\beta_3 - \frac{m_2m_4}{M}l^2(\dot{\gamma}+\dot{\alpha}_2)(\dot{\gamma}+\dot{\alpha}_4)\sin\beta_2\cos\beta_4 \\
&\quad - \mu R_c^{-3}\frac{1}{M}l^2\begin{bmatrix} -2m_1m_2[\cos\alpha_1\cos\alpha_2\sin(\beta_1-\beta_2)] \\ +2m_2m_3[\cos\alpha_2\cos\alpha_3\sin(\beta_2-\beta_3)] \\ +2m_2m_4[\cos\alpha_2\cos\alpha_4\sin(\beta_2-\beta_4)] \end{bmatrix} = Q_{\beta_2}
\end{aligned} \tag{A.5}$$

$$\begin{aligned}
&\frac{m_3}{M}(m_1+m_2+m_4)l^2\ddot{\alpha}_3\cos^2\beta_3 - \frac{m_1m_3}{M}l^2\ddot{\alpha}_1\cos\beta_1\cos\beta_3 \\
&\quad - \frac{m_2m_3}{M}l^2\ddot{\alpha}_2\cos\beta_2\cos\beta_3 - \frac{m_3m_4}{M}l^2\ddot{\alpha}_4\cos\beta_3\cos\beta_4 \\
&\quad - 2\frac{m_3}{M}(m_1+m_2+m_4)l^2(\dot{\gamma}+\dot{\alpha}_3)\dot{\beta}_3\sin\beta_3\cos\beta_3 \\
&\quad + \frac{m_1m_3}{M}l^2(\dot{\gamma}+\dot{\alpha}_1)\dot{\beta}_1\sin\beta_1\cos\beta_3 + \frac{m_1m_3}{M}l^2(\dot{\gamma}+\dot{\alpha}_1)\dot{\beta}_3\cos\beta_1\sin\beta_3 \\
&\quad + \frac{m_2m_3}{M}l^2(\dot{\gamma}+\dot{\alpha}_2)\dot{\beta}_2\sin\beta_2\cos\beta_3 + \frac{m_2m_3}{M}l^2(\dot{\gamma}+\dot{\alpha}_2)\dot{\beta}_3\cos\beta_2\sin\beta_3 \\
&\quad + \frac{m_3m_4}{M}l^2(\dot{\gamma}+\dot{\alpha}_4)\dot{\beta}_3\sin\beta_3\cos\beta_4 + \frac{m_3m_4}{M}l^2(\dot{\gamma}+\dot{\alpha}_4)\dot{\beta}_4\cos\beta_3\sin\beta_4 \\
&\quad - \mu R_c^{-3}\frac{1}{M}l^2\begin{bmatrix} 2m_1m_3[\cos\alpha_1\sin\alpha_3\cos(\beta_1-\beta_3) - \sin\alpha_1\cos\alpha_3] \\ 2m_2m_3[\cos\alpha_2\sin\alpha_3\cos(\beta_2-\beta_3) - \sin\alpha_2\cos\alpha_3] \\ 2m_3m_4[\sin\alpha_3\cos\alpha_4\cos(\beta_3-\beta_4) - \cos\alpha_3\sin\alpha_4] \end{bmatrix} = Q_{\alpha_3}
\end{aligned} \tag{A.6}$$

$$\begin{aligned}
&\frac{m_3}{M}(m_1+m_2+m_4)l^2\ddot{\beta}_3 - \frac{m_1m_3}{M}l^2\ddot{\beta}_1 - \frac{m_2m_3}{M}l^2\ddot{\beta}_2 - \frac{m_3m_4}{M}l^2\ddot{\beta}_4 \\
&\quad + \frac{m_3}{M}(m_1+m_2+m_4)l^2(\dot{\gamma}+\dot{\alpha}_3)^2\sin\beta_3\cos\beta_3 - \frac{m_1m_3}{M}l^2(\dot{\gamma}+\dot{\alpha}_1)(\dot{\gamma}+\dot{\alpha}_3)\cos\beta_1\sin\beta_3 \\
&\quad - \frac{m_2m_3}{M}l^2(\dot{\gamma}+\dot{\alpha}_2)(\dot{\gamma}+\dot{\alpha}_3)\cos\beta_2\sin\beta_3 - \frac{m_3m_4}{M}l^2(\dot{\gamma}+\dot{\alpha}_3)(\dot{\gamma}+\dot{\alpha}_4)\sin\beta_3\cos\beta_4 \\
&\quad + \mu R_c^{-3}\frac{1}{M}l^2\begin{bmatrix} 2m_1m_3\cos\alpha_1\cos\alpha_3\sin(\beta_1-\beta_3) \\ 2m_2m_3\cos\alpha_2\cos\alpha_3\sin(\beta_2-\beta_3) \\ -2m_3m_4\cos\alpha_3\cos\alpha_4\sin(\beta_3-\beta_4) \end{bmatrix} = Q_{\beta_3}
\end{aligned} \tag{A.7}$$

$$\begin{aligned}
&\frac{m_4}{M}(m_1+m_2+m_3)l^2\ddot{\alpha}_4\cos^2\beta_4-\frac{m_1m_4}{M}l^2\ddot{\alpha}_1\cos\beta_1\cos\beta_4\\
&-\frac{m_2m_4}{M}l^2\ddot{\alpha}_2\cos\beta_2\cos\beta_4-\frac{m_3m_4}{M}l^2\ddot{\alpha}_3\cos\beta_3\cos\beta_4\\
&-2\frac{m_4}{M}(m_1+m_2+m_3)l^2(\dot{\gamma}+\dot{\alpha}_4)\dot{\beta}_4\cos\beta_4\sin\beta_4\\
&+\frac{m_1m_4}{M}l^2(\dot{\gamma}+\dot{\alpha}_1)\dot{\beta}_1\sin\beta_1\cos\beta_4+\frac{m_1m_4}{M}l^2(\dot{\gamma}+\dot{\alpha}_1)\dot{\beta}_4\cos\beta_1\sin\beta_4\\
&+\frac{m_2m_4}{M}l^2(\dot{\gamma}+\dot{\alpha}_2)\dot{\beta}_2\sin\beta_2\cos\beta_4+\frac{m_2m_4}{M}l^2(\dot{\gamma}+\dot{\alpha}_2)\dot{\beta}_4\cos\beta_2\sin\beta_4\\
&+\frac{m_3m_4}{M}l^2(\dot{\gamma}+\dot{\alpha}_3)\dot{\beta}_3\sin\beta_3\cos\beta_4+\frac{m_3m_4}{M}l^2(\dot{\gamma}+\dot{\alpha}_3)\dot{\beta}_4\cos\beta_3\sin\beta_4\\
&-\mu R_c^{-3}\frac{1}{M}l^2\begin{bmatrix}2m_1m_4[\cos\alpha_1\sin\alpha_4\cos(\beta_1-\beta_4)-\sin\alpha_1\cos\alpha_4]\\2m_2m_4[\cos\alpha_2\sin\alpha_4\cos(\beta_2-\beta_4)-\sin\alpha_2\cos\alpha_4]\\2m_3m_4[\cos\alpha_3\sin\alpha_4\cos(\beta_3-\beta_4)-\sin\alpha_3\cos\alpha_4]\end{bmatrix}=Q_{\alpha_4}
\end{aligned}\tag{A.8}$$

$$\begin{aligned}
&\frac{m_4}{M}(m_1+m_2+m_3)l^2\ddot{\beta}_4-\frac{m_1m_4}{M}l^2\ddot{\beta}_1-\frac{m_2m_4}{M}l^2\ddot{\beta}_2-\frac{m_3m_4}{M}l^2\ddot{\beta}_3\\
&+\frac{m_4}{M}(m_1+m_2+m_3)l^2(\dot{\gamma}+\dot{\alpha}_4)^2\sin\beta_4\cos\beta_4-\frac{m_1m_4}{M}l^2(\dot{\gamma}+\dot{\alpha}_1)(\dot{\gamma}+\dot{\alpha}_4)\cos\beta_1\sin\beta_4\\
&-\frac{m_2m_4}{M}l^2(\dot{\gamma}+\dot{\alpha}_2)(\dot{\gamma}+\dot{\alpha}_4)\cos\beta_2\sin\beta_4-\frac{m_3m_4}{M}l^2(\dot{\gamma}+\dot{\alpha}_3)(\dot{\gamma}+\dot{\alpha}_4)\cos\beta_3\sin\beta_4\\
&+\mu R_c^{-3}\frac{1}{M}l^2\begin{bmatrix}2m_1m_4\cos\alpha_1\cos\alpha_4\sin(\beta_1-\beta_4)\\+2m_2m_4\cos\alpha_2\cos\alpha_4\sin(\beta_2-\beta_4)\\+2m_3m_4\cos\alpha_3\cos\alpha_4\sin(\beta_3-\beta_4)\end{bmatrix}=Q_{\beta_4}
\end{aligned}\tag{A.9}$$

Appendix B: Graph Theory

In this research, we use a weighted undirected simple graph to describe the communication topology between the MUs. A weighted undirected simple graph $G\triangleq(\upsilon,\varepsilon,\boldsymbol{A})$ consists of a finite non-empty node set $\upsilon=\{\upsilon_1,\upsilon_2,\ldots,\upsilon_n\}$, an edge set $\varepsilon\subseteq\upsilon\times\upsilon$, and a weighted adjacency matrix $\boldsymbol{A}=\left[a_{ij}\right]\in\mathbb{R}^{n\times n}$. If the ith node can get information from the jth node, then there is an edge from the jth node to the ith node, denoted as $(\upsilon_i,\upsilon_j)\in\varepsilon$. For the undirected graph, the adjacency matrix $\boldsymbol{A}=\left[a_{ij}\right]\in\mathbb{R}^{n\times n}$ is defined as $a_{ij}=a_{ji}>0$ if $(\upsilon_i,\upsilon_j)\in\varepsilon$ and $i\neq j$, otherwise $a_{ij}=0$, since $(\upsilon_i,\upsilon_j)\in\varepsilon$ implies $(\upsilon_j,\upsilon_i)\in\varepsilon$. An undirected graph is connected if there exists a path between any distinct pair of nodes. Further, the degree matrix of a graph is defined as $\boldsymbol{D}=\mathrm{diag}(d(\upsilon_1),d(\upsilon_2),\ldots,d(\upsilon_n))$ where $d(\upsilon_i)=\sum_{j\neq i}a_{ij}$. The Laplacian matrix $\boldsymbol{L}=\left[l_{ij}\right]\in\mathbb{R}^{n\times n}$ of undirected graph G is defined as $\boldsymbol{L}=\boldsymbol{D}-\boldsymbol{A}$, which has the property that it is symmetric positive semi-definite. Define a matrix $\boldsymbol{B}=\mathrm{diag}(b_1,b_2,\ldots,b_n)$ where $b_i>0$ if the ith MU can get information from the platform satellite, otherwise $b_i = 0$. Further, to express the relationship of communication between the platform satellite and the MUs, we define a matrix $\boldsymbol{H}$, where $\boldsymbol{H} = \boldsymbol{L} + \boldsymbol{B}$.

In the undirected case, because $\boldsymbol{L}$ has zero row sums, 0 is an eigenvalue of $\boldsymbol{L}$ with the associated eigenvector $\mathbf{1}_n$, the $n \times 1$ column vector of ones. Note that $\boldsymbol{L}$ is diagonally dominant and has nonnegative diagonal entries. Following from the matrix theory, for an undirected graph, all nonzero eigenvalues of $\boldsymbol{L}$ are positive. For an undirected graph, 0 is a simple eigenvalue of $\boldsymbol{L}$ if and only if the undirected graph is connected. For an undirected graph, let $\lambda_i(\boldsymbol{L})$ be the ith eigenvalue of $\boldsymbol{L}$ with $\lambda_1(\boldsymbol{L}) \leq \lambda_2(\boldsymbol{L}) \leq \cdots \leq \lambda_n(\boldsymbol{L})$, so that $\lambda_1(\boldsymbol{L}) = 0$. For an undirected graph, $\lambda_2(\boldsymbol{L})$ is the algebraic connectivity, which is positive if and only if the undirected graph is connected. The algebraic connectivity quantifies the convergence rate of consensus algorithms.

Matrix Theory

Theorem 1 Let $\boldsymbol{A} = \left[a_{ij}\right] \in \mathbb{R}^{n \times n}$, and let

$$R_i'(\boldsymbol{A}) \equiv \sum_{j=1, j \neq i}^{n} \left|a_{ij}\right|, \quad i = 1, 2, \ldots, n$$

denote the deleted absolute row sums of $\boldsymbol{A}$. Then all eigenvalues of $\boldsymbol{A}$ are located in the union of n discs

$$\cup_{i=1}^{n} \left\{z \in C : |z - a_{ii}| \leq R_i'(\boldsymbol{A})\right\} \equiv G(\boldsymbol{A})$$

Furthermore, if a union of k of these n discs forms a connected region that is disjoint from all of the remaining $n - k$ discs, then there are precisely k eigenvalues of $\boldsymbol{A}$ in this region.

Lemma 1 Suppose that $\boldsymbol{A} \in R^{m \times n}$, $\boldsymbol{B} \in R^{p \times q}$, $\boldsymbol{C} \in R^{n \times s}$, and $\boldsymbol{D} \in R^{q \times t}$. The following arguments are valid.

(1) $(\boldsymbol{A} \otimes \boldsymbol{B})(\boldsymbol{C} \otimes \boldsymbol{D}) = \boldsymbol{AC} \otimes \boldsymbol{BD}$.
(2) $(\boldsymbol{A} \otimes \boldsymbol{B})^{\mathrm{T}} = \boldsymbol{A}^{\mathrm{T}} \otimes \boldsymbol{B}^{\mathrm{T}}$.
(3) If $\boldsymbol{A}$ and $\boldsymbol{B}$ are symmetrical, so is $\boldsymbol{A} \otimes \boldsymbol{B}$.
(4) If $\boldsymbol{A}$ and $\boldsymbol{B}$ are symmetrical positive definite (respectively, positive semi-definite), so is $\boldsymbol{A} \otimes \boldsymbol{B}$.
(5) Suppose that $\boldsymbol{A} \in R^{m \times m}$ and $\boldsymbol{B} \in R^{n \times n}$ are invertible. Then $(\boldsymbol{A} \otimes \boldsymbol{B})^{-1} = \boldsymbol{A}^{-1} \otimes \mathbf{B}^{-1}$.
(6) If all of the eigenvalues of matrix $\boldsymbol{A} \in R^{m \times m}$ are $\lambda_1, \lambda_2, \ldots, \lambda_m$, and the eigenvalues of matrix $\boldsymbol{B} \in R^{n \times n}$ are $\mu_1, \mu_2, \ldots, \mu_n$, then the eigenvalues of matrices $\boldsymbol{A} \otimes \boldsymbol{B}$ and $\boldsymbol{A} \otimes \boldsymbol{I}_n + \boldsymbol{I}_m \otimes \boldsymbol{B}^{\mathrm{T}}$ are $\lambda_i \mu_j$ and $\lambda_i + \mu_j$ $(i = 1, 2, \ldots, m; j = 1, 2, \ldots, n)$, respectively.

CPSIA information can be obtained
at www.ICGtesting.com
Printed in the USA
LVHW081036081219
639811LV00001B/86/P

9 789811 503863